T0335985

Studies in Fuzziness and Soft Computing

Volume 336

Series editor

Janusz Kacprzyk, Polish Academy of Sciences, Warsaw, Poland
e-mail: kacprzyk@ibspan.waw.pl

About this Series

The series "Studies in Fuzziness and Soft Computing" contains publications on various topics in the area of soft computing, which include fuzzy sets, rough sets, neural networks, evolutionary computation, probabilistic and evidential reasoning, multi-valued logic, and related fields. The publications within "Studies in Fuzziness and Soft Computing" are primarily monographs and edited volumes. They cover significant recent developments in the field, both of a foundational and applicable character. An important feature of the series is its short publication time and world-wide distribution. This permits a rapid and broad dissemination of research results.

More information about this series at http://www.springer.com/series/2941

Susanne Saminger-Platz · Radko Mesiar
Editors

On Logical, Algebraic, and Probabilistic Aspects of Fuzzy Set Theory

Editors
Susanne Saminger-Platz
Department of Knowledge-Based
Mathematical Systems
Johannes Kepler University Linz
Linz
Austria

Radko Mesiar
Department of Mathematics and Descriptive
Geometry
Slovak University of Technology
Bratislava
Slovakia

ISSN 1434-9922 ISSN 1860-0808 (electronic)
Studies in Fuzziness and Soft Computing
ISBN 978-3-319-28807-9 ISBN 978-3-319-28808-6 (eBook)
DOI 10.1007/978-3-319-28808-6

Library of Congress Control Number: 2015960233

Printed on acid-free paper

This Springer imprint is published by SpringerNature
The registered company is Springer International Publishing AG Switzerland

Dedicated to Erich Peter Klement

Copyright OÖN/Weihbold

Preface

Erich Peter Klement got interested in fuzzy set theory already in the 1970s while being a young Assistant Professor at Johannes Kepler University in Linz, Austria. In 1979, he stayed with Lotfi A. Zadeh at Berkeley University as a Visiting Research Associate, several other research visits to universities in Europe and the United States followed. It was also in 1979 when, together with Ulrich Höhle and Robert Lowen, he first organized the "International Seminar on Fuzzy Set Theory" in Linz. At that time, he could not know that there would be more than 35 seminars to follow, well-established and widely known to the scientific community as the "Linz Seminars on Fuzzy Set Theory".

Organizing and hosting the seminars for so many years has not only been a great service to the community but also had a big impact on the evolvement of science dedicated to this field. The philosophy of the seminar has always been to encourage critical discussions on mathematical aspects of fuzzy set theory by bringing together researchers from different fields. Those, sometimes even controversial, discussions have helped to develop a common understanding of the treatment of fuzzy sets and fuzzy logic. And it happened more than once, in particular after the fall of the Iron Curtain, that researchers from different countries first met in person on the occasion of one of the Linz seminars.

It has also been during the early 1990s that Peter Klement founded the Fuzzy Logic Laboratorium Linz Hagenberg (FLLL). Through industrial, applied and basic research projects, he provided a working place and research perspectives for (young) colleagues from different countries and disciplines. As the head of the FLLL, and also the Department of Knowledge-Based Mathematical Systems at Johannes Kepler University, he has hosted numerous international researchers within, but also independently of, many research actions such as CEEPUS and COST encouraging again discussions on the theory and application of fuzzy set theory and beyond.

Besides his activities for the scientific community, Peter Klement has always been active as a researcher himself. His early research interests have mainly been devoted to (fuzzy) measures and integrals as several of his articles from the 1980s prove. Fuzzy measures have also been the background of his close collaboration with the Dan Butnariu leading to the publication of the joint monograph entitled "Triangular Norm-Based Measures and Games with Fuzzy Coalitions" published in 1993.

In 1992, as a result of the first visit of Radko Mesiar and Endre Pap to Linz with an original intention to work on fuzzy measures and integrals, another long time research cooperation, namely on the triangular norms and triangular conorms was established resulting in a lot of journal articles, but in particular in the publication of the joint monograph entitled "Triangular Norms" in 2000. By the intensified work on triangular norms, Peter Klement's attention had also been drawn to copulas so that since the early years of the new millennium also copulas and quasi-copulas had appeared in the titles of his articles, as well as topics related to aggregation functions leading back to his original research interests in (generalized) integration.

It is therefore not by chance that the current edited volume reflects and covers several aspects of Peter Klement's research acitivities. Among the authors one can find former Ph.D. students, former colleagues from the Department of Knowledge-Based Mathematical Systems, as well as colleagues, co-authors, friends of Peter Klement with more than 30 years of experience in fuzzy set theory. Some of the chapters included reflect personal views on traditional topics of the Linz seminar—some of which containing even controversial aspects and fostering a discussion on the mathematics behind. Other chapters deal with deep mathematical theory of the algebraic and logical foundations of fuzzy set theory and fuzzy logic. Several chapters approach topics related to Peter Klement's personal research interests in copulas, measures and integrals, as well as aggregation problems.

We briefly summarize the single chapters included in this volume:

Siegfried Gottwald has contributed to chapter discussing the main developments in the field of mathematically oriented fuzzy logics and how they found their representation over the years in the Linz Seminars on Fuzzy Set Theory. Let us acknowledge that Siegfried Gottwald had been a regular and active participant to the Linz Seminars since 1990 and we are deeply sorrow that he passed away before the finalization of this edited volume.

Enric Trillas has provided a very individual view on fuzzy sets and their personal and scientific perception since their introduction by Lotfi A. Zadeh in his seminal paper in 1965.

In his contribution "Modules in the Category Sup" Ulrich Höhle explains basic properties of left modules on unital quantales with perspectives towards fuzzy set theory and contributes to the clarification of mathematical, in particular the algebraic, basis of fuzzy set theory inside mathematics.

Daniele Mundici elaborates in his chapter a geometric approach to MV-algebras and relates algebraic aspects to the basis of fuzzy resp. many-valued logics.

Francesc Esteva and Lluis Godo discuss the equational characterization of continuous t-norms being an indispensable tool for modelling the semantic interpretation of the intersection in fuzzy logics in narrow sense.

Also Thomas Vetterlein and Milan Petrík focus on the semantics of fuzzy logics by discussing two different ways of investigating totally ordered monoids as an interpretation of the conjunction in fuzzy logics.

Andrea Mesiarová-Zemánková's chapter provides a characterization of the structure of uninorms with continuous diagonal functions. Uninorms may be seen as generalizations of t-norms and t-conorms, as they are associative and commutative increasing operations on the unit interval whose neutral element can be, in contrast to t-norms, respectively t-conorms, any interior element of the unit interval and allow to model also bipolar behaviour in aggregation problems.

Humberto Bustince, Edurne Barrenechea, Miguel Pagola and Javier Fernandez provide an overview on concepts of overlap and grouping functions generalizing ideas of connectives from fuzzy set theory for the aggregation of information in fuzzy classification systems.

Fabrizio Durante and Elisa Perrone in their chapter focus on asymmetric copulas and their application in the design of experiments. The importance of copulas stems from Sklar's theorem clarifying that the dependence of a multivariate distribution function of its univariate marginal distributions is, in case of continuity, completely captured by a unique copula. The asymmetry of a copula therefore reflects the non-exchangeability of the underlying random variables.

Carlo Sempi elaborates in his chapter the relationship between copulas and stochastic processes, in particular the Brownian motion.

Anna Kolesárová and Andrea Stupňanová discuss extensions of capacities to n-ary aggregation functions with relationships to the discrete Choquet and Sugeno integral stressing the role of n-ary copulas when generalizing Lovász and Owen extensions.

Ronald R. Yager approaches a more recent problem in aggregation, namely the problem of multi-source information fusion by using measure representations. The concepts of assurance and opportunity in the measure framework are also discussed.

Michel Grabisch focusses on bases and transforms of set functions on a finite set. The basic duality between bases and invertible linear transforms is established, covering, among others, the case of the Moebius transform, the Fourier transform and interaction transforms.

Siegfried Weber in his chapter deals with conditioning for Boolean subsets, indicator functions and fuzzy subsets. It introduces and discusses two types of iteration.

Endre Pap discusses the integration of multivalued functions from additive to arbitrary non-negative set functions. In particular, a set-valued Gould-type integral of multifunctions is introduced and discussed.

Our special thanks go to our authors for their willingness to contribute to this comprehensive volume. And we hope that the readers will enjoy reading all or part of the chapters.

We congratulate Peter Klement for his scientific achievements and we are thankful for the support he has given to us and to the scientific community in fuzzy set theory throughout so many years. We are happy to witness that, although being retired from being a university professor, he still enjoys being an active researcher.

We wish him all the best, in particular healthiness, for pursuing his goals in the future.

We have been supported by our universities, the Johannes Kepler University in Linz and the Slovak University of Technology in Bratislava. We also gratefully acknowledge the support of the grants APVV-14-0013 and the support in the framework of the Technologie-Transfer-Förderung Wi-2014-200710/3KX/Kai of the Upper Austrian Government, as well as the encouragement and the help of Prof. Janusz Kacprzyk for the preparation of this edited volume.

Linz, Bratislava
November 2015

Susanne Saminger-Platz
Radko Mesiar

Contents

Fuzzy Logic and the Linz Seminar: Themes and Some Personal Reminiscences

Siegfried Gottwald

Abstract The paper discusses concisely the main developments in the field of mathematically oriented fuzzy logics and how they found their representation over the years in the Linz Seminars on Fuzzy Set Theory.

1 Introduction

The *Linz Seminar on Fuzzy Set Theory*, first organized by Peter Klement in 1979, soon became famous among mathematicians interested in fuzzy set topics, pure as well as applied ones.

Personally, I first met Peter Klement in 1983 at the *Polish Symposium on Interval & Fuzzy Mathematics* [1] in Poznań, and then again in 1985 at an *International Workshop on Fuzzy Sets Applications* [2] in Eisenach. We soon had a quite friendly relationship—and he told me that I'd have a standing invitation to the Linz Seminar as soon as I was able to "tunnel" the iron curtain. This, however, proved to be impossible till 1989. So I first attended the Linz Seminar in 1990—and ever since with only a few exceptions. And already in 1990 Peter asked me kindly to join the Program Committee, what I accepted with pleasure.

2 Logic and Fuzzy Sets—The Early Years

As I had graduated in 1969 at Leipzig University under the supervision of Dieter Klaua, from the very beginning it was clear to me that fuzzy sets should be discussed within the framework of many-valued, particularly Łukasiewicz-like logics. Throughout the 1970s I did this, particularly in my Habilitation Thesis (published

S. Gottwald (✉)
Universität Leipzig, Abt. Logik Am Institut Für Philosophie, Beethovenstr. 15, 04107 Leipzig, Germany
e-mail: gottwald@uni-leipzig.de

S. Saminger-Platz and R. Mesiar (eds.), *On Logical, Algebraic, and Probabilistic Aspects of Fuzzy Set Theory*, Studies in Fuzziness and Soft Computing 336,
DOI 10.1007/978-3-319-28808-6_1

as) [3–5], using (only) Łukasiewicz logic—but with a strong feeling for the fact that the whole approach would naturally work within a more general logical framework.

It was the contact and cooperation with E. Czogała from Gliwice in the beginning 1980s, and soon also with his graduate student W. Pedrycz, and particularly a stay of W. Pedrycz at Delft University of Technology, that brought to me the information of the t-norm framework. And it was this cooperation together with a 1983 Summer School on fuzzy set matters in Bulgaria where I learned much about the importance of fuzzy relation equations for fuzzy modeling, and of the solution approach by E. Sanchez.

As a result I started to consider t-norm based logics [6] and fuzzy set theory in this realm [7], and I generalized also solvability considerations for systems of fuzzy relation equations to this framework [8, 9].

From the Linz circle then I learned that the Linz Seminar was at its early meetings instrumental in establishing the fundamental role of the t-norms as suitable candidates for generalized, interactive intersections of fuzzy sets, and hence of generalized, non-idempotent conjunction connectives for systems of many-valued logics. Core members of the Linz Seminar crew had contributed to the discussions around t-norms, e.g. at the 2nd Linz Seminar of 1980 with t-norm related contributions by D. Dubois and E.P. Klement.

Additionally, what I learned from the Linz Seminars in the early 1990s was the central role that algebraic considerations, even category theoretic ones, should play in the fuzzy sets context to structure the theory in a suitable way. And I realized with pleasure the open-mindedness of most of the Linz participants regarding a large diversity of mathematical topics.

Besides the t-norm topic, themes from formal logic have been considered only occasionally in the early years of the Linz Seminar: U. Cerutti discussed in 1981 category theoretic aspects, L. Valverde problems of generalized connectives in 1981 and 1982. Logical operators had also been the topic of L. Kohout in 1984 and 1985. A whole day was devoted to logic, for the first time, in 1988, with U. Höhle and L. Kohout as speakers.

As times went on and the mathematical discussions in the fuzzy sets field became more mature, the Linz Seminar was for the first time mainly devoted to a particular topic in 1988: *Measures and Integrals*, followed by the topic of *Applications of Category Theory to Fuzzy Subsets* in 1989.

3 A New Time—Open Borders in Europe

With the breakdown of the iron curtain in 1989 immediately a much wider audience was reached by and interested in the Linz Seminar, and new funding possibilities became available which Peter Klement was able to use in a quite effective manner. Devoted mainly to *Applications of Logical and Algebraic Aspects of Fuzzy Relations* the 1990 seminar considerably enlarged its number of participants and joined researchers from the former East and West. With my personal focus on logic, let me mention F. Esteva, D. Mundici, and V. Novák, who for the first time attended the Linz Seminar in 1990.

Since, logical aspects of fuzzy sets and mathematical aspects of suitable non-classical logics have quite regularly been in the focus of the Linz Seminar. This was e.g. the case of the Linz Seminars of 1992 on *Non-Classical Logics and Their Applications*, of 1996 on *Fuzzy Sets, Logics, and Artificial Intelligence*, and again of 1997 on *Enriched Lattice Structures for Many-Valued and Fuzzy Logics*. These last two seminars did have P. Hájek among the participants—who, in those years, was in the main phase of his research for the logic BL of continuous t-norms presented 1998 in the seminal monograph [10].

This series of logic related Linz Seminars continued in 2000 with the topic of *Mathematical Aspects of Non-Classical Logics and Fuzzy Inference*, in 2005 with *Fuzzy Logics and Related Structures*, in 2010 with *Lattice-Valued Logic and its Applications*, and in 2014 with *Graded Logical Approaches and their Applications*. But also the Linz Seminars in the years 1993, 2004, and 2009 had logic as one of the fields of particular interest.

Thus, from the mid-1990 the Linz Seminar well mirrored the fact of increasingly strong relationships between fuzzy logics and mathematics.

4 Core Topics of the Development

What happened in fuzzy logic in that time, from a mathematical point of view?

4.1 T-Norm Based Logics

As already mentioned, it was clear from the mid-1980s that t-norm based logics offer a suitable framework for fuzzy set theory, and to identify the membership degrees with the truth degrees of such logics. And the standard understanding of the membership degrees of fuzzy sets also supported the choice of the truth degree 1 as the only designated one. There were, thus, two prominent particular cases for such logics: the infinite-valued versions L_∞ of the Łukasiewicz and G_∞ of the Gödel families of many-valued logics, cf. [11].

The context which was given by the t-norms as truth degree functions for non-idempotent conjunction connectives and offered, however, some different pathways for the introduction of further related connectives. Thus, t-conorms became natural candidates to determine non-idempotent disjunction connectives. Furthermore the fuzzy community agreed to have idempotent conjunction $\wedge$ and disjunction $\vee$ connectives with min, max, respectively, as truth degree functions, too.

There was, yet, no such standard choice neither for a negation nor for an implication connective. The matter is, however, not too difficult regarding candidates for generalized, many-valued negation functions $\mathbf{n}$. Standard properties should only be

$\mathbf{n}(0) = 1$ and $\mathbf{n}(1) = 0$ together with antitonicity. It seems, essentially, to be a matter of choice whether one additionally likes to have e.g. involutiveness $\mathbf{n}(\mathbf{n}(x)) = x$, or strict antitonicity, or continuity.

This situation triggered research which started at the one hand from de Morgan algebras to study their algebraic properties, and on the other hand from functional equations which reflect central logical laws of classical logic and study solutions in the context of de Morgan algebras. This area of research was and is carried through essentially by Spanish researchers like E. Trillas or L. Valverde, and has only occasionally been presented at the Linz Seminars (e.g. in 1981, 1982, and 1987). But these considerations have often been restricted to implication free fragments of propositional languages.

For suitably generalized implication connectives the matter proved to be more intricate.

The context of conjunction, disjunction, and negation connectives—provided by t-norms, t-conorms, and suitable negation functions—offers various possibilities to define implication connectives like in classical logic. The easiest way is to define an implication $\rightarrow$ by $\varphi \rightarrow \psi =_{df} \neg\varphi \vee \psi$ with the connectives $\vee$, $\neg$ characterized by a t-conorm and an involutive negation function, respectively. Such implication connectives are called S-implications.[1] The standard implication of the Łukasiewicz logic L_∞ has such a characterization.

But in the area of fuzzy relation equations another type of implication connective proved to be important to describe the inclusion-maximal solutions: so called R-implications[2] characterized for a t-norm $*$ by the *adjointness condition*

$$a * b \leq c \quad \text{iff} \quad a \leq b \Rightarrow c\,. \tag{1}$$

The standard implication of the Gödel logic G_∞ has such a characterization. In this case one has $* = \min$ and the R-implication coincides with the relative pseudo-complement in the complete lattice ([0, 1], min, max, 0, 1).

Interestingly the standard implication of the Łukasiewicz logic L_∞ is also an R-implication. And because of the connections between R-implications and fuzzy relation equations, those R-implications had also been used in the previously mentioned naive approaches toward t-norm based logics and related fuzzy set theoretic developments [6, 9].

Nevertheless it seems to be a kind of competitive situation between t-norm based logics with R-implications, and such ones with S-implications. And indeed, the latter situation was studied in [12]. However, the mainstream development has focussed on t-norm based logics with R-implications. It seems that the crucial point for this decision was the fact that for R-implications the rule of detachment has a simple and convincing argument for its correctness in the fact that the formula

[1] This comes from the fact that in early phases of these considerations t-conorms also had been discussed under the name "S-norms".

[2] The name derives from the algebraic operation of residuation.

$$\varphi \,\&\, (\varphi \Rightarrow \psi) \Rightarrow \psi \tag{2}$$

is logically valid. With $\Rightarrow$ read as an S-implication, however, this formula fails to be logically valid.

The t-norm based residuated logics had in the beginning 1990s been developed only on an intuitive level: there existed axiomatizations only for two particular cases: Łukasiewicz logic L_∞ and Gödel logic G_∞. What was additionally known was the fact that the choice of an R-implication, i.e. the acceptance of the adjointness condition (1) forced the t-norm T involved in this condition to be left-continuous, i.e. to have all its unary parameterizations $T_a(x) = T(a, x)$ as left-continuous functions.

A first breakthrough came from U. Höhle who gave in [13, 14] an adequate axiomatization of a logic ML which was characterized by an algebraic semantics constituted by the class of all integral commutative residuated lattice ordered monoids, and who claimed that this should be the formalization of fuzzy logic. Important results on these algebraic structures U. Höhle had presented at the 1992 Linz Seminar. These integral commutative residuated lattice ordered monoids proved to be an important specification of the commutative lattice ordered semigroups which Goguen [15] had proposed as suitable algebraic structures for membership degrees of fuzzy sets.

In the beginning 1990s also P. Hájek got interested in the topic of logics related to fuzzy sets. I remember that, after a colloquium talk he had given at Leipzig University, he asked me with reference to the German language forerunner [16] of [11] whether I had ever thought about an approach toward a product-based infinite-valued logic similar to C.C. Chang's approach [17] toward (a completeness proof for) Łukasiewicz logic L_∞ using MV-algebras.

Such a product logic Π, i.e. a residuated t-norm based logic with the arithmetic product as basic t-norm, was presented in [18]. Its algebraic characterization by the class of all product algebras proved to be prototypical for P. Hájek's later approach toward the logic BL of all continuous t-norms.

Because nobody saw a possibility to axiomatize residuated t-norm based logics in general, the idea of P. Hájek was to axiomatize the common logic of all continuous t-norms. P. Hájek's restriction to continuous t-norms came from his conviction that for application only continuous t-norms should be relevant.

For this approach it was substantial that an algebraic characterization of the continuity of t-norms by the *divisibility* condition

$$a \wedge b = a * (a \Rightarrow b) \tag{3}$$

was known from [14]. And it was equally important that P. Hájek restricted the class of integral commutative residuated lattice ordered monoids to those ones which additionally satisfied the *prelinearity* condition

$$(a \Rightarrow b) \vee (b \Rightarrow a) = 1\,, \tag{4}$$

called *algebraic strong de Morgan law* in [14]. Since then, these structures are known as *BL-algebras*.

In the Linz Seminars of 1996 and 1997 P. Hájek presented core ideas and results of his approach.

4.2 Graded Notions of Consequence

Another stream in the area of fuzzy sets related logics started in 1979 with J. Pavelka's discussion [19] of many-valued propositional logics with graded notions of consequence, i.e. of logics which allowed to consider syntactic as well as semantic consequence hulls of fuzzy sets of formulas. The general context was the one offered by Goguen's commutative lattice ordered semigroups with residuation added.

The semantic consequence hull $\mathsf{Cn}^*_{\models}(\Sigma)$ of a fuzzy set Σ of formulas is defined in a rather standard way with reference to models of Σ. In this context, a [0, 1]-evaluation $\mathbf{e}$ is called a *model* of a fuzzy set Σ of formulas iff for each formula φ one has

$$\text{membership degree of } \varphi \text{ in } \Sigma \;\leq\; \mathbf{e}(\varphi) \tag{5}$$

for the truth degree $\mathbf{e}(\varphi)$ of φ under $\mathbf{e}$.

Similar as in classical model theory one can connect with each [0, 1]-evaluation $\mathbf{e}$ a fuzzified theory $\mathsf{Th}(\mathbf{e})$ of $\mathbf{e}$ in choosing for each formula φ the truth degree $\mathbf{e}(\varphi)$ as the membership degree of φ in $\mathsf{Th}(\mathbf{e})$. These notions allow to define in a natural way

$$\mathsf{Cn}^*_{\models}(\Sigma) \;=_{\text{df}}\; \bigcap \{\mathsf{Th}(\mathbf{e}) \mid \mathbf{e} \text{ model of } \Sigma\}\,. \tag{6}$$

A corresponding syntax for such a graded notion of consequence has to cope with membership degrees. Therefore the propositional language of [19] was enriched with truth degree constants for each possible membership/truth degree, i.e. with constants for each real out of [0, 1]. Furthermore, derivations with a fuzzy set Σ of premisses have to treat formulas and degrees in parallel, and so have to act inference rules. As a result, each derivation is a derivation of a formula φ to some degree a, the *proof degree* of φ from Σ for that particular derivation. Because formulas may have different derivations, they may have different proof degrees in this context. The *provability degree* $\mathsf{pr}_{\Sigma}(\varphi)$ of φ from Σ then is the supremum of all possible proof degrees of φ from Σ.

And the syntactic consequence hull $\mathsf{Cn}^*_{\vdash}(\Sigma)$ of a fuzzy set Σ of formulas is defined by the condition

$$\text{membership degree of } \varphi \text{ in } \mathsf{Cn}^*_{\vdash}(\Sigma) \;=_{\text{df}}\; \mathsf{pr}_{\Sigma}(\varphi)\,. \tag{7}$$

A general *completeness theorem* then is the statement $\mathsf{Cn}^*_{\models}(\Sigma) = \mathsf{Cn}^*_{\vdash}(\Sigma)$ and could be proved by J. Pavelka [19] only for Łukasiewicz logic L_∞ as basic system. The reason is that the completeness proof needs the continuity of the residuation operation, and only the Łukasiewicz t-norm and its isomorphic versions have continuous R-implications [20].

This line of research was extended to first-order logic by V. Novák [21]. He attended the Linz Seminar, beginning in 1990, quite often and presented various of his theoretical achievements, e.g. about model theoretic results, but also applications e.g. to problems in theoretical linguistics like the modeling of intermediate quantifiers [22].

Also in this first-order case a general completeness theorem $\mathsf{Cn}^*_{\models}(\Sigma) = \mathsf{Cn}^*_{\vdash}(\Sigma)$ can be proved only for Łukasiewicz logic L_∞ as basic system–the proof again needs the continuity of the R-implication.

The whole problem of graded notions of consequence can also be treated more algebraically, having in mind that for classical logic there is a strong relationship between consequence relations and closure operators in the class of all sets of formulas.

Such a treatment needs the reference to closure operators in the class of all fuzzy sets of formulas. They are defined in the standard way, i.e. via increasingness, monotonicity, and idempotency and need only the reference to the (binary) inclusion relation for fuzzy sets. To a large extent this approach was studied by G. Gerla [23] who attended the Linz Seminar e.g. in 1996 and presented basic ideas of this approach.

However, this more algebraic way toward graded notions of consequence does not give more general results: a completeness theorem results again only for the case of Łukasiewicz logic L_∞ as basic system.

4.3 *Lattice-Valued Structures and Category Theory*

The focus of the considerations in the field of mathematical fuzzy logics toward residuated lattice-ordered monoids forced also other investigations into the theory of fuzzy sets with membership degrees in such structures or in similar ones.

The approaches toward lattice-valued mathematics brought together earlier investigations on fuzzy topologies and on category theoretic treatments of fuzzy set matters which had been main topics e.g. of the 1989 Linz Seminar on *Applications of Category Theory to Fuzzy Subsets* [24]. They have repeatedly been core topics for the Linz Seminars, e.g. in 1993, 1997, 2004, and 2010, and they somehow culminated in the 2012 Seminar on *Enriched Category Theory and Related Topics* documented in [25]. Enriched categories have hom-sets which themselves are structured, e.g. as residuated lattices. Enriched category theory can be used to understand the underlying structure of fuzzy set theory as a particular monoidal closed category. From this point of view, the grading membership functions become enriched presheaves.

The understanding of fuzzy sets as particular presheaves works, however, also in the simpler context of usual categories. Also this point of view was quite often discussed in the Linz Seminar and gave rise to M-valued sets as early as in [26], and in this context to the understanding of the graded self-identity as measuring an extent of existence [27, 28].

This notion of M-valued set touches a further important problem for many-valued and fuzzy logics: *graded identities*. Such an M-valued set is an ordered pair $A = (|A|, \delta_A)$ consisting of a crisp set $|A|$ and a graded, i.e. M-valued (local) equality relation δ_A satisfying

(E1): $\delta_A(x, y) \leq \delta_A(x, x) \wedge \delta_A(y, y)$, *strictness*
(E2): $\delta_A(x, y) = \delta_A(y, x)$, *symmetry*
(E3): $\delta_A(x, y) * (\delta_A(y, y) \to \delta_A(y, z)) \leq \delta_A(x, z)$. *transitivity*

The degree of self-identity $\delta_A(x, x)$ is understood as *degree of extent* for x in A, or also as *degree of existence*, as was done for the Heyting algebra valued case in [29].

The problem of graded identities is a general problem in many-valued first-order logics. It is not yet solved and has important philosophical aspects which, from this author's point of view, are still quite incompletely understood.

For the case of the finitely-valued Łukasiewicz logics it had already in 1958 by H. Thiele [30] been shown that the usual Leibniz principle of substitutivity of equals forces the equality relation to be crisp. A weakening of that principle, however, allows for graded identities, as shown by this author in [31], cf. also [11].

Also for fuzzy sets a graded identity defined with respect to a graded membership predicate ε by

$$x \equiv y =_{\mathrm{df}} \forall z(z\varepsilon x \to z\varepsilon y) \ \& \ \forall z(z\varepsilon y \to z\varepsilon x) \tag{8}$$

appears quite naturally and is used e.g. in [7, 32].

But there is also a competing possibility. P. Hájek [33] offered, expanding an early approach [34] by Th. Skolem, an "almost naive" axiomatic set theory in the realm of Łukasiewicz logic, having unrestricted comprehension as its sole axiom schema. This Cantor-Łukasiewicz fuzzy set theory **CŁ** gets two identity relations, extensional equality $\approx$ and Leibniz equality $=$ defined by

$$x \approx y =_{\mathrm{df}} \forall z(z\varepsilon x \leftrightarrow z\varepsilon y)\,, \tag{9}$$

$$x = y =_{\mathrm{df}} \forall u(x\varepsilon u \leftrightarrow y\varepsilon u)\,. \tag{10}$$

From those identity relations, Leibniz equality is crisp and extensional equality really graded.

To a large extent these results can also be generalized to the setting with the logic MTL of all left-continuous t-norms as background logic, as discussed in [35].

This topic of graded identities occasionally occurred during the Linz Seminars, but never was a core topic there. It is not only a mathematical problem.

5 Concluding Remarks

Of course, it is nearly impossible to cover with such a series like that of the Linz Seminars all the main developments. However, if one tries to find rather important gaps, only proof theory comes into the focus. For long, proof theory–understood here as the field of sequent and tableau calculi–for many-valued and fuzzy logics was not a well developed and really active field. Besides some isolated research results often restricted to finitely-valued logics, explained e.g. in [11], prior to the end of the 1990s there had not been published important papers.

The situation changed as hypersequents, i.e. finite multisets of sequents as introduced by A. Avron [36], became known as a useful tool for fuzzy logics. M. Baaz et al. [37] indicate the first steps into this new field, and the monograph [38] documents the amount of work done in the first decade of this century.

There are other areas, related to mathematical fuzzy logics but more application oriented, which have not been discussed here, but have been present at the Linz Seminars to at least some degree. Here I have in mind the field of approximate reasoning, and particularly possibility theory, which was often discussed in Linz Seminars devoted to more applied fields of mathematics.

Summing up and reconsidering all the 35 *Linz Seminars on Fuzzy Set Theory* from 1979 till 2014, all of them locally well organized by Peter Klement, it is remarkable that most of the essential topics in the development of mathematically oriented fuzzy logics have been presented and actively supported by those seminars.

References

1. Albrycht, J., Wiśniewski, H. (eds.): Proceedings of the Polish Symposium on Interval & Fuzzy Mathematics, Poznań. Wydawn Politech Pozn (1983)
2. Bocklisch, S., Orlovski, S., Peschel, M., Nishiwaki, Y. (eds.): Fuzzy sets applications, methodological approaches, and results. In: Mathematische Forschung, vol. 30. Akademie-Verlag, Berlin (1986)
3. Gottwald, S.: A cumulative system of fuzzy sets. In: Marek, W., Srebrny, M., Zarach, A. (eds.) Set Theory Hierarchy Theory, Mem. Tribute A. Mostowski, Bierutowice (1975) Lecture Notes In Mathematics, vol. 537, pp. 109–119. Springer, Berlin (1976)
4. Gottwald, S.: Set theory for fuzzy sets of higher level. Fuzzy Sets Syst. **2**, 125–151 (1979)
5. Gottwald, S.: Fuzzy uniqueness of fuzzy mappings. Fuzzy Sets Syst. **3**, 49–74 (1980)
6. Gottwald, S.: T-Normen und φ-Operatoren als Wahrheitswertfunktionen mehrwertiger Junktoren. In: Wechsung, G. (ed.) Frege Conference 1984 (Schwerin, 1984). Mathematical Research, vol. 20, pp. 121–128. Akademie-Verlag, Berlin (1984)
7. Gottwald, S.: Fuzzy set theory with t-norms and φ-operators. In: Di Nola, A., Ventre, A.G.S. (eds.) The Mathematics of Fuzzy Systems. Interdisciplinary Systems Research, vol. 88, pp. 143–195. TÜV Rheinland, Köln (Cologne) (1986)
8. Gottwald, S.: Generalized solvability criteria for fuzzy equations. Fuzzy Sets Syst. **17**, 285–296 (1985)
9. Gottwald, S.: Characterizations of the solvability of fuzzy equations. Elektron. Informationsverarbeitung Kybernetik **22**, 67–91 (1986)

10. Hájek, P.: Metamathematics of Fuzzy Logic. In: Trends in Logic, vol. 4. Kluwer Acad. Publ, Dordrecht (1998)
11. Gottwald, S.: A Treatise on Many-Valued Logics. In: Studies in Logic and Computation, vol. 9. Research Studies Press, Baldock (2001)
12. Butnariu, D., Klement, E.P., Zafrany, S.: On triangular norm-based propositional fuzzy logics. Fuzzy Sets Syst. **69**, 241–255 (1995)
13. Höhle, U.: Monoidal logic. In: Kruse, R., Gebhard, J., Palm, R. (eds.) Fuzzy Systems in Computer Science. Artificial Intelligence, pp. 233–243. Verlag Vieweg, Wiesbaden (1994)
14. Höhle, U.: Commutative, residuated l-monoids. In: Höhle, U., Klement, E.P. (eds.) Non-Classical Logics and Their Applications to Fuzzy Subsets. Theory and Decision Library Series B, vol. 32, pp. 53–106. Kluwer Acad. Publ., Dordrecht (1995)
15. Goguen, J.A.: The logic of inexact concepts. Synthese **19**, 325–373 (1968–69)
16. Gottwald, S.: Mehrwertige Logik. Logica Nova. Akademie-Verlag, Berlin (1989)
17. Chang, C.C.: Algebraic analysis of many valued logics. Trans. Am. Math. Soc. **88**, 476–490 (1958)
18. Hájek, P., Lluís, G., Francesc, E.: A complete many-valued logic with product-conjunction. Arch. Math. Log. **35**, 191–208 (1996)
19. Pavelka, J.: On fuzzy logic. I–III. Zeitschr. math. Logik Grundl. Math. **25**, 45–52, 119–134, 447–464 (1979)
20. Jayaram, B.: On the continuity of residuals of triangular norms. Nonlinear Anal. **72**, 1010–1018 (2010)
21. Novák, V.: On the syntactico-semantical completeness of first-order fuzzy logic. I: Syntax and semantics. II: Main results. Kybernetika **26**, 47–66, 134–154 (1990)
22. Novák, V.: A formal theory of intermediate quantifiers. Fuzzy Sets Syst. **159**(10), 1229–1246 (2008)
23. Gerla, G.: Fuzzy logic. Mathematical Tools for Approximate Reasoning. In: Trends in Logic, vol. 11. Kluwer Academic Publishers (2001)
24. Rodabaugh, S.E., Klement, E.P., Höhle, U. (eds.): Applications of Category Theory to Fuzzy Subsets. Kluwer Acad. Publ, Dordrecht (1992)
25. Höhle, U., Klement, E.P. (eds.): Fuzzy Sets and Systems, vol. 256 (2014)
26. Höhle, U.: M-valued sets and sheaves over integral commutative CL-monoids. In: Rodabaugh, S.E., et al. (eds.) Applications of Category Theory to Fuzzy Subsets, Theory and Decision Library Series B, vol. 14, pp. 34–72. Kluwer Acad. Publ., Dordrecht (1992)
27. Höhle, U.: Many valued logic and sheaf theory. Sci. Math. Japon. **68**(3), 417–433 (2008)
28. Höhle, U.: Many-valued equalities and their representations. In: Klement, E.P., Mesiar, R. (eds.) Logical, Algebraic, Analytic, and Probabilistic Aspects of Triangular Norms, pp. 301–319. Elsevier, Dordrecht (2005)
29. Scott, D.S.: Identity and existence in intuitionistic logic. In: Fourman, M.P., Mulvey, C.J., Scott, D.S. (eds.) Applications of Sheaves. Lecture Notes in Mathematics, vol. 753, pp. 660–696. Springer, New York (1979)
30. Thiele, H.: Theorie der endlichwertigen Łukasiewiczschen Prädikatenkalküle der ersten Stufe. Zeitschr. math. Logik Grundl. Math **4**, 108–142 (1958)
31. Gottwald, S.: A generalized Łukasiewicz-style identity logic. In: de Alcantara, L.P. (ed.) Mathematical Logic and Formal Systems. Lecture Notes Pure Applied Mathematics, vol. 94, pp. 183–195. Marcel Dekker, New York (1985)
32. Gottwald, S.: Fuzzy Sets and Fuzzy Logic. Artificial Intelligence. Verlag Vieweg, Wiesbaden, and Tecnea, Toulouse (1993)
33. Hájek, P.: On equality and natural numbers in Cantor-Łukasiewicz set theory. Log. J. IGPL **21**(3), 91–100 (2013)
34. Skolem, Th.: Bemerkungen zum Komprehensionsaxiom. Zeitschr. math. Logik Grundl. Math. **3**, 1–17 (1957)
35. Běhounek, L., Haniková, Z.: Set theory and arithmetic in fuzzy logic. In: Montagna, F. (ed.) Petr Hájek on Mathematical Fuzzy Logic. Outstanding Contributions to Logic, vol. 6, pp. 63–89. Springer, Switzerland (2015)

36. Arnon, A.: Hypersequents, logical consequence and intermediate logics for concurrency. Ann. Math. Log. AI **4**, 225–248 (1991)
37. Baaz, M., Ciabattoni, A., Fermüller, C., Veith, H. (eds.): Proof theory of fuzzy logics: urquhart's C and related logics. In: Mathematical Foundations of Computer Science. Lecture Notes in Computer Science, vol. 1450, pp. 203–212. Springer, Berlin (1998)
38. Metcalfe, G., Olivetti, N., Gabbay, D.: Proof theory for product logics. Neural Netw. World **13**, 549–558 (2003)

How I Saw, and How I See Fuzzy Sets

Enric Trillas

Abstract This paper does not pretend a 'technical' presentation of a particular topic with an exhausting list of references; it just would like to contain some reflections of the author concerning how he sees, or better, he wishes, the future of current fuzzy logic that, in his view and at the risk of stagnation, cannot lie on any kind of 'logicism' but on 'scienticism'.

1 Introduction

It can be said that as it was originally introduced by Zadeh [16], fuzzy set theory mainly deals with two important linguistic phenomena, imprecision and non-random uncertainty, and that fuzzy sets can be applied, among others, to the study of dynamical systems whose behavior can be described by sets of imprecise linguistic rules, and to the random uncertainty associated to some linguistic statements [7, 13]. For instance, the theory of possibility can deal with non-random uncertainty, fuzzy control with dynamical systems, and fuzzy probability with random fuzzy events.

The ground of fuzzy set theory lies in the, historically not surprising, fact that predicates acting in a universe of discourse generate linguistic collectives in it; collectives [14], except when they degenerate in just a single classical set, are cloudy linguistic entities neither well known, nor easy to specify virtual or 'thought' entities whose appearances, or states, are just membership functions, fuzzy sets allowing to see their projections inside the fog of ordinary language. Hence, fuzzy sets can be seen as a starting point for the currently non existing scientific study of linguistic collectives. In sum and grossly speaking, fuzzy sets deal with ordinary language; they are mathematical entities contextually reflecting collectives, and modeled by their membership functions. They meant to pass from an old world of exact thinking

To Professor Peter Klement, with deep affection.

E. Trillas (✉)
European Centre for Soft Computing, Mieres, Asturias, Spain
e-mail: enric.trillas@softcomputing.es

S. Saminger-Platz and R. Mesiar (eds.), *On Logical, Algebraic, and Probabilistic Aspects of Fuzzy Set Theory*, Studies in Fuzziness and Soft Computing 336,
DOI 10.1007/978-3-319-28808-6_2

represented by sets, to a new world of approximate thinking represented by them. The future of fuzzy sets can be seen around a new mathematical study of ordinary, or common sense, reasoning in which the central idea is, instead of 'deducing' from precise premises reflecting totally known information even if not fully describing something, to that of 'conjecturing' [9] from imprecise premises reflecting information partially known and also able to reach creative conclusions. That is, to increase the informative content of the premises or previous information; in sum, to be in touch with creativity.

Fuzzy sets have to do with both the representation of information, and to how new one can be obtained by just a 'previous thinking' as it is done always in searching for a new aspect of a problem, and that, latter on, should be either formalized, or checked against some reality to acquire the status of 'new' knowledge. Of course, in these processes of conjecturing those of deducing, abducing, and also lucubrating are included [11].

Without no doubt, it can also be said that the idea of fuzzy sets was born in the 'cultural' neighborhood of cybernetics, where analogical computers [1] were seriously taken into account. Fuzzy sets can be seen indeed as 'analogical entities' in contrast to the 'digital crisp sets' and, since most of the human knowledge is essentially analogical, it is not at all surprising that fuzzy sets can be suitable for representing, at least, expert knowledge. In fact, the first application of fuzzy sets to the control of machines, introduced in 1972 by the late Abe Mamdani [5], can be considered as a method for the management of imprecise expert knowledge, and who knows if, in a future, and provided analogical quantum computers [15] were actually constructed, fuzzy sets will not play some new role in their functioning. If from a philosophical and scientific point of view fuzzy sets are but measures, from a technological one they are just analogical tool constructs representing knowledge.

2 My First Steps into Fuzzy Logic

I entered into fuzzy logic by chance. It was through an interview, in a French newspaper, with the late professor Arnold Kaufmann in which he spoke on his then recently appeared book 'Ensembles flous'. The subject interested me since I was doing my research work on the probabilistic metrics introduced by Karl Menger, and knew his paper entitled 'Ensembles flous', a new concept that he translated into English by hazy sets. I bought Kaufmann's book, read a good part of it, and to some extent I was actually disappointed; my first glance at fuzzy sets make me to believe that they were just a simple generalization of sets. Nevertheless, since some of the examples in the book called upon my attention and made me curious, I decided to read the 1965 paper where fuzzy sets were originally introduced, whose title is 'Fuzzy Sets' and was written by Zadeh [16]. Before reading this paper, and as a consequence of both my mathematical formation, and the reading of Menger's paper, I was unable to see fuzzy sets unlinked with probability; indeed, hazy sets represent something like the probability of an element belonging to a set.

But the reading of Zadeh's paper suddenly changed my view. The subject was towards representing the multitude of imprecise predicates of which language is full; it was, for me, the first mathematical model for taking into account the imprecision that, permeating language, affects ordinary reasoning with concepts that are not definable like those managed in mathematics by 'if and only if' conditions, but only describable from their use in several contexts as they appear in dictionaries. It was for me a new land to be explored, and I was captivated as I see the possibility of building up mathematical models of common reasoning. I decide to start with such an exploration! At the end, in that time I was unhappy with the 'bourbakism' of which mathematics was full in Spain, and I was also worried by the giving up that, from time ago, logicians kept on ordinary reasoning.

Although the only references in the first Zadeh's paper on fuzzy sets are the purely mathematical books by Birkhoff, Halmos, and Kleene, and since from very young I kept a deep interest in Bertrand Russell's philosophical writings, I remembered the Russell's paper 'On Vagueness' and believed that there should be some links between fuzzy sets and vagueness. This idea conducted me to the 1972 paper by Aldo De Luca and Settimo Termini, where they established the then new concept of a 'fuzzy entropy', and I thought it is nothing else than a measure of the vagueness or, by duality of its classicality, booleanity, or crispness, the linguistic label of a fuzzy set presents. This idea make me to think that 'fuzziness' is just a restriction of the vagueness of a predicate whenever it can be represented by a fuzzy set, and this was for me a challenging philosophical idea that, many years ahead, conducted me to see that fuzzy sets are nothing else than measures of the meaning of predicates. My first papers on fuzzy sets dealt, between 1976 and 1978, with trying to find functionally expressible mathematical formulas able to represent fuzzy entropies, but different and more general than the unique logarithmic fuzzy entropy shown by De Luca and Termini in his paper. In addition, I also tried to relate them with the Sugeno's fuzzy integral since, in the meantime, I was acquainted with Michio Sugeno in Toulouse, and with his 1974 Ph.D. Thesis.

Early after these worries, I began to be interested in the subject of fuzzy connectives and fuzzy inference. For the first I was essentially motivated by the fact that fuzzy connectives can't show the same properties in all contexts, and that distributivity is, for instance, a very constraining and crisp property. What conducted me in such direction were the papers by Bellman and Giertz [2], and that on negation by Lowen [4]. My dedication to Probabilistic Metric Spaces, that bring me to know Bert Schweizer and Able Sklar after meeting Karl Menger in Chicago, introduced me to solve Functional Equations, and I got the idea of characterizing the (continuous) strong negations by just solving an easy functional equation. Since I was familiar with Schweizer and Sklar's t-norms (a restriction by adding associativity to those introduced by Menger [6]), I can introduce in fuzzy logic these ordered semi-groups in the unit interval. Finally, the fact that as examples of his Compositional Rule of Inference, Zadeh did show some that were non-preserving the classical Modus Ponens when the input is just the antecedent of the rule, I tried to study this 'Modus' in fuzzy logic by formulating it as Hardegree did in Orthomodular lattices [3].

To end this section, that corresponds with the time in which I met for the first time professor Peter Klement and in Barcelona, let me remember again that, as another consequence of Menger's work, but due to the idea of mathematically modeling the breaking of synonymous chains, I introduced in fuzzy logic the T-indistinguishabilities, or T-equivalences, allowing to relating such problem with that of Poincaré concerning the physical continuum. I would like to say that Menger's trace in fuzzy logic or, at least, in my contributions to it, is certainly of some relevance.

3 Zadeh's Fuzzy Sets Are but Measures of Meaning

For a lot of time after 1965, the mathematical nature of fuzzy sets in relation with the meaning of their linguistic label, was not clearly explained. They were simply viewed as membership functions generalizing the characteristic function of crisp sets and, supposedly, representing its meaning in the universe of discourse but without counting with a meaning's operational description [8]. If philosophers largely debated on the meaning of 'meaning', they never attended the representation of meaning, and it lacked a scientific study that today can be considered started with the work of Zadeh, and in a form close to the Wittgenstein of the 'Philosophical Investigations', when he states that almost always 'the meaning of a word is its use in language'. How can even if not defined, the use or management in language of a linguistic label be mathematically described?

If P is a linguistic label, or predicate, acting in a universe of discourse X through the elemental statements 'x is P', for a suitably management of P the two binary relations in X, that empirically come from linguistic perception, from its use,

- $x =_P y \Leftrightarrow x$ shows the property named P equally than y shows it $\Leftrightarrow x$ is equally P than y,
- $x \leq_P y \Leftrightarrow x$ is less P than y,

should be known [2, 14]. When both relations $\leq_P$ and $=_P$ do coincide, it is said that the use of P in X is *precise*, *rigid*, or *crisp*, $=_P$ is an equivalence, and X is partitioned in the equivalence classes in the quotient set $X/=$, $[x] = \{y \in X; y =_P x\}$. Instead and when $\leq_P \neq =_P$ that, provided it can be supposed $=_P = \leq_P \cap \leq_P^{-1}$, implies that it is not $\leq_P \subseteq \leq_P^{-1}$, it is said that the use of P in X is *imprecise, flexible,* or *fuzzy*. In any case, the graph $(X, \leq_P)$ represents the *qualitative*, or *primary*, meaning of P in X. In this way, the previously amorphous universe of discourse X, is softly structured thanks to the use of P in it. The simple and usual act of 'speaking' on a property recognizable in the elements of X, endows X with the arcs of this graph; an idea corresponding with the intuitive one that rational speech tries to introduce some kind of 'ordering' in the universe of discourse, also corresponding to the establishment of some necessary link between ordering and understanding. Nevertheless, the graph does not exhaust the 'full meaning' of P in X, and when it is $\leq_P = \emptyset$, it can be said that P is metaphysically used in X, that P is *metaphysical*, or *meaningless* in X [11]. Notice that it is thanks to the relation $\leq_P$ that can be seen the variability of the property named P along the elements of X; that it is $\leq_P \neq =_P$, is what permits to say that the use of P is imprecise in X.

If P is not metaphysically used in X, that is, if $\leq_P \neq \emptyset$, then a *measure* of the extent of P in X, is a mapping μ_P: $X \to [0, 1]$, such that

(1) $x \leq_P y \Rightarrow \mu_P(x) \leq \mu_P(y)$
(2) z maximal for $\leq_P \Rightarrow \mu_P(z) = 1$
(3) z minimal for $\leq_P \Rightarrow \mu_P(z) = 0$.

Once the graph $(X, \leq_P)$ is known it can be said that P is *measurable* in X, and once a measure μ_P is known that it is *effectively measurable* in X [14].

The three former properties are not sufficient, in general, to specify a measure, but there is only a single one if the predicate is precise; to specify a measure either more information on the use of P, or to establish a reasonable hypothesis on it, is necessary.

In any case, each measure μ_P is the membership function of a fuzzy set in X labeled P. Fuzzy sets are defined by the measures of the extent up to which the elements in X are P show the property named P; shortly speaking it can be said that fuzzy sets are measures of meaning, like probabilities are measures of random uncertainty, and fuzzy entropies are measures of fuzziness. It should be noticed that each quantity $(X, \leq_P, \mu_P)$ represents a good enough knowledge on the meaning of P in X for its scientific consideration; it can be said that such quantities are the typically scientific *domestication* of meaning [10], and can offer a new perspective for studying both fuzzy sets and fuzzy logic.

It should be noticed that if the use of P in X is rigid, it is $x =_P y \Leftrightarrow x \leq_P y$ & $y \leq_P x \Rightarrow \mu_P(x) = \mu_P(y)$, and hence, μ_P is constant in the classes modulo P, the only values μ_P can take are 0 or 1, $\mu_P^{-1}(1)$ is the crisp subset specified by P in X, $\mu_P^{-1}(0)$ its classical complement, and one of them can be empty.

In praxis, a fuzzy set is designed by means of the information on its linguistic label that is available and that, most of the times, is not the full relation $\leq_P$, but a part of it; there are cases in which obtaining $\leq_P$ can be very difficult. Hence and very often, neither it is always $\leq_P$ completely known, nor it can be stated that the designed membership function μ_P^* is truly a measure, but some unknown approximation to it. Consequently, the designer cannot work with $\leq_P$ but only with the total order defined by x $\leq_{\mu_P^*} y \Leftrightarrow \mu_P^*(x) \leq \mu_P^*(y)$, called the *working meaning* of P in X. Provided μ_P^* were actually a measure or, at least, it can be supposed it verifies property (1), and since then, $x \leq_P y \Rightarrow \mu_P^*(x) \leq \mu_P^*(y) \Leftrightarrow x \leq_{\mu_P^*} y$, implies $\leq_P \subseteq \leq_{\mu_P^*}$, that is, the working meaning extends the qualitative meaning of P. The act of measuring P, modifies its qualitative meaning by adding more arcs to it [14].

Notice that since in most cases the relation $\leq_P$ has not a total, or linear, character, it cannot coincide with the linear orders $\leq_{\mu_P}$. When there is coincidence, it is said that the measure *perfectly reflects* the qualitative meaning of P. It is easy to proof that, provided $\leq_P$ is reflexive and transitive, then $=_P$ is an equivalence relation, and that the mapping $C : X \to X/=_P$, assigning to each x the equivalence class $[x] = C(x)$, verifies $x \leq_P y \Leftrightarrow [x] \leq_P^* [y] \Leftrightarrow C(x) \leq_P^* C(y)$. Hence, not only C

perfectly reflects the qualitative meaning of P in X, but this idea opens the door for defining 'qualitative measures' of a predicate by taking, instead of the unit interval, some non-numerical posets with which more possibilities of perfectly reflecting the qualitative meaning could appear.

4 On the Other Types of Fuzzy Sets

In those cases in which the measure does not perfectly reflect the qualitative meaning, and since in science is not at all rare to manage measures with complex values, it could be suitable to substitute the real interval [0, 1] by the complex one $\{a + bi; a, b \in [0, 1]\}$, the complex circle, endowed with the usual partial order $a_1 + b_1 i \leq a_2 + b_2 i \Leftrightarrow a_1 \leq a_2$ & $b_1 \leq b_1$, and with analogous properties [14] to the former (1), (2), and (3). This substitution cannot guarantee that a complex-valued measure will perfectly reflect the qualitative meaning, but just that it can offer more possibilities for it, since the working order will be not linear. This substitution that can be equivalently seen by taking an interval-valued measure, just changing a+bi by the interval $[a, b]$, and corresponding to a particular type of the so-called type-2 fuzzy sets reflecting that the value of the measure carries with the uncertainty coming from only being sure that it is in the interval $[a, b]$.

Analogously, and instead of the real or the complex unit intervals, it can be taken the set $[0, 1]^{[0,1]}$, of the fuzzy sets in the unit interval (type-2 fuzzy sets) that contains images isomorphic to both the unit interval and the complex unit interval, and for those cases in which the only that can be asserted is that the value of the measure is, for instance, either 'around 0.7', or 'high' [12]. In this form, all the types of fuzzy sets currently considered, are integrated thanks to the quantities, either numerical or functional, representing the meaning of its linguistic label.

The full meaning of a linguistic label P is not unique, but it is actually context-dependent and purpose-driven. Each quantity $(X, \leq_P, \mu_P)$, real, complex or fuzzy valued, is obtained through what the designer can know, in a given context, of the use, action or behavior of P in X, or through some reasonable hypothesis he could be able to make on such behavior. This last is the often considered case in most applications, in which the real-valued measure, the membership function, is supposed to be trapezoidal, or just triangular.

Once seen that the membership functions of fuzzy sets mean nothing else than a 'measure of the meaning' of its linguistic label, it can be remembered the famous words of Lord Kelvin shortened to 'If you cannot measure it, it is not science'. There are, notwithstanding and at least, two aspects introducing important differences between Lord Kelvin's times and ours. In the first place, it is the fact that if, lets it say, science is essentially concerned with matter and energy, fuzzy set theory is concerned with knowledge and information, and directly related with the so called Information Technologies. In a second place, in Lord Kelvin's science there were and are known systematic procedures and laboratory methods, to measure the basic parameters of the studied things, but now and for what concerns, for instance, the design of membership functions, the situation is different and more linked to some analogy with virtual objects, than with physically real objects. It is not the same to study the chemical composition of an organic product, or the movement of a star, than to study the meaning of a written piece, or the control of a machine whose behavior is known by the knowledge of the experts in their functioning and once linguistically described. Nevertheless, this is the kind of problems currently worrying Artificial Intelligence.

5 The Evoluation of Fuzzy Logic

Anyway, the evolution of fuzzy logic towards Zadeh's Computing with Words and Perceptions, CwW for short [17], is conducting towards the mathematical representation of statements larger and more complex than the more or less simple rules considered in fuzzy control [8]. This will mean to face with the necessity of considering different ways of expressing conditional statements, and the already known linguistic connectives 'and', 'or', 'not', etc., since there is not a universal form of expressing them in language, like it is in classical logic and set theory, but respectively represented in fuzzy logic by residuated implications, S-implications, conjunctive implications, t-norms, t-conorms, negation functions, etc. For all that there are a lot of mathematical models facilitating to fuzzy logic a remarkable armamentarium for the representation of statements, and for doing deductive inferences with them, but what is not yet clear enough are the linguistic subjects to which such armamentarium is applicable, and to which is not. For instance, fuzzy logic only considers functionally expressible connectives, but no suitable criteria are known for recognizing this hypothesis in concrete cases, and, analogously, the use of non strong but continuous negations is not yet spread into fuzzy logic applications to represent language. Even more, almost always the used connectives are min, max, prod, and 1-id; there exists a big separation between what is employed by practitioners of fuzzy logic, and what is kept in the theoretical armamentarium generated by mathematicians.

In sum, it seems that fuzzy logic is approaching the time in which it should face a turning point. The great subjects fuzzy logic deals with are linguistic imprecision and non-random uncertainty, not to say anything on the very important but scientifically almost pending subjects of ambiguity, the presence of multiple meanings, and common sense non-deductive reasoning [11] with imprecise, non-randomly uncertain, and ambiguous words.

The only way to properly afford it is, in the author's view, the transformation of fuzzy logic in a kind of 'physics' of imprecision, non-random uncertainty and ambiguity. That is, in a new experimental science that, based in Natural Language, can count with mathematical models able to give important parameters to be experimentally computed at each case once their can be found in the same study of language, and not by abstract mathematical thought considerations. What is needed is to transform the study of language from a logic one in a scientific one.

When fuzzy logic was initially developed in the past Century's seventies and eighties, almost the only back referents for its study were classical and multiple-valued logics, but now it should be centered in Natural Language. If current fuzzy logic already meant an important progress in the way asked by John von Neumann of introducing mathematical analysis in the study of those subjects without a just 'yes' or 'not' hypothesis for its validity, it can be the right moment to go a step ahead and turning towards the Artificial Intelligence's 'Gordian Knot' of trying to reach computers thinking like people usually do.

6 Conclusion

Up to some point, and although many papers of a mathematical character, even with some of them of a true mathematical quality, are being continuously published in the setting of 'theoretic fuzzy logic', its evolution seems to be actually stagnated because of some moving away of what is the essence of fuzzy logic. By one side, those papers remain practically unknown or, at least, not considered, for those who devote their efforts to the applications of fuzzy logic, and by the other the motivation of their authors is almost always purely abstract; in them, it rarely appears a 'real fuzzy problem' to which either their results could be applied to, or just it can be suggested by the paper's content. It seems as if in current fuzzy logic it were two streams, that of mathematicians and that of engineers, but fuzzy logic should be an integrated study of what is 'fuzzy' and, in principle, that practitioners ignore the obtained mathematical results, marks a limit in their capability of designing fuzzy systems. There is, perhaps, some kind of isolation between both types of researchers, the most relevant of the ones not mixed with the most relevant of the others. This and to some extent, goes against the cross-fertilization of both groups and can contribute to the closing of the first in their own mathematical interest. Anyway, and in the last years, I hopefully heard on mixed groups working in some specific projects. Notwithstanding, and as far as I know, such projects are on very specific topics not directly related with CwW.

In the author's view, and by looking at what is fuzzy and what is for its study, the great challenge for the best continuation of theoretic fuzzy logic lies in the problems that are in the back of the new Zadeh's 'Computing with Words and Perceptions', where the problems that were essential for the introduction of fuzzy sets could acquire all their relevance as clearly dealing with Natural Language's complex phrases, and with the non-deductive varieties of Commonsense Reasoning. Nevertheless, the natural and dynamic characters of both language and reasoning, seems to suggest that a new, and scientific, study of them cannot be completely afforded by only counting with the abstract reasoning reached through mathematical theorems that only can be successfully applied provided all what is being supposed for their proofs is actually verified in a concrete and actual situation. Something that is, usually, very difficult to check as, it happens, for instance, when trying to use an S-implication function for linguistic rules in which the representation of the negation of their antecedents is actually unknown.

It is in the thought of the author that the main subjects of 'fuzzy logic' or, by extension CwW, are both the representation and technical management of the imprecision and the uncertainty pervading natural language and commonsense reasoning in non-trivial statements. In some cases, for instance, the meaning of the components of a large statement is only captured after having captured the full meaning of the full statement, something different of what is done in logic where always it is done by departing from the meaning of the components.

To afford those subjects it seems recommendable to face them as they are, natural phenomena of which, and in addition, we have a scarce knowledge that, notwithstanding, should be increased by the only way it can be followed for any natural phenomena, namely, by experimenting in controlled forms as it is typical of science. It is with the conjunction of experimentation and mathematical modeling how measurable parameters can be obtained and deep conclusions attained. Science always needs to count with suitable frames for representing what it deals with, thanks to which some mathematical models could be established and that, at its turn, facilitates some numerical parameters necessary to going on with more experimentation. À la Popper, research is always an un-ended quest.

A new experimental science dealing with linguistic imprecision and uncertainty, both random and not random, seems to appear in the horizon and into the complex knitting of language. It is an enterprise that jointly with, and close to, fully knowing the brain's functioning, could contribute to capture what is rationality by going far from old metaphysical ideas, and by means of the single way mankind has for acquiring safe knowledge, the scientific method.

Would young researchers in the XXI Century devote their efforts to such a challenging enterprise!

Acknowledgments This paper is partially funded by the 'Foundation for the Advancement of Soft Computing', Mieres (Asturias), Spain.

References

1. Basáñez, L., Batle, N., Ferraté, G., Grané, J., Trillas, E.: A first approach to sigma-transform. J. Math. Anal. Appl. **92**(1), 224–233 (1983)
2. Bellman, R., Giertz, M.: On the analytic formalism of fuzzy sets. Inf. Sci. **5**, 149–156 (1973)
3. Hardegree, G.M.: The conditional in quantum logic. Synthese **29**, 63–80 (1974)
4. Lowen, R.: On fuzzy complements. Inf. Sci. **14**, 107–113 (1978)
5. Mamdani, E.H., Assilian, S.: An experiment in linguistic synthesis with a fuzzy logic controller. Int. J. Hum. Comput. Stud. **7**(1), 1–13 (1975)
6. Menger, K.: Statistical metries. Proc. Nat. Acad. Sci. USA **28**(12), 535–537 (1942)
7. Nguyen, H.T., Walker, E.A.: A First Course in Fuzzy Logic. Chapman & Hall, Boca Raton (2000)
8. Trillas, E., Guadarrama, S.: Fuzzy representations need a careful design. Int. J. Gen. Sys. **39**(3), 329–346 (2010)

9. Trillas, E.: A model for 'crisp reasoning' with fuzzy sets. Int. J. Intell. Syst. **27**, 859–872 (2012)
10. Trillas, E.: En defensa del razonamiento creativo, Universidad Pública de Navarra (2014)
11. Trillas, E.: Razonamiento; significado, incertidumbre y borrosidad; Ed. Upna, Pamplona (2015)
12. Trillas, E.: An algebraic model of reasoning to support Zadeh's CwW (2015)
13. Trillas, E., Eciolaza, L.: Fuzzy logic: an introductory course for engineering students. Springer (2015)
14. Trillas, E., Termini, S., Moraga, C.: A naïve way of looking at fuzzy sets. Fuzzy Sets Syst. In press (2015)
15. Williams, C.P., Clearwater, S.H.: Explorations in Quantum Computing. Springer, NY (1988)
16. Zadeh, L.A.: Fuzzy sets. Inf. Control **8**, 338–353 (1965)
17. Zadeh, L.A.: Computing with Words: Principal Concepts and Ideas. Springer, Berlin (2012)

Modules in the Category Sup

Ulrich Höhle

Abstract This chapter explains basic properties of left modules on unital quantales with the perspective towards fuzzy set theory. Typical constructions such as the fuzzy power set, Zadeh's forward operator or binary operations defined according to Zadeh's extension principle are constructions in the symmetric monoidal closed category of complete lattices and join preserving maps. Moreover, involutive left modules play a significant role in the representation theory of C^*-algebras.

1 Introduction

The motivation of this chapter is to make a contribution to the mathematical foundations of fuzzy set theory and to describe the place where fuzzy set theory is residing inside mathematics. Let Sup be the category of complete lattices and join preserving maps. Our thesis is that module theory in Sup is the algebraic basis of fuzzy set theory. We justify this thesis by the following observations:

- In Zadeh's pioneering paper on fuzzy sets (cf. [30]), the real unit interval provided with the bounded sum appears as underlying mathematical structure. Since the bounded sum is the t-conorm of the Łukasiewicz arithmetic conjunction,

$$\alpha * \beta = \max(\alpha + \beta - 1, 0), \quad \alpha, \beta \in [0, 1],$$

  L.A. Zadeh has tacitly used the unital quantale given by the canonical MV-algebra as algebraic basis.
- Let $\mathfrak{Q}$ be a unital quantale and X be a set. Then the $\mathfrak{Q}$-valued (i.e. $\mathfrak{Q}$-fuzzy) power set of X is the free left $\mathfrak{Q}$-module generated by X. This result appears for the first time in [12] and describes the *fuzzy power set* by a universal property in the language of module theory in Sup.

U. Höhle (✉)
Fachbereich C Mathematik und Naturwissenschaften, Bergische Universität, Wuppertal, Germany
e-mail: uhoehle@uni-wuppertal.de

S. Saminger-Platz and R. Mesiar (eds.), *On Logical, Algebraic, and Probabilistic Aspects of Fuzzy Set Theory*, Studies in Fuzziness and Soft Computing 336,
DOI 10.1007/978-3-319-28808-6_3

- Left $\mathfrak{Q}$-modules and complete $\mathfrak{Q}$-valued (i.e. $\mathfrak{Q}$-fuzzy) ordered sets are equivalent concepts—a result which goes back to I. Stubbe in a more general context given by quantaloid enriched categories (cf. [29]).
- Binary operations constructed according to Zadeh's extension principle are the tensor product of the respective Minkowski multiplications with the multiplication of the underlying unital quantale (cf. [8]).

Because of the previous observations we believe that it is meaningful to put together various important aspects of module theory in Sup with the perspective towards fuzzy set theory. In this sense this paper is a survey on some basic properties of left modules on unital quantales.

We begin with the construction of the tensor product in Sup which goes back to Z. Shmuely 1974 (cf. [27]). Since there exist various approaches to the tensor product (cf. [3, 12]), we prefer here the understanding of the tensor product as "function space"—i.e. as a complete lattice of join reversing maps (cf. Proposition 1 on p. 6 in [12]). Subsequently, we view unital quantales as monoids in Sup and develop the usual module theory in Sup including complete many-valued ordered sets which represent a typical phenomenon of left-modules in Sup. In this context it is interesting to realize that involutive left modules play a significant role in the representation theory of C^*-algebras (cf. [13, 14, 16, 24]).

2 The Tensor Product in the Category Sup

Let Sup be the category of complete lattices and arbitrary join preserving maps. It is well known that Sup is a complete and cocomplete category. In particular, regular quotients of complete lattices L can be identified with closure operators on L—these are isotone self-maps $L \xrightarrow{c} L$ provided with the properties $1_L \le c$ and $c \circ c \le c$.

With regard to algebraic considerations the most important property of Sup is the existence of a tensor product transforming Sup into a symmetric monoidal closed category (cf. [17]). The purpose of this section is to recall these fundamental properties and to explain the role of the tensor product for the construction of fuzzy power sets.

Let L and M be complete lattices. A map $L \xrightarrow{f} M$ is *join reversing* if for all subsets A of L the relation $f(\bigvee A) = \bigwedge f(A)$ holds. Obviously, join reversing maps $L \xrightarrow{f} M$ are always part of a Galois connection (f, g) between L and M (cf. [4, 7]) where the second polarity $M \xrightarrow{g} L$ is determined by

$$g(m) = \bigvee \{\ell \in L \mid m \le f(\ell)\} \tag{1}$$

Therefore we denote the set of all join reserving maps $L \xrightarrow{f} M$ by $\mathcal{G}(L, M)$ and consider the following partial order on $\mathcal{G}(L, M)$:

$$f_1 \leq f_2 \quad \Leftrightarrow \quad \forall \ell \in L : \; f_1(\ell) \leq f_2(\ell). \tag{2}$$

Obviously, $(\mathcal{G}(L, M), \leq)$ is a complete lattice in which meets (but in general not joins) are computed pointwisely. In this context, the formula (1) determines an order isomorphism between $\mathcal{G}(L, M)$ and $\mathcal{G}(M, L)$.

Let N be a further complete lattice. Referring to [3] a map $L \times M \xrightarrow{b} N$ is a *bimorphism* (in Sup) iff b is join preserving in each variable separately—i.e. for all $m \in M$ and $\ell \in L$ the correspondences $\ell \mapsto b(\ell, m)$ and $m \mapsto b(\ell, m)$ are join preserving.

Example 1 Let L and M be complete lattices. For every pair $(\ell, m) \in L \times M$ we construct a join reversing map $L \xrightarrow{f_{(\ell,m)}} M$ as follows (cf. Lemma 1.5 in [27])

$$f_{(\ell,m)}(z) = \begin{Bmatrix} \bot, \; z \not\leq \ell, \\ m, \; z \leq \ell, z \neq \bot, \\ \top, \; z = \bot, \end{Bmatrix} \qquad z \in L. \tag{3}$$

where $\bot$ $(\top)$ denotes the respective universal lower (upper) universal bound in L and M. Obviously, the following properties hold for all $\ell, \ell' \in L$ and $m, m' \in M$:

- $f_{(\bot,m)}$ and $f_{(\ell,\bot)}$ coincide with the universal lower bound in $\mathcal{G}(L, M)$.
- If $\ell \neq \bot$ and $m \neq \bot$, then $f_{(\ell,m)} \leq f_{(\ell',m')}$ iff $\ell \leq \ell'$ and $m \leq m'$.

On this background we introduce a bimorphism $L \times M \xrightarrow{\beta} \mathcal{G}(L, M)$ by:

$$\beta(\ell, m) = f_{(\ell,m)}, \qquad (\ell, m) \in L \times M. \tag{4}$$

It is evident that β is join preserving in its second argument. In order to verify that β is also join preserving in its first argument, it is sufficient to consider a non empty subset $\{\ell_i \mid i \in I\}$ of L. Then $f \in \mathcal{G}(L, M)$ is an upper bound of $A := \{\beta(\ell_i, m) \mid i \in I\}$ iff $m \leq f(\ell_i)$ holds for all $i \in I$. Since f is join reversing, f is an upper bound of A iff $m \leq f\big(\bigvee_{i \in I} \ell_i\big)$. Hence, $\beta\big(\bigvee_{i \in I} \ell_i, m\big)$ is the smallest upper bound of A—i.e. $\bigvee_{i \in I} \beta(\ell_i, m) = \beta\big(\bigvee_{i \in I} \ell_i, m\big)$.

Definition 1 Let L and M be complete lattices. A pair (β, X) is called the *tensor product* of L and M, if X is a complete lattice and $L \times M \xrightarrow{\beta} X$ is a bimorphism such that the following universal property holds:

For every bimorphism $L \times M \xrightarrow{b} N$ there exists a unique join preserving map $X \xrightarrow{h} N$ making the following diagram commutative:

$$\begin{array}{ccc} L \times M & \xrightarrow{\beta} & X \\ & \searrow^{b} & \vdots h \\ & & N \end{array} \tag{5}$$

In the special case of Sup the next theorem is a version of the Corollary of Proposition 5 in [3].

Theorem 1 *For every pair (L, M) of complete lattices L and M the tensor product (β, X) exists and is unique up to an isomorphism in the sense of* Sup.

Proof The uniqueness of the tensor product is an immediate corollary from its universal property. In order to verify the existence of the tensor product, we consider the pair $\big(\beta, \mathcal{G}(L, M)\big)$ constructed in Example 1.

Let $L \times M \xrightarrow{b} N$ be a bimorphism. Since every join reversing map $L \xrightarrow{f} M$ is the join of $\{\beta(\ell, f(\ell)) \mid \ell \in L\}$ in $\mathcal{G}(L, M)$—i.e.

$$f = \bigvee_{\ell \in L} \beta(\ell, f(\ell)), \tag{6}$$

there exists at most a join preserving map $\mathcal{G}(L, M) \xrightarrow{h} N$ making the diagram

$$\begin{array}{ccc} L \times M & \xrightarrow{\beta} & \mathcal{G}(L, M) \\ & \searrow^{b} & \vdots h \\ & & N \end{array} \tag{7}$$

commutative.

In order to establish the existence of a join preserving map $\mathcal{G}(L, M) \xrightarrow{h} N$ s.t. the diagram (7) commutes, we proceed as follows. We define a map $\mathcal{G}(L, M) \xrightarrow{h} N$ by

$$h(f) = \bigvee_{(\ell,m) \in A} b(\ell, m), \qquad A \subseteq L \times M, \quad \bigvee_{(\ell,m) \in A} \beta(\ell, m) = f. \tag{8}$$

Provided that we can show that h is well defined, then formula (8) shows immediately that h is join preserving and makes the diagram (7) commutative. Therefore we only verify that h is well defined. For this purpose we choose a further subset B of $L \times M$ with $\bigvee_{(\ell,m) \in B} \beta(\ell, m) = f$. Since N is complete, it is sufficient to verify that the respective sets of upper bounds of $\{b(\ell, m) \mid (\ell, m) \in A\}$ and $\{b(\ell, m) \mid (\ell, m) \in B\}$ coincide.

First we apply the property that b is join preserving in each variable separately and define a join reversing map $L \xrightarrow{f_z} M$ for all $z \in N$ as follows

$$f_z(\ell) = \bigvee \{m \in M \mid b(\ell, m) \le z\}, \quad \ell \in L.$$

If $z \in N$ be an upper bound of $\{b(\ell, m) \mid (\ell, m) \in A\}$, then $m \leq f_z(\ell)$ holds for all $(\ell, m) \in A$. Since f is the join of $\{\beta(\ell, m) \mid (\ell, m) \in A\}$, we obtain $f \leq f_z$. Since f is also the join of $\{\beta(\ell, m) \mid (\ell, m) \in B\}$, the relation

$$m \leq f_z(\ell), \quad (\ell, m) \in B \tag{9}$$

follows. Because of $b(\ell, f_z(\ell)) \leq z$ we infer from (9) that z is also an upper bound of $\{b(\ell, m) \mid (\ell, m) \in B\}$. Interchanging the role of A and B the assertion follows. □

Because of the previous theorem we fix the following notation and terminology. The tensor product of two complete lattices L and M is denoted by $L \otimes M$. By abuse of notation the bimorphism $L \times M \xrightarrow{\beta} L \otimes M$ is also denoted by $\otimes$, and instead of $\beta(\ell, m)$ we also write $\ell \otimes m$. Elements of $L \otimes M$ are called *tensors*, and tensors of the special type $\ell \otimes m$ are called *elementary tensors*. It follows immediately from (6) that every tensor is a join of an appropriate family of elementary tensors. Sometimes, in the case of $\ell \neq \bot$ and $m \neq \bot$, the following equivalence is useful:

$$\ell \otimes m \leq \ell' \otimes m' \quad \Longleftrightarrow \quad \ell \leq \ell' \text{ and } m \leq m'.$$

Let L be a complete lattice. We show that the *L-fuzzy power set* is the tensor product of L with the ordinary power set. For this purpose we choose a set X and consider the partial ordering $\leq$ on L^X defined by:

$$g_1 \leq g_2 \quad \Leftrightarrow \quad \forall x \in X : \; g_1(x) \leq g_2(x).$$

Then L^X is a complete lattice. In the literature on fuzzy sets L^X is called the *L-fuzzy* or *L-valued power set of X*.

Theorem 2 ([8]) *Let L be a complete lattice, X be a set, and let $\mathcal{P}(X)$ be the ordinary power set of X. There exists an order isomorphism $L^X \xrightarrow{\Phi_L} L \otimes \mathcal{P}(X)$ defined by:*

$$\Phi_L(g) = \bigvee_{x \in X} g(x) \otimes \{x\}, \quad g \in L^X.$$

Proof It is easily seen that every element $g \in L^X$ can be identified with a join reversing map $\mathcal{P}(X) \xrightarrow{F_g} L$

$$F_g(A) = \bigwedge_{x \in A} g(x), \qquad A \in \mathcal{P}(X),$$

and vice versa. Hence the map $L^X \xrightarrow{\Phi_L} \mathcal{G}\big(L, \mathcal{P}(X)\big) = L \otimes \mathcal{P}(X)$ defined by

$$[\Phi_L(g)](\ell) = \{x \in X \mid \ell \leq g(x)\}, \qquad \ell \in L \tag{10}$$

is bijective and isotone. Since the inverse map of Φ_L has the form

$$[\Phi_L^{-1}(f)](x) = \bigvee\{\ell \in L \mid x \in f(\ell)\}, \quad f \in \mathcal{G}(L, \mathcal{P}(X)),$$

Φ_L is an order isomorphism. The proof is complete, if we can show that $\Phi_L(g)$ is the smallest upper bound of

$$A := \{g(x) \otimes \{x\} \mid x \in X\}.$$

An element $k \in L \otimes \mathcal{P}(X)$ is an upper bound of A iff $\{x\} \subseteq k(g(x))$ holds for all $x \in X$ iff $\{x \in X \mid \ell \leq g(x)\} \subseteq k(\ell)$ holds for all $\ell \in L$. Hence $\Phi_L(g)$ (cf. (10)) is the smallest upper bound of A. □

It follows from the universal property of the tensor product that the tensor product in Sup induces a bifunctor (also denoted by $\otimes$) from Sup $\times$ Sup to Sup. In particular, the tensor product of join preserving maps $L_1 \xrightarrow{h_1} L_2$ and $M_1 \xrightarrow{h_2} M_2$ is determined by the commutativity of the following diagram:

$$\begin{array}{ccc} L_1 \times M_1 & \xrightarrow{\otimes} & L_1 \otimes M_1 \\ {\scriptstyle h_1 \times h_2}\downarrow & & \vdots\, {\scriptstyle h_1 \otimes h_2} \\ L_2 \times M_2 & \xrightarrow[\otimes]{} & L_2 \otimes M_2 \end{array}$$

Before we show that $\otimes$ induces a symmetric monoidal closed structure on Sup we apply the tensor product of join preserving maps to the construction of variable-basis fuzzy power set operators (cf. [25]). For this purpose we recall the following notation. Let $\mathsf{P}\colon \mathtt{Set} \to \mathtt{Set}$ be the covariant power set functor—i.e.

$$\mathsf{P}(X) = \mathcal{P}(X), \qquad X \xrightarrow{\varphi} Y, \quad \mathsf{P}(\varphi)(A) = \varphi(A), \ A \in \mathcal{P}(X).$$

Further, let $L \xrightarrow{h} M$ be a join preserving map. Then for any map $X \xrightarrow{\varphi} Y$ the variable-basis power set forward operator $L^X \xrightarrow{(\varphi,h)^{\rightarrow}} M^Y$ has the form (cf. 3.9(6) in [25]):

$$[(\varphi, h)^{\rightarrow}(g)](y) = \bigvee\{h \circ g(x) \mid \varphi(x) = y\}, \qquad y \in Y.$$

Corollary 1 *Let $X \xrightarrow{\varphi} Y$ be a map and $L \xrightarrow{h} M$ be a join preserving map. Then the following diagram is commutative:*

$$\begin{array}{ccc} L^X & \xrightarrow{(\varphi,h)^{\rightarrow}} & M^Y \\ {\scriptstyle \Phi_L}\downarrow & & \downarrow{\scriptstyle \Phi_M} \\ L \otimes \mathcal{P}(X) & \xrightarrow[h \otimes \mathsf{P}(\varphi)]{} & M \otimes \mathcal{P}(Y) \end{array}$$

Proof Referring to Theorem 2 we obtain the following relation for all $g \in L^X$:

$$
\begin{aligned}
h \otimes \mathsf{P}(\varphi)\big(\Phi_L(g)\big) &= \bigvee_{x \in X} h(g(x)) \otimes \{\varphi(x)\} \\
&= \bigvee_{y \in Y} \big(\bigvee \{h(g(x)) \mid \varphi(x) = y\}\big) \otimes \{y\} \\
&= \Phi_M\big((\varphi, h)^{\rightarrow}(g)\big).
\end{aligned}
$$

□

Because of the previous corollary the variable-basis power set forward operator coincides with the *tensor product* of the traditional power set forward operator (i.e. the operator taking direct images) with the join preserving map performing the *change of basis*. In this sense the tensor product of Sup plays a fundamental role in the mathematical foundations of fuzzy set theory.

The next important observation is the property that for every complete lattice M the endofunctor $\mathsf{F}\colon \mathtt{Sup} \to \mathtt{Sup}$ defined by

$$
\mathsf{F}(L) = L \otimes M, \quad L_1 \xrightarrow{h} L_2, \quad \mathsf{F}(h) = h \otimes 1_M
$$

has a right adjoint functor G. This result follows from the general constructions in [3]. But we give here a direct proof.

For complete lattices M and N the set $[M, N]$ of all join preserving maps $M \xrightarrow{h} N$ is a complete lattice w.r.t. the pointwisely defined order. Then $\mathsf{G}\colon \mathtt{Sup} \to \mathtt{Sup}$ is defined by:

$$
\mathsf{G}(N) = [M, N], \qquad N_1 \xrightarrow{k} N_2, \quad [\mathsf{G}(k)](h) = k \circ h, \quad h \in [M, N_1].
$$

Further, the evaluation map $[M, N] \times M \xrightarrow{\mathrm{ev}_N} N$ with $\mathrm{ev}_N(h, m) = h(m)$ is a bimorphism. Because of the universal property of the tensor product there exists a unique join preserving map $[M, N] \otimes M \xrightarrow{\varepsilon_N} N$ making the diagram

$$
\begin{array}{ccc}
[M, N] \times M & \xrightarrow{\otimes} & [M, N] \otimes M \\
& \searrow^{\mathrm{ev}_N} & \Big\downarrow \varepsilon_N \\
& & N
\end{array}
$$

commutative. It is not difficult to see that $\varepsilon = (\varepsilon_N)_{N \in |\mathtt{Sup}|}$ is a natural transformation from $\mathsf{F} \circ \mathsf{G}$ to $\mathrm{id}_{\mathtt{Sup}}$.

Theorem 3 *Let L, M and N be complete lattices and $L \otimes M \xrightarrow{h} N$ be a join preserving map. Then there exists a unique join preserving map $L \xrightarrow{\ulcorner h \urcorner} [M, N]$ making the following diagram commutative:*

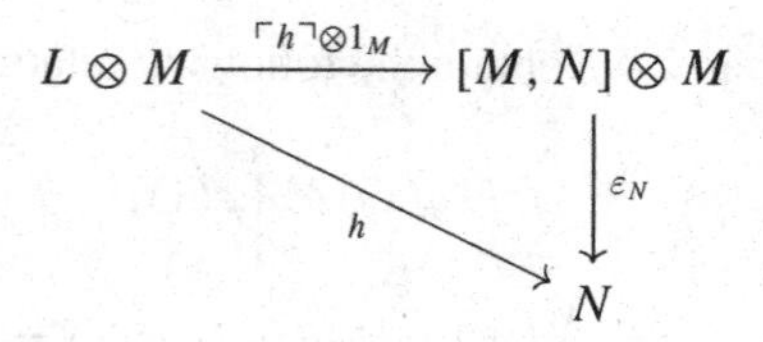

Proof The bimorphism $L \times M \xrightarrow{b_h} N$ corresponding to $L \otimes M \xrightarrow{h} N$ is given by

$$b_h(\ell, m) = h(\ell \otimes m), \quad (\ell, m) \in L \times M.$$

Then we conclude from the commutativity of the following diagram

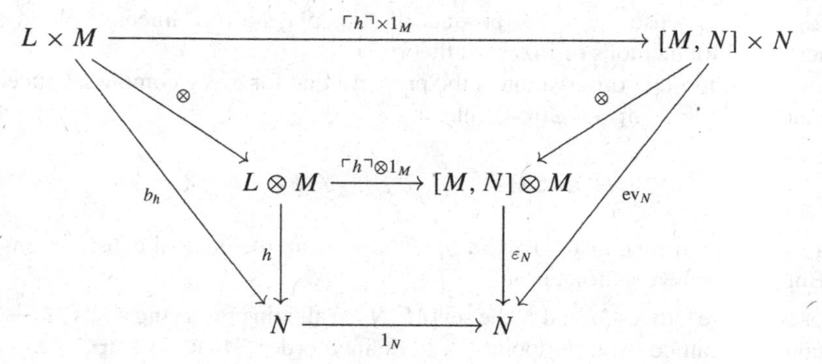

that $\ulcorner h \urcorner$ is unique and given by:

$$[\ulcorner h \urcorner(\ell)](m) = b_h(\ell, m), \qquad \ell \in L, \ m \in M. \tag{11}$$

Since joins in $[M, N]$ are computed pointwisely, the map $\ulcorner h \urcorner$ is in fact join preserving. □

The previous theorem motivates the following terminology. Let $L \otimes M \xrightarrow{h} N$ be a join preserving map. Then the join preserving map $L \xrightarrow{\ulcorner h \urcorner} [M, N]$ determined by (11) is called the *monoidal adjoint map* of h.

The correspondence $h \mapsto \ulcorner h \urcorner$ induces an order isomorphism from $[L \otimes M, N]$ onto $[L, [M, N]]$ (Proposition 5 in [3]). As a corollary of this fact we obtain the associativity of the tensor product.

Corollary 2 *Let L, M and N be complete lattices. There exists a unique order isomorphism $(L \otimes M) \otimes N \xrightarrow{\mathfrak{a}_{LMN}} L \otimes (M \otimes N)$ satisfying the following property for all $(\ell, m, n) \in L \times M \times N$:*

$$\mathfrak{a}_{LMN}\big((\ell \otimes m) \otimes n\big) = \ell \otimes (m \otimes n). \tag{12}$$

Proof By X^{op} we denote the dual lattice of X. Hence the tensor product $L \otimes M$ has the form $L \otimes M = [L, M^{op}]^{op}$.

If we choose $(\ell \otimes m) \otimes n \in (L \otimes M) \otimes N = [L \otimes M, N^{op}]^{op}$, then it is not difficult to conclude from (11) that the relation $\ulcorner(\ell \otimes m) \otimes n\urcorner = \ell \otimes (m \otimes n)$ holds. Hence the formation of taking monoidal adjoint maps is the desired order isomorphism from $(L \otimes M) \otimes N = [L \otimes M, N^{op}]^{op}$ to $[L, [M, N^{op}]]^{op} = L \otimes (M \otimes N)$. □

It follows immediately from (12) that $\mathfrak{a}_{LMN}$ is a component of a natural isomorphism making the pentagonal diagram (cf. [15, 17]) commutative. Hence the bifunctor determined by the tensor product is associative.

Comment The associativity of the tensor product appears already as Theorem 1.7 in [27], but without an explicit statement of property (12).

The commutativity of the tensor product—i.e. the symmetry $\mathfrak{c}$ of the bifunctor $\otimes$—goes back to Z. Shmuely 1974 (cf. [27, Theorem1.5, Lemma1.5]).

Lemma 1 *Let L and M be complete lattices. Then there exist a unique order isomorphism $L \otimes M \xrightarrow{\mathfrak{c}_{LM}} M \otimes L$ satisfying the following condition for all pairs $(\ell, m) \in L \times M$:*

$$\mathfrak{c}_{LM}(\ell \otimes m) = m \otimes \ell. \tag{13}$$

Proof We fix $\ell \in L$ and $m \in M$ with $\ell \neq \bot$ and $m \neq \bot$. Because of the relation

$$\bigvee \{\ell' \in L \mid m' \leq (\ell \otimes m)(\ell')\} = (m \otimes \ell)(m'), \quad m' \in M$$

the pair $(l \otimes m, m \otimes \ell)$ is a Galois connection between L and M. Hence the correspondence determined by (1) is the desired order isomorphism $\mathfrak{c}_{LM}$ satisfying (13). □

The next lemma explains that the 2-chain is the unit object w.r.t. the tensor product in Sup (cf. [12, p. 6, Proposition 2(ii)]).

Lemma 2 *Let L and M be complete lattices and let $\mathbb{1} = \{0, 1\}$ be the lattice with two different elements. Then there exist order isomorphisms $\mathbb{1} \otimes M \xrightarrow{l_M} M$ and $L \otimes \mathbb{1} \xrightarrow{r_L} L$ such that the following diagrams are commutative:*

$$\begin{array}{ccc} (L \otimes \mathbb{1}) \otimes M & \xrightarrow{a_{X\mathbb{1}Y}} & L \otimes (\mathbb{1} \otimes M) \\ & {\scriptstyle r_L \otimes 1_M} \searrow \quad \swarrow {\scriptstyle 1_L \otimes l_M} & \\ & L \otimes M & \end{array} \qquad \begin{array}{ccc} L \otimes \mathbb{1} & \xrightarrow{\mathfrak{c}_{L\mathbb{1}}} & \mathbb{1} \otimes L \\ & {\scriptstyle r_L} \searrow \quad \swarrow {\scriptstyle l_L} & \\ & L & \end{array} \tag{14}$$

In the case of $L = M = \mathbb{1}$ the relation $r_{\mathbb{1}} = l_{\mathbb{1}}$ holds.

Proof Let 0 be the bottom element of two element lattice $\mathbb{1} = \{0, 1\}$. For $f \in L \otimes \mathbb{1}$ and $g \in \mathbb{1} \otimes M$ we define $\ell_0 \in L$ and $m_0 \in M$ by

$$\ell_0 = \bigvee \{\ell \in L \mid f(\ell) = 1\} \quad \text{and} \quad y_0 = g(1)$$

and observe $f = \ell_0 \otimes 1$ and $g = 1 \otimes m_0$. Hence the desired order isomorphisms r_X and l_Y are defined by $r_L(\ell \otimes 1) = \ell$ and $l_Y(1 \otimes m) = m$. It is easily seen that the diagrams in (14) are commutative and the property $r_{\mathbb{1}} = l_{\mathbb{1}}$ holds. □

We can summarize the results of Theorem 3, Corollary 2 and Lemmas 1, 2 as follows.

Fact. The septuple $(\mathtt{Sup}, \otimes, \mathfrak{a}, \mathfrak{c}, l, r, \mathbb{1})$ is a symmetric monoidal closed category (cf. [15, 17]).

3 Unital Quantales as Monoids in Sup

A pair $(X, *)$ is a *prequantale* (cf. [26]) if X is a complete lattice and $X \times X \xrightarrow{*} X$ is a bimorphism of Sup. Instead of $*(x, y)$ we also write $x * y$ for $x, y \in X$. If $(X, *)$ is a prequantale, then $*$ is called the *multiplication of* X.

Because of the universal property of the tensor product in Sup the multiplication $*$ of a prequantale X can be identified with a binary operation in the sense of Sup—i.e. a join preserving map $X \otimes X \xrightarrow{\circledast} X$ s.t. $* = \circledast \circ \otimes$ holds. Hence *prequantales* and *magmas in* Sup are equivalent concepts.

Notation If $*$ is the multiplication of a prequantale X with its corresponding binary operation $\circledast$, then instead of $\circledast(x \otimes y)$ we also write $x \circledast y$ with $x, y \in X$. Since $x * y$ and $x \circledast y$ coincide, it will depend on the context which kind of notation we will prefer.

A *homomorphism* between prequantales is a join preserving map $X \xrightarrow{h} Y$ which also preserves the respective multiplications—i.e. $h(x * z) = h(x) * h(z)$. Hence a homomorphism $(X, *) \xrightarrow{h} (Y, *)$ is characterized by the commutativity of the following diagram:

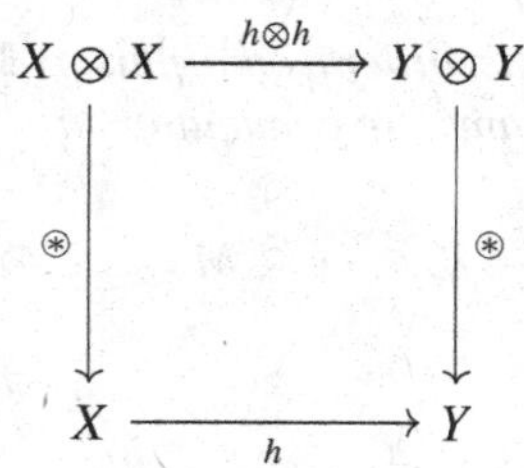

Since $(\mathtt{Sup}, \otimes, \mathfrak{a}, \mathfrak{c}, l, r, \mathbb{1})$ is bi-closed (cf. Fact and [15]) and Sup is cocomplete, we conclude from the free algebra algorithm (cf. [1] and p. 186 in [2]) that the algorithm converges for the endofunctor $\mathsf{T}\colon \mathtt{Sup} \to \mathtt{Sup}$ determined by

$$\mathsf{T}(X) = X \otimes X, \qquad X \xrightarrow{h} Y, \quad \mathsf{T}(h) = h \otimes h.$$

Hence every complete lattice generates a free magma (i.e. free prequantale) (cf. [6]).

A prequantale $(X, *)$ is a *quantale* (cf. [26]) if $*$ is associative—i.e. $(x * y) * z = x * (y * z)$. A quantale is *unital*, if $*$ has a unit e—i.e. $x * e = e * x$ for all $x \in X$. A homomorphism h is unital, if h preserves the respective units—i.e. $h(e) = e$.

Let $(X, *, e)$ be a unital quantale. If we identify the multiplication with its corresponding binary operation and the unit with the join preserving map $\mathbb{1} = \{0, 1\} \to X$ sending 1 to e, then $(X, \circledast, e)$ is a *monoid* in the sense of the symmetric monoidal category $(\mathtt{Sup}, \otimes, \mathfrak{a}, \mathfrak{c}, l, r, \mathbb{1})$ (cf. Corollary 2, Lemmas 1 and 2). Since $(\mathtt{Sup}, \otimes, \mathfrak{a}, \mathfrak{c}, l, r, \mathbb{1})$ is bi-closed and Sup is cocomplete, every complete lattice generates a free monoid (i.e. a free unital quantale) (cf. Theorem 2 on p. 172 in [17]).

Example 2 Let $\mathbb{N}_0$ be the set of natural numbers including 0, and let $\mathbb{1}$ be the unit object of the tensor product in Sup. Then the free unital quantale generated by $\mathbb{1}$ coincides with the power set $\mathcal{P}(\mathbb{N}_0)$ of $\mathbb{N}_0$ equipped with the *Minkowski addition* $\boxplus$—i.e.

$$A \boxplus B = \{a + b \mid a \in A,\ b \in B\}, \qquad A, B \in \mathcal{P}(\mathbb{N}_0).$$

In fact, if $\mathbb{1} = \{0, 1\} \xrightarrow{\eta_{\mathbb{1}}} \mathcal{P}(\mathbb{N}_0)$ is the join preserving embedding with $\eta_{\mathbb{1}}(1) = \{1\}$ (and $\eta_{\mathbb{1}}(0) = \varnothing$), then for every further unital quantale $(X, *, e)$ and for every further join preserving map $\mathbb{1} \xrightarrow{h} X$ (e.g. $h(1) = x$) there exists a unique unital homomorphism $\mathcal{P}(\mathbb{N}_0) \xrightarrow{h^\sharp} X$ with the property $h^\sharp \circ \eta_{\mathbb{1}} = h$. In particular, $h^\sharp$ is given by:

$$h^\sharp(A) = \bigvee \{x^n \mid n \in A\}, \quad A \in \mathcal{P}(\mathbb{N}_0)$$

where $x^0 = e$ and $x^n = x * x^{n-1}$ for $n \in \mathbb{N}$.

As always in monoidal categories the tensor product of monoids exist. In $(\mathtt{Sup}, \otimes, \mathfrak{a}, \mathfrak{c}, l, r, \mathbb{1})$ the situation is as follows. Let $(X_1, *_1, e_1)$ and $(X_2, *_2, e_2)$ be unital quantales. First we infer from the associativity and commutativity of the tensor product that there exists a unique order isomorphism

$$(X_1 \otimes X_2) \otimes (X_1 \otimes X_2) \xrightarrow{\Phi} (X_1 \otimes X_1) \otimes (X_2 \otimes X_2)$$

satisfying the condition $\Phi\big((x_1 \otimes x_2) \otimes (y_1 \otimes y_2)\big) = (x_1 \otimes y_1) \otimes (x_2 \otimes y_2)$ for all $x_1, y_1 \in X_1$ and $x_2, y_2 \in X_2$. Hence, if $X_1 \otimes X_1 \xrightarrow{\circledast_1} X_1$ and $X_2 \otimes X_2 \xrightarrow{\circledast_2} X_2$ denote the binary operations corresponding to $*_1$ and $*_2$, then there exists a binary operation on $X_1 \otimes X_2$ defined by the composition of Φ with the tensor product of $\circledast_1$ and $\circledast_2$—i.e. $(\circledast_1 \otimes \circledast_2) \circ \Phi$. Obviously, the corresponding bimorphism $(X_1 \otimes X_2) \times (X_1 \otimes X_2) \xrightarrow{\star} X_1 \otimes X_2$ is uniquely determined by the property

$$(x_1 \otimes x_2) \star (y_1 \otimes y_2) = (x_1 *_1 y_1) \otimes (x_2 *_2 y_2), \quad x_1, y_1 \in X_1, \ x_2, y_2 \in X_2. \tag{15}$$

The associativity of $\star$ follows immediately from the associativity of $*_1$ and $*_2$ and the fact that every tensor is the join of an appropriate family of elementary tensors. Hence the triple $(X_1 \otimes X_2, \star, e_1 \otimes e_2)$ is a unital quantale and is called the *tensor product* of $(X_1, *_1, e_1)$ and $(X_2, *_2, .e_2)$.

A special case of the previous situation is the construction of binary operations on many-valued power sets according to *Zadeh's extension principle* (cf. [8]) where L.A. Zadeh considers only the special case $* = \wedge$ in his original papers [31–33].

Example 3 Let $(X, \cdot, e)$ be a monoid in $\mathtt{Set}$ and $\mathcal{P}(X)$ be the power set of X provided with the set inclusion as partial order and the Minkowski multiplication $\boxdot$ w.r.t. the multiplication $\cdot$ on X. Then $(\mathcal{P}(X), \boxdot, \{e\})$ is a unital quantale.

Let $(\mathfrak{Q}, *, e)$ be a further unital quantale. The binary operation $\odot$ on the $\mathfrak{Q}$-valued power set $\mathfrak{Q}^X$ defined according to Zadeh's extension principle is given by:

$$(g_1 \odot g_2)(x) = \bigvee_{x_1 \cdot x_2 = x} g_1(x_1) * g_2(x_2), \qquad g_1, g_2 \in \mathfrak{Q}^X, \ x \in X. \tag{16}$$

If we now identify the tensor product $\mathfrak{Q} \otimes \mathcal{P}(X)$ with $\mathfrak{Q}^X$ (cf. Theorem 2), then $\odot$ coincides with the tensor product of $*$ with $\boxdot$. In fact, because of (15) the relation

$$\begin{aligned}
\bigvee_{x \in X} (g_1 \odot g_2)(x) \otimes \{x\} &= \bigvee_{x_1, x_2 \in X} \big(g_1(x_1) * g_2(x_2)\big) \otimes \{x_1 \cdot x_2\} \\
&= \bigvee_{x_1, x_2 \in X} \big(g_1(x_1) * g(x_2)\big) \otimes \big(\{x_1\} \boxdot \{x_2\}\big) \\
&= \bigvee_{x_1, x_2 \in X} \big(g_1(x_1) \otimes \{x_1\}\big) \star \big(g(x_2)\big) \otimes \{x_2\}\big) \\
&= \Big(\bigvee_{x \in X} g_1(x) \otimes \{x\}\Big) \star \Big(\bigvee_{x \in X} g_2(x) \otimes \{x\}\Big).
\end{aligned}$$

holds. Hence the identification of the multiplication $\odot$ (defined according to Zadeh's extension principle) with the tensor product $\star$ of the respective multiplications follows from Theorem 2.

For further details on the tensor product of unital quantales the reader is referred to [8].

A unital quantale $(X, *, e)$ is called a *unital quantale with involution* or briefly *involutive* if there exists an involutive anti-automorphism $X \xrightarrow{\iota} X$—i.e. ι is an order preserving involution on X satisfying the property $\iota(x * y) = \iota(y) * \iota(x)$ for all $x, y \in X$. Instead of $\iota(x)$ we also write x'. An element $x \in X$ is *self-adjoint* if $x = x'$ holds. The unit and the universal bounds are always self-adjoint.

A unital homomorphism $X \xrightarrow{h} Y$ between involutive and unital quantales is *involutive* if $h(x') = h(x)'$ holds for all $x \in X$.

The next lemma shows that every order preserving involution on a complete lattice X can be extended to an involutive anti-automorphism on the free unital quantale $(X^\sharp, *, e^\sharp)$ generated by X.

Lemma 3 *Let X be a complete lattice and $(X^\sharp, *, e^\sharp)$ be the free unital quantale generated by X with the corresponding embedding $X \xrightarrow{\eta_X} X^\sharp$. For every order preserving involution $X \xrightarrow{\ell} X$ there exists a unique involutive anti-automorphism $\iota^\sharp$ on $(X^\sharp, *, e^\sharp)$ satisfying the condition $\iota^\sharp \circ \eta_X = \eta_X \circ \iota$.*

Proof Since the transposition of the multiplication preserves the associativity law, the triple $(X^\sharp, *^{op}, e^\sharp)$ with $x *^{op} y = y * x$ is again a unital quantale. Then $\iota^\sharp$ is given by the extension of $X \xrightarrow{\eta_X \circ \iota} X^\sharp$ to a unique unital homomorphism $(X^\sharp, *, e^\sharp) \xrightarrow{\iota^\sharp} (X^\sharp, *^{op}, e^\sharp)$. □

Because of the previous lemma there exist a plenty of involutive and unital quantales. We finish this section with two prominent examples. (cf. [21]).

Example 4 Let $\mathfrak{A}$ be a unital C^*-algebra with involution * (cf. Sect. 6.3), and let $\mathrm{Max}(\mathfrak{A})$ be the set of all closed linear subspaces of $\mathfrak{A}$. The adjoint of a closed, linear subspace M has the form $M' = \{a^* \mid a \in M\}$. If M and N are closed, linear subspaces, then the product $M * N$ is given by the closure of the set of all finite sums $\sum a_i \cdot b_i$ with $a_i \in M$ and $b_i \in N$—i.e. $M * N = \overline{MN}$. Hence the restriction of $*$ to closed ideals coincides with the usual ideal multiplication. In particular, $(M * N)' = N' * M'$ holds for all $M, N \in \mathrm{Max}(\mathfrak{A})$.

On $\mathrm{Max}(\mathfrak{A})$ we consider the partial order determined by the set inclusion. Then $\mathrm{Max}(\mathfrak{A})$ is a complete lattice, and the multiplication $*$ of closed linear subspaces is join preserving in each variable separately. Hence $(\mathrm{Max}(\mathfrak{A}), *, ')$ is an involutive and unital quantale. It is well known that $\mathrm{Max}(\mathfrak{A})$ characterizes $\mathfrak{A}$ up to an $*$-isomorphism (cf. Theorem 3.3 in [16]). Sometimes $\mathrm{Max}(\mathfrak{A})$ is also called the *spectrum* of $\mathfrak{A}$.

Example 5 A *complete De Morgan algebra* is a complete lattice L equipped with an order reversing involution ^{0}—this means that the self-map $\lambda \longmapsto \lambda^0$ of L is an involution on L satisfying the condition

$$\lambda_1 \le \lambda_2 \quad \Longrightarrow \quad \lambda_2^0 \le \lambda_1^0, \qquad \lambda_1, \lambda_2 \in L.$$

Hence $\lambda \longmapsto \lambda^0$ is an anti-automorphism of L.

We show that every complete De Morgan algebra $(L, {}^0)$ gives rise to an involutive and unital quantale. For this purpose we consider the complete lattice $[L, L]$ of all join preserving self-maps of L (cf. Sect. 2) provided with the composition as multiplication—i.e.

$$f_1 \circ f_2(\lambda) = f_1(f_2(\lambda)), \quad \lambda \in L.$$

Then $([L, L], \circ, 1_L)$ is a unital quantale. Further, the order reversing involution ^{0} on L induces an order preserving involution $'$ on $[L, L]$ by:

$$f'(\lambda) = \big(f^{\vdash}(\lambda^0)\big)^0, \quad \lambda \in L$$

where $f^{\vdash}$ is the right adjoint map of f—i.e.

$$f^{\vdash}(\mu) = \bigvee \{\lambda \in L \mid f(\lambda) \leq \mu\}, \quad \mu \in L.$$

Since adjoint situations can be composed, the relation $(f_1 \circ f_2)' = f_2' \circ f_1'$ follows. Hence $([L, L], \circ, 1_L, ')$ is an involutive and unital quantale. (cf. [20], Example (5) in [22]).

4 Modules on Unital Quantales

In any monoidal category C the concept of left-modules is available—i.e. objects of C provided with a left action w.r.t. a monoid in C (cf. [17]). Here we recall the axioms of a left-module in the special case that the monoidal category coincides with $(\mathsf{Sup}, \otimes, \mathfrak{a}, \mathfrak{c}, \ell, r, \mathbb{1})$ (cf. Fact in Sect. 2).

Let $(\mathfrak{Q}, *, e)$ be a unital quantale with the corresponding binary operation $\mathfrak{Q} \otimes \mathfrak{Q} \xrightarrow{\circledast} \mathfrak{Q}$ determined by the multiplication $*$. A *left action* on a complete lattice M w.r.t. $(\mathfrak{Q}, *, e)$ is a join preserving map $\mathfrak{Q} \otimes M \xrightarrow{\odot} M$ such that the following diagrams commute (cf. [17]):

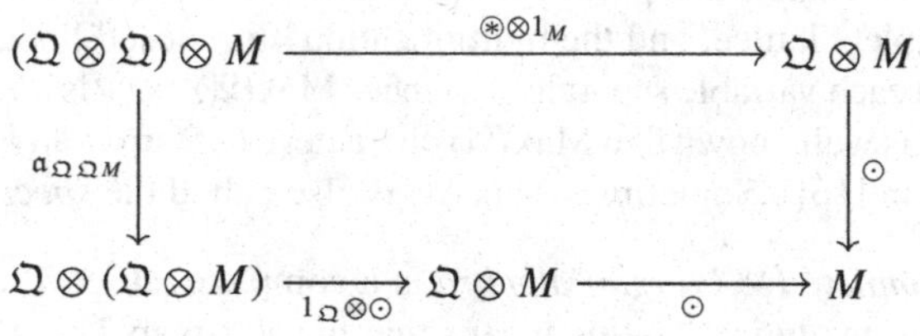

$$\begin{array}{ccccc}
(\mathfrak{Q} \otimes \mathfrak{Q}) \otimes M & & \xrightarrow{\circledast \otimes 1_M} & & \mathfrak{Q} \otimes M \\
\downarrow{\scriptstyle \mathfrak{a}_{\mathfrak{Q}\mathfrak{Q}M}} & & & & \downarrow{\scriptstyle \odot} \\
\mathfrak{Q} \otimes (\mathfrak{Q} \otimes M) & \xrightarrow[1_{\mathfrak{Q}} \otimes \odot]{} & \mathfrak{Q} \otimes M & \xrightarrow[\odot]{} & M
\end{array} \tag{17}$$

$$\begin{array}{ccc}
\mathbb{1} \otimes M & \xrightarrow{e \otimes 1_M} & \mathfrak{Q} \otimes M \\
& {\scriptstyle \ell_M} \searrow & \downarrow{\scriptstyle \odot} \\
& & M
\end{array} \tag{18}$$

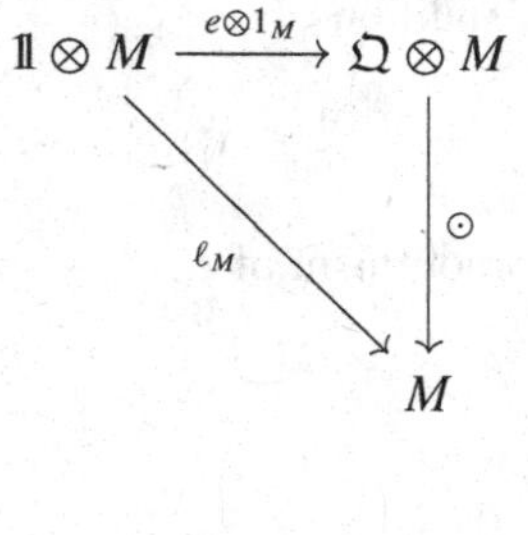

A pair $(M, \odot)$ is a *left $\mathfrak{Q}$-module* if M is a complete lattice and $\odot$ is a left action on M w.r.t. $(\mathfrak{Q}, *, e)$.

Let $(\alpha, x) \longmapsto \odot(\alpha \otimes x)$ be the bimorphism $\mathfrak{Q} \times M \to M$ corresponding to the left action $\odot$ on M. Then instead of $\odot(\alpha \otimes x)$ we also write $\alpha \odot x$.

Since every tensor is the join of an appropriate family of elementary tensors, the commutativity of the diagrams (17) and (18) is equivalent to the following axioms:

(M1) If $\alpha, \beta \in \mathfrak{Q}$ and $m \in M$, then $\alpha \odot (\beta \odot m) = (\alpha * \beta) \odot m$.
(M2) If e denotes the unit of $\mathfrak{Q}$, then $e \odot m = m$ for all $m \in M$.

Proposition 1 *Let $\mathfrak{Q}$ be a unital quantale, M be a complete lattice, and let $[M, M]$ be the unital quantale of all join preserving self-maps of M provided with the composition as multiplication. Further, let $[M, M] \times M \xrightarrow{\mathrm{ev}_M} M$ be the evaluation map—i.e. $\mathrm{ev}_M(f, m) = f(m)$. Then there exists a bijective map between the set of all left actions $\odot$ on M in the sense of* Sup *and the set of all unital homomorphisms $\mathfrak{Q} \xrightarrow{h} [M, M]$ making the following diagram commutative:*

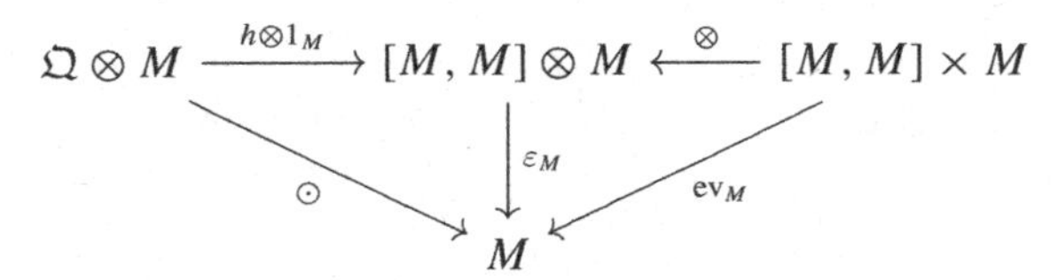

where ε_M is the join preserving map determined by the evaluation map ev_M.

Proof Since Sup is monoidal closed (cf. Theorem 3), the homomorphism h coincides with the monoidal adjoint of the left-action $\odot$. The fact that h preserves the algebraic structure follows from a chase of diagrams or by a simple calculation using the axioms (M1) and (M2) directly. □

The previous proposition says that left $\mathfrak{Q}$-modules M can simply be characterized by unital homomorphisms from $\mathfrak{Q}$ to $[M, M]$.

Before we continue, we first give a simple example of a left $\mathfrak{Q}$-module.

Example 6 Let X be a set and $\mathfrak{Q}^X$ be the set of all maps $X \xrightarrow{f} \mathfrak{Q}$ provided with the partial order defined pointwisely—i.e.

$$f \le g \;\Leftrightarrow\; f(x) \le g(x) \quad \text{for all } x \in X.$$

Then $\mathfrak{Q}^X$ is the $\mathfrak{Q}$-valued power set of X (cf. Sect. 2), and the left action $\odot$ on $\mathfrak{Q}^X$ is determined by

$$(\alpha \odot f)(x) = \alpha * f(x), \qquad \alpha \in \mathfrak{Q},\ f \in \mathfrak{Q}^X,\ x \in X. \tag{19}$$

Hence $\mathfrak{Q}^X$ is a left $\mathfrak{Q}$-module.

Left $\mathfrak{Q}$-module homomorphisms are join preserving maps $M \xrightarrow{h} N$ preserving the respective left actions—i.e. the commutativity of the following diagram:

$$\begin{array}{ccc} \mathfrak{Q} \otimes M & \xrightarrow{1_{\mathfrak{Q}} \otimes h} & \mathfrak{Q} \otimes N \\ \big\downarrow{\scriptstyle \odot} & & \big\downarrow{\scriptstyle \odot} \\ M & \xrightarrow[h]{} & N \end{array}$$

As usually left $\mathfrak{Q}$-modules and left $\mathfrak{Q}$-module homomorphisms form a category which we denote by $\mathtt{Mod}(\mathfrak{Q})$.

It is well known that the forgetful functor $\mathsf{U}_1 : \mathtt{Mod}(\mathfrak{Q}) \to \mathtt{Sup}$ has a left adjoint functor $\mathsf{F}_1 : \mathtt{Sup} \to \mathtt{Mod}(\mathfrak{Q})$ sending a complete lattice X to $\mathfrak{Q} \otimes X$ where the left action $\odot$ on $\mathfrak{Q} \otimes X$ is defined by the commutativity of the following diagram (cf. [17]):

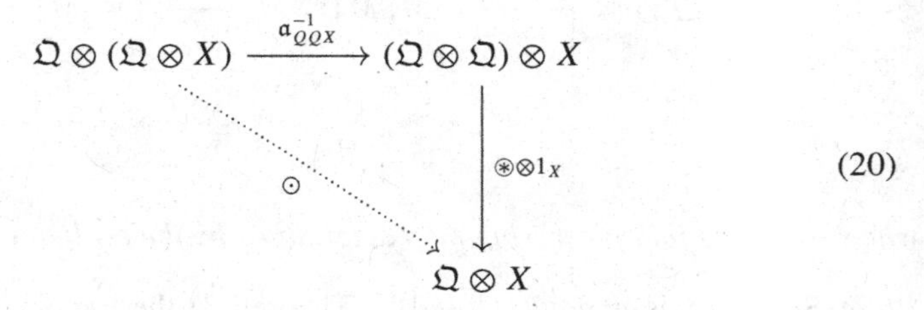

Since in [17] the commutativity of the diagrams (17) and (18) has not been verified in the general case of monoidal categories, we insert here this information in the special case of $\mathtt{Sup}$ for the convenience of the reader.

Lemma 4 *Let X be a complete lattice. Then the join preserving map $\mathfrak{Q} \otimes (\mathfrak{Q} \otimes X) \xrightarrow{\odot} \mathfrak{Q} \otimes X$ determined by (20) is a left action on $\mathfrak{Q} \otimes X$.*

Proof Since tensors in $\mathfrak{Q} \otimes X$ are joins of elementary tensors, we restrict our interest to elementary tensors and choose $x \in X$ and $\alpha, \beta, \gamma \in \mathfrak{Q}$. Since $\circledast$ is associative, we obtain:

$$\begin{aligned} \alpha \odot (\beta \odot (\gamma \otimes x)) &= \alpha \odot \big((\beta \circledast \gamma) \otimes x\big) \\ &= (\alpha \circledast (\beta \circledast \gamma)) \otimes x \\ &= ((\alpha \circledast \beta) \circledast \gamma) \otimes x \\ &= (\alpha \circledast \beta) \odot (\gamma \otimes x). \end{aligned}$$

Hence (M1) is verified. The axiom (M2) is evident. □

Addition. For $\alpha \in \mathfrak{Q}$ and $f \in \mathfrak{Q} \otimes X$ the expression $\alpha \odot f$ is explicitly given by $\alpha \odot f = \bigvee_{\beta \in \mathfrak{Q}} (\alpha * \beta) \otimes f(\beta)$ where we have made use of (6).

A combination of Lemma 4 with Example 6 leads to the following corollary of Theorem 2.

Corollary 3 ([12]) *Let X be a set and $\mathcal{P}(X)$ be the power set of X. Then the order isomorphism $\mathfrak{Q}^X \xrightarrow{\Phi_{\mathfrak{Q}}} \mathfrak{Q} \otimes \mathcal{P}(X)$ specified in Theorem 2 is a left $\mathfrak{Q}$-module isomorphism.*

Proof We maintain the notation from Example 6 and Lemma 4. In order to verify the relation $\alpha \odot \Phi_{\mathfrak{Q}}(g) = \Phi_{\mathfrak{Q}}(\alpha \odot g)$ for all $\alpha \in \mathfrak{Q}$ and $g \in \mathfrak{Q}^X$ we refer to Theorem 2 and obtain:

$$\begin{aligned} \alpha \odot \Phi_{\mathfrak{Q}}(g) &= \alpha \odot \Big(\bigvee_{x \in X} g(x) \otimes \{x\} \Big) \\ &= \bigvee_{x \in X} \alpha \odot (g(x) \otimes \{x\}) \\ &= \bigvee_{x \in X} (\alpha \circledast g(x)) \otimes \{x\} \\ &= \bigvee_{x \in X} (\alpha * g(x)) \otimes \{x\} \\ &= \Phi_{\mathfrak{Q}}(\alpha \odot g). \end{aligned}$$

□

Theorem 4 *Let X be a complete lattice, $(\mathfrak{Q}, *, e)$ be a unital quantale and $X \xrightarrow{\eta_X^M} \mathfrak{Q} \otimes X$ be the join preserving map determined by $\eta_X^M(x) = e \otimes x$ for all $x \in X$. Then for every left $\mathfrak{Q}$-module N and for every join preserving map $X \xrightarrow{h} N$ there exists a unique left $\mathfrak{Q}$-module homomorphism $\mathfrak{Q} \otimes X \xrightarrow{h^\sharp} N$ making the following diagram commutative:*

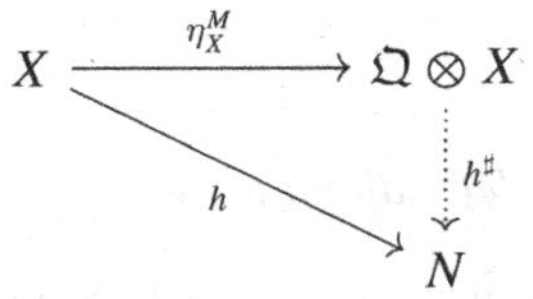

$$\begin{array}{ccc} X & \xrightarrow{\eta_X^M} & \mathfrak{Q} \otimes X \\ & {\scriptstyle h} \searrow & \vdots\, {\scriptstyle h^\sharp} \\ & & N \end{array} \qquad (21)$$

Proof (a) (Uniqueness). Let $h^\sharp$ be a left $\mathfrak{Q}$-module homomorphism making the diagram (21) commutative. Then for all $x \in X$ the relation $h^\sharp(e \otimes x) = h(x)$ follows. Because of (20) (cf. Lemma 4) we obtain for $\gamma \in \mathfrak{Q}$ and $x \in X$:

$$h^\sharp(\gamma \otimes x) = h^\sharp((\gamma * e) \otimes x) = h^\sharp(\gamma \odot (e \otimes x)) = \gamma \odot h^\sharp(e \otimes x) = \gamma \odot h(x).$$

Since every tensor in $\mathfrak{Q} \otimes X$ is a join of elementary tensors, $h^\sharp$ is uniquely determined by the commutativity of the diagram (21).

(b) (Existence). Let $\odot$ be the left action on N. We define a join preserving map $\mathfrak{Q} \otimes X \xrightarrow{h^\sharp} N$ by

$$h^\sharp = \odot \circ (1_{\mathfrak{Q}} \otimes h). \tag{22}$$

Then for every $\alpha \in \mathfrak{Q}$ and for every elementary tensor $\gamma \otimes x$ the relation

$$\begin{aligned} h^\sharp(\alpha \odot (\gamma \otimes x)) &= h^\sharp((\alpha \circledast \gamma) \otimes x) \\ &= [\odot \circ (1_{\mathfrak{Q}} \otimes h)]\big((\alpha \circledast \gamma) \otimes x\big) \\ &= (\alpha \circledast \gamma) \odot h(x) \\ &= \alpha \odot (\gamma \odot h(x)) \\ &= \alpha \odot h^\sharp(\gamma \otimes x). \end{aligned}$$

holds. Since $h^\sharp$ is join preserving and every tensor is a join of elementary tensors, $h^\sharp$ is obviously a left $\mathfrak{Q}$-module homomorphism.

Finally, since e is the unit of $\mathfrak{Q}$, we conclude from (22) that the relation

$$h^\sharp(e \otimes x) = e \odot h(x) = h(x)$$

holds for all $x \in X$. Hence $h^\sharp$ makes the diagram (21) commutative. □

Referring to Lemma 4 and Theorem 4 there exists a functor

$$\mathsf{F}_1 \colon \mathtt{Sup} \to \mathtt{Mod}(\mathfrak{Q})$$

acting on objects and morphisms as follows:

$$\mathsf{F}_1(X) = \mathfrak{Q} \otimes X \quad \text{and for } X \xrightarrow{\varphi} Y, \qquad \mathsf{F}_1(\varphi) = (\eta_Y^M \circ \varphi)^\sharp. \tag{23}$$

Corollary 4 *The functor* F_1 *is left adjoint to* U_1.

Proof The assertion follows immediately from Theorem 4. □

Corollary 5 *The forgetful functor from* $\mathsf{U}_2 \colon \mathtt{Mod}(\mathfrak{Q}) \to \mathtt{Set}$ *has a left adjoint.*

Proof Let X be a set, $\mathcal{P}(X)$ be the ordinary power set of X, and let L be a complete lattice. Since every map $X \xrightarrow{f} L$ has a unique extension to a join preserving map $\mathcal{P}(X) \xrightarrow{f^\sharp} L$ with

$$f^\sharp(A) = \bigvee_{x \in A} f(x), \qquad A \in \mathcal{P}(X),$$

the forgetful functor $\mathsf{U}_0 \colon \mathtt{Sup} \to \mathtt{Set}$ has a left adjoint functor $\mathsf{F}_0 \colon \mathtt{Set} \to \mathtt{Sup}$ which sends a set X to its power set $\mathcal{P}(X)$. Since adjoint situations compose, we conclude from Corollary 4 that $\mathsf{F}_2 = \mathsf{F}_1 \circ \mathsf{F}_0$ is left adjoint to $\mathsf{U}_2 = \mathsf{U}_0 \circ \mathsf{U}_1$. □

It follows from Corollaries 3, 4 and 5 that the monad corresponding to the adjoint situation $\mathsf{F}_2 \dashv \mathsf{U}_2$ is the $\mathfrak{Q}$-valued power set monad $(\mathsf{P}_{\mathfrak{Q}}, \eta^{\mathfrak{Q}}, \mu^{\mathfrak{Q}})$ (on $\mathtt{Set}$) where

$$\mathsf{P}_{\mathfrak{Q}}(X) = \mathfrak{Q}^X, \quad X \xrightarrow{\varphi} Y, \quad [\mathsf{P}_{\mathfrak{Q}}(\varphi)(g)](y) = \bigvee \{g(x) \mid \varphi(x) = y\}, \quad y \in Y,$$

$$\eta^{\mathfrak{Q}}_X(x) = \Phi^{-1}_{\mathfrak{Q}}(e \otimes \{x\}) = 1_{\{x\}}, \quad 1_{\{x\}}(z) = \begin{cases} e, \ z = x \\ \bot, \ z \neq x, \end{cases} \quad z, x \in X,$$

$$[\mu^{\mathfrak{Q}}_X(G)](x) = \bigvee_{g \in \mathfrak{Q}^X} G(g) * g(x), \quad x \in X, \ G \in \mathfrak{Q}^{\mathfrak{Q}^X}.$$

The Eilenberg-Moore category of the $\mathfrak{Q}$-valued power set monad is isomorphic to $\mathtt{Mod}(\mathfrak{Q})$ (cf. [28]). The proof of this result is a generalization of the standard proof that $\mathtt{Sup}$ is isomorphic to the Eilenberg-Moore category of the ordinary power set monad (see Example I.5.15 in [18]). What is important here for us is the special relationship between ALGEBRA and FUZZY SET THEORY expressed by following statement:

The $\mathfrak{Q}$-valued power set $\mathfrak{Q}^X$ is the free left $\mathfrak{Q}$-module *generated by the set X.*

This observation goes back to Joyal and Tierney (cf. p. 10 in [12]) and means from a mathematical point of view that *fuzzy set theory* is *module theory on unital quantales*. The next section is a confirmation of this insight.

At the end of this section we would like to draw the attention of the reader to the interesting fact that submodules of $\mathfrak{Q}^X$ play a strategic role in the study of *stratified $\mathfrak{Q}$-valued topological spaces* (cf. p. 180 in [9]).

5 Complete $\mathfrak{Q}$-Valued Order Sets and Left $\mathfrak{Q}$-Modules

First we recall the axioms of a many-valued preorder (cf. [5, 11, 23]). Let $(\mathfrak{Q}, *, e)$ be a unital quantale and X be a set. A map $X \times X \xrightarrow{p} \mathfrak{Q}$ is a *$\mathfrak{Q}$-valued preorder* (*$\mathfrak{Q}$-preorder* for short) if p satisfies the following properties for all $x, y, z \in X$:

$$e \leq p(x, x), \qquad \text{(Reflexivity)}$$
$$p(x, y) * p(y, z) \leq p(x, z). \qquad \text{(Transitivity)}$$

Every $\mathfrak{Q}$-preorder p has an underlying ordinary preorder $\leq$ defined by

$$x \leq y \quad \Leftrightarrow \quad e \leq p(x, y). \tag{24}$$

A $\mathfrak{Q}$-preorder p is *skeletal* or *antisymmetric* iff its underlying preorder is antisymmetric. An antisymmetric $\mathfrak{Q}$-preorder is also called a *$\mathfrak{Q}$-valued order* (or *$\mathfrak{Q}$-order* for short).

A pair (X, p) is a *$\mathfrak{Q}$-preordered set* if X is a set and p is a $\mathfrak{Q}$-preorder on X. The same applies to $\mathfrak{Q}$-orders. It is well known that $\mathfrak{Q}$-preordered sets are $\mathfrak{Q}$-enriched categories where $\mathfrak{Q}$ is viewed as a monoidal biclosed category (cf. [15]).

Let (X, p) and (Y, q) be $\mathfrak{Q}$-preordered sets. A *$\mathfrak{Q}$-homomorphism* $X \xrightarrow{h} Y$ is a map satisfying the following condition

$$p(x, z) \leq q(h(x), h(z))$$

for all $x, z \in X$. Obviously, $\mathfrak{Q}$-homomorphisms are always isotone w.r.t. the underlying preorders. The class of $\mathfrak{Q}$-preordered sets and the class of $\mathfrak{Q}$-homomorphisms form a category denoted by $\mathtt{Pre}(\mathfrak{Q})$.

Let (X, p) be a $\mathfrak{Q}$-preordered set. A *covariant $\mathfrak{Q}$-presheaf* on (X, p) is a $\mathfrak{Q}$-fuzzy set $X \xrightarrow{f} \mathfrak{Q}$ which is *right-extensional*—i.e.

$$f(x) * p(x, y) \leq f(y), \qquad x, y \in X.$$

On the set $P(X, p)$ of all covariant $\mathfrak{Q}$-presheaves on (X, p) we introduce a $\mathfrak{Q}$-order $P(X, p) \times P(X, p) \xrightarrow{d} \mathfrak{Q}$ as follows. First, we recall the left-implication $\swarrow$ of $\mathfrak{Q}$

$$\alpha \swarrow \beta = \bigvee \{\gamma \in \mathfrak{Q} \mid \gamma * \beta \leq \alpha\}, \qquad \alpha, \beta \in \mathfrak{Q}.$$

Then $\swarrow$ is an antisymmetric $\mathfrak{Q}$-preorder on $\mathfrak{Q}$ and its underlying partial order coincides with the dual order of $\mathfrak{Q}$. Now the $\mathfrak{Q}$-order d on $P(X, p)$ is defined by

$$d(f, g) = \bigwedge_{x \in X} f(x) \swarrow g(x), \qquad f, g \in P(X, p), \tag{25}$$

Hence $\big(P(X, p), d\big)$ is a $\mathfrak{Q}$-ordered set.

Lemma 5 *Let (X, p) be a $\mathfrak{Q}$-preordered set and $P(X, p) \xrightarrow{\xi} X$ be a $\mathfrak{Q}$-homomorphism satisfying the condition $\xi\big(p(x, \_)\big) = x$ for all $x \in X$. Then p is antisymmetric, and every further $\mathfrak{Q}$-homomorphism $P(X, p) \xrightarrow{\overline{\xi}} X$ with $\overline{\xi}\big(p(x, \_)\big) = x$ coincides with ξ—i.e. $\xi = \overline{\xi}$.*

Proof Let us choose $x, y \in X$ with $e \leq p(x, y)$ and $e \leq p(y, x)$. Then we conclude from the transitivity of p that $p(x, z) = p(y, z)$ holds for all $z \in X$. Hence $x = \xi(p(x, \_)) = \xi(p(y, \_)) = y$ follows—i.e. p is antisymmetric.

Moreover, since every covariant $\mathfrak{Q}$-presheaf is right-extensional, the relation

$$d(f, p(x, \_)) = \bigwedge_{z \in X} f(z) \swarrow p(x, z) = f(x)$$

holds for all $x \in X$. Hence we obtain:

$$f(x) \leq p\big(\xi(f), \xi(p(x, \_))\big) = p\big(\xi(f), x\big), \quad x \in X. \tag{26}$$

If $P(X, p) \xrightarrow{\overline{\xi}} X$ is a further $\mathfrak{Q}$-homomorphism with $\overline{\xi}\big(p(x, \_)\big) = x$ for all $x \in X$, then we infer from (26) that $\xi(f) \leq \overline{\xi}(f)$ holds where $\leq$ is the underlying order in (X, p). Interchanging now the role of ξ and $\overline{\xi}$ the relation $\xi = \overline{\xi}$ follows from the antisymmetry of $\leq$. □

Motivated by Lemma 5 we introduce the following terminology.

Definition 2 A triple (X, p, ξ) is a called a *complete $\mathfrak{Q}$-ordered set* if (X, p) is a $\mathfrak{Q}$-ordered set and $P(X, p) \xrightarrow{\xi} X$ is a $\mathfrak{Q}$-homomorphism provided with the property:

$$\xi\big(p(x, \_)\big) = x, \qquad x \in X. \tag{27}$$

Since ξ is unique, ξ is also called the *formation of arbitrary meets* in (X, p).

Theorem 5 *Let M be a left $\mathfrak{Q}$-module with the left action $\odot$. There exists a $\mathfrak{Q}$-preorder p on M provided with the following properties:*

(i) $p(x, y) = \bigvee\{\alpha \in \mathfrak{Q} \mid \alpha \odot y \leq x\}, \quad x, y \in M.$

(ii) The map $P(X, p) \xrightarrow{\xi} X$ defined by

$$\xi(f) = \bigvee_{x \in M} f(x) \odot x, \qquad f \in P(M, p). \tag{28}$$

is a $\mathfrak{Q}$-homomorphism and satisfies (27)—i.e. (M, p, ξ) is a complete $\mathfrak{Q}$-ordered set.

Proof We define p by (i) and show that p is a $\mathfrak{Q}$-preorder. Because of (M2) the reflexivity of p is evident. With regard to the transitivity of p we use (M1) and observe:

$$\big(p(x, y) * p(y, z)\big) \odot z = p(x, y) \odot (p(y, z) \odot z) \leq p(x, y) \odot y \leq x.$$

Hence $p(x, y) * p(y, z) \leq p(x, z)$ follows. Since the underlying preorder of p coincides with the dual order of M, p is even antisymmetric—i.e. p is $\mathfrak{Q}$-order on M.

In order to verify (ii) we proceed as follows. Because of $p(x, z) \odot z \leq x$ it follows immediately from (28) that $\xi\big(p(x, \_)\big) = \bigvee_{z \in M} p(x, z) \odot z = x$ holds for all $x \in X$.

Hence ξ satisfies (27). It remains to show that ξ is a $\mathfrak{Q}$-homomorphism. For this purpose we choose $f, g \in P(M, p)$ and observe:

$$\begin{aligned} d(f,g) \odot \xi(g) &= \bigvee_{x \in X} d(f,g) \odot (g(x) \odot x) \\ &= \Big(\Big(\bigvee_{x\in X} f(x) \swarrow g(x)\Big) * g(x)\Big) \odot x \\ &\leq \bigvee_{x \in X} f(x) \odot x \\ &= \xi(f). \end{aligned}$$

Hence $d(f,g) \leq p(\xi(f), \xi(g))$ follows—i.e. ξ is in fact a $\mathfrak{Q}$-homomorphism. □

Corollary 6 *Let X be a set. Then the complete $\mathfrak{Q}$-ordered set $(\mathfrak{Q}^X, p, \xi)$ induced by the free left $\mathfrak{Q}$-module $\mathfrak{Q}^X$ has the following form:*

$$p(f,g) = \bigwedge_{x \in X} f(x) \swarrow g(x), \qquad f, g \in \mathfrak{Q}^X,$$
$$P(\mathfrak{Q}^X, p) \xrightarrow{\xi} \mathfrak{Q}^X, \quad \xi(F)(x) = \bigvee_{f \in \mathfrak{Q}^X} F(f) * f(x), \quad F \in P(\mathfrak{Q}^X, p).$$

Proof Let $\odot$ be the left action on $\mathfrak{Q}^X$ determined by (19). Then

$$\bigwedge_{x \in X} f(x) \swarrow g(x) = \bigvee \{\alpha \in \mathfrak{Q} \mid \alpha \odot g \leq f\}$$

follows. Hence the assertion follows immediately from Theorem 5, (28) and (19). □

Theorem 5 shows that every left $\mathfrak{Q}$-module gives rise to a complete $\mathfrak{Q}$-ordered set. In the following considerations we show that also the converse holds. For this purpose we first complete the object function $(X, p) \mapsto P(X, p)$ to an endofunctor P of $\mathtt{Pre}(\mathfrak{Q})$:

$$(X,p) \xrightarrow{h} (Y,q), \quad P(X,p) \xrightarrow{\mathsf{P}(h)} P(Y,q),$$
$$[\mathsf{P}(h)(g)](y) = \bigvee_{x \in X} g(x) * q(h(x), y), \quad y \in Y.$$

Further, we need a natural transformation $\mu\colon \mathsf{P} \circ \mathsf{P} \to \mathsf{P}$ determined by

$$\mu = (\mu_{(X,p)})_{(X,p) \in |\mathtt{Pre}(\mathfrak{Q})|}, \quad \text{where} \quad P(P(X,p), d) \xrightarrow{\mu_{(X,p)}} P(X,p),$$
$$[\mu_{(X,p)}(F)](x) = \bigvee_{f \in P(X,p)} F(f) * f(x), \quad F \in P\big(P(X,p), d\big), \; x \in X.$$

Lemma 6 *Let (X, p, ξ) be a complete $\mathfrak{Q}$-ordered set. Then the following diagram is commutative:*

$$\begin{array}{ccc} P(P(X,p),d) & \xrightarrow{\mathsf{P}(\xi)} & P(X,p) \\ \Big\downarrow{\scriptstyle \mu_{(X,p)}} & & \Big\downarrow{\scriptstyle \xi} \\ P(X,p) & \xrightarrow[\xi]{} & X \end{array} \tag{29}$$

Proof Let $\leq^{op}$ be the dual partial order of the underlying partial order of p—i.e.

$$x \leq^{op} y \quad \Leftrightarrow \quad e \leq p(y, x).$$

Since ξ is a $\mathfrak{Q}$-homomorphism, for $f, g \in P(X, p)$ we derive the following implication from the definition of d:

$$f \leq g \quad \Rightarrow \quad \xi(f) \leq^{op} \xi(g). \tag{30}$$

Further, because of the right-extensionality of covariant presheaves $f \in P(X, p)$ the relation

$$f(x) = d\big(f, p(x, \_)\big) = \bigwedge_{z \in X} f(z) \swarrow d(x, z).$$

holds for all $x \in X$. Now we apply (27) and again the property that ξ is a $\mathfrak{Q}$-homomorphism and obtain

$$f(x) \leq p(\xi(f), x) \qquad f \in P(X, p), \ x \in X \tag{31}$$

which implies $\mu_{(X,p)}(F) \leq \mathsf{P}(\xi)(F)$ for all $F \in P\big(P(X, p), d\big)$. Because of (30) the following relation holds:

$$\xi\big(\mu_{(X,p)}(F)\big) \leq^{op} \xi\big(\mathsf{P}(\xi)(F)\big). \tag{32}$$

On the other hand, for all $f_0 \in P(X, p)$ we observe $F(f_0) \leq d\big(\mu_{(X,p)}(F), f_0\big)$. Hence the relation

$$F(f_0) * p\big(\xi(f_0), z\big) \leq p\big(\xi(\mu_{(X,p)}(F)), \xi(f_0)\big) * p\big(\xi(f_0), z\big) \leq p\big(\xi(\mu_{(X,p)}(F)), z\big)$$

is valid—i.e. $\mathsf{P}(\xi)(F) \leq p\big(\xi(\mu_{(X,p)}(F)), \_\big)$. Now we apply (27) and (30) and obtain:

$$\xi\big(\mathsf{P}(\xi)(F)\big) \leq^{op} \xi\big(\mu_{(X,p)}(F)\big). \tag{33}$$

Finally, since $\leq^{op}$ is antisymmetric, $\xi\big(\mathsf{P}(\xi)(F)\big) = \xi\big(\mu_{(X,p)}(F)\big)$ follows from (32) and (33). Thus the diagram (29) is commutative. □

Theorem 6 *Let (X, p, ξ) be a complete $\mathfrak{Q}$-ordered set and $\leq^{op}$ be the dual order w.r.t. the underlying partial order of p. Then $(X, \leq^{op})$ is a complete (ordinary) lattice, and the map $(\alpha, x) \longmapsto \alpha \cdot x = \xi\big(\alpha * p(x, \_)\big)$ is a bimorphism. Moreover, the join preserving map $\mathfrak{Q} \otimes X \xrightarrow{\odot} X$ determined by*

$$\alpha \odot x = \xi\big(\alpha * p(x, \_)\big), \quad \alpha \in \mathfrak{Q},\ x \in X \tag{34}$$

is a left action on X—i.e. $(X, \odot)$ is a left $\mathfrak{Q}$-module, and the following relation holds:

$$\xi(g) = \bigvee_{x \in X} g(x) \odot x, \qquad g \in P(X, p). \tag{35}$$

Proof Let $\bot$ be the universal lower bound of $P(X, p)$ in the sense of the dual order of the underlying partial order of d—i.e.

$$\bot(z) \leq g(z), \quad g \in P(X, p), \quad z \in X$$

where $\leq$ is the partial order on $\mathfrak{Q}$. Because of (30) the element $\xi(\bot)$ is the universal lower bound of $(X, \leq^{op})$. If A is a non empty subset of X, then it follows immediately from (30) that $\xi\big(\bigvee_{x \in A} p(x, \_)\big)$ is the join of A w.r.t. $\leq^{op}$—this means that $(X, \leq^{op})$ is a complete lattice.

(a) We define a map $\mathfrak{Q} \times X \to X$ by

$$\alpha \cdot x = \xi\big(\alpha * p(x, \_)\big), \quad \alpha \in \mathfrak{Q},\ x \in X \tag{36}$$

and show that $\cdot$ is a bimorphism (in Sup). Because of the definition of $\cdot$ the relation $\bot \cdot x = \bot$ is evident for all $x \in X$. On the other hand, let us consider the universal lower bound $\bot = \xi(\bot)$ in $(X, \leq^{op})$. Then we define a right-extensional map $P(X, p) \xrightarrow{F} \mathfrak{Q}$ by

$$F(f) = \alpha * d(\bot, f), \qquad f \in P(X, p),\ \alpha \in \mathfrak{Q}.$$

Obviously, $\mu_{(X,p)}(F)$ coincides with the universal lower bound in $P(X, p)$—i.e. $\mu_{(X,p)}(F) = \bot$. Further, we observe:

$$[\mathsf{P}(\xi)(F)](z) = \bigvee_{f \in P(X,p)} \alpha * d(\bot, f) * p(\xi(f), z) = \alpha * p(\xi(\bot), z), \quad z \in X.$$

Hence the relation

$$\bot = \xi \circ \mu_{(X,p)}(F) = \xi \circ \mathsf{P}(\xi)(F) = \xi\big(\alpha * p(\xi(\bot), \_)\big) = \alpha \cdot \xi(\bot) = \alpha \cdot \bot$$

follows from Lemma 6 and (36).

Further, for any non empty subset $\{g_i \mid i \in I\}$ of $P(X, p)$ we define an element F of $P(P(X, p), d)$ by:

$$F(f) = \bigvee_{i \in I} d(g_i, f), \quad f \in P(X, p).$$

Referring again to Lemma 6 we obtain:

$$\xi(\bigvee_{i \in I} g_i) = \xi(\mu_{(X,p)}(F)) = \xi(\mathsf{P}(\xi)(F)) = \xi(\bigvee_{i \in I} p(\xi(g_i), \_)).$$

Hence ξ is join preserving w.r.t. $\leq^{op}$.

Because of the previous observation the relation $(\bigvee_{i \in I} \alpha_i) \cdot x = \bigvee_{i \in I} \alpha_i \cdot x$ follows immediately from the definition of $\cdot$. Now we consider a non empty subset $\{x_i \mid i \in I\}$ of X and define $g = \bigvee_{i \in I} p(x_i, \_)$. Hence $\xi(g)$ is the join of $\{x_i \mid i \in I\}$ w.r.t. $\leq^{op}$. Further, for $\alpha \in \mathfrak{Q}$ we define $F \in P(P(X, p), d)$ by:

$$F(f) = \alpha * d(g, f), \quad f \in P(X, p).$$

Obviously, $\mu_{(X,p)}(F) = \alpha * g$ holds. Now we apply Lemma 6 and use again the fact that ξ is join preserving:

$$\begin{aligned} \bigvee_{i \in I} \alpha \cdot x_i &= \xi(\bigvee_{i \in I} \alpha * p(x_i, \_)) \\ &= \xi(\alpha * g) \\ &= \xi(\mu_{(X,p)}(F)) \\ &= \xi(\mathsf{P}(\xi)(F)) \\ &= \xi\big(\alpha * p(\xi(g), \_)\big) \\ &= \alpha \cdot (\bigvee_{i \in I} x_i). \end{aligned}$$

Hence we have verified that $\cdot$ is join preserving in each variable separately w.r.t. $\leq^{op}$.

(b) Because of the universal property of the tensor product there exists a unique join preserving map $\mathfrak{Q} \otimes X \xrightarrow{\odot} X$ making the diagram

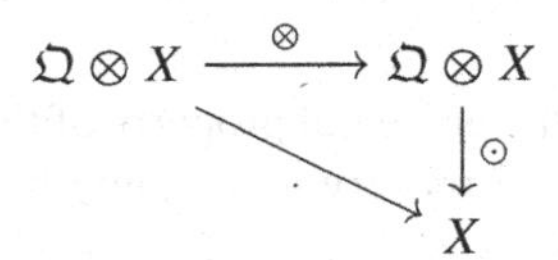

commutative. Obviously, $\odot$ satisfies (34) because of (36). Therfore we only show that $\odot$ is a left action on $(X, \leq^{op})$.

Because of (27) and (34) the axiom (M2) is evident. In order to verify (M1) we fix α, $\beta \in \mathfrak{Q}$ and $x \in X$. Then we define $F \in P(P(X, p), d)$ as follows:

$$g(z) = \alpha * p(x, z), \quad z \in X, \qquad F(f) = \beta * d(g, f), \quad f \in P(X, p).$$

Obviously, $\mu_{(X,p)}(F) = (\beta * \alpha) * p(x, \_)$ and $\xi(g) = \alpha \odot x$ hold. Now we apply again Lemma 6 and obtain:

$$(\beta * \alpha) \odot x = \xi(\mu_{(X,p)}(F)) = \xi(\mathsf{P}(\xi)(F)) = \xi(\beta * p(\xi(g), \_)) = \beta \odot (\alpha \odot x).$$

(c) In order to verify (35) we fix $g \in P(X, p)$ and use its right-extensionality—i.e.

$$g(z) = \bigvee_{x \in X} g(x) * p(x, z), \quad z \in X.$$

Hence (35) follows form (34) and the property that ξ is join preserving. □

From Theorems 5 and 6 it follows that left $\mathfrak{Q}$-modules and complete $\mathfrak{Q}$-ordered sets are equivalent concepts—a result which goes back to Stubbe in the more general context of quantaloid enriched categories (cf. [29], see also Remark 5.6 in [10]). In this context, complete $\mathfrak{Q}$-ordered sets emphasizes the *many-valued* (i.e. enriched categorical) aspect, while left $\mathfrak{Q}$-modules refer to the *algebraic* properties of this theory. Since in the fuzzy community pre-singletons of the form $\alpha * p(x, \_)$ are viewed as *$\mathfrak{Q}$-fuzzy points*, it is important to realize that the *left action of α on x* in the sense of Sup means the *join* of $\alpha * p(x, \_)$ w.r.t. the dual order determined by p (cf. Theorem 5(i)).

Finally, we mention the fact that complete $\mathfrak{Q}$-ordered sets coincide with algebras of the monad of covariant $\mathfrak{Q}$-presheaves (cf. Remark 5.7 in [10]).

6 Left $\mathfrak{Q}$-Modules on Involutive Quantales

Let $\mathfrak{Q}$ be an involutive and unital quantale with unit e and an order preserving involution $'$. Further, let $(M, \odot)$ be a left $\mathfrak{Q}$-module, and M^{op} be the complete lattice provided with the dual order of M. It is easily seen that the right implication $\mathfrak{Q} \times M^{op} \xrightarrow{\searrow} M^{op}$ defined by

$$\alpha \searrow m = \bigvee \{n \in M \mid \alpha \cdot n \leq m\}, \qquad \alpha \in \mathfrak{Q}, \, m \in M$$

is a bimorphism. Referring to the universal property of the tensor product in Sup the right implication $\searrow$ and the order preserving involution $'$on $\mathfrak{Q}$ induce a left action $\odot$ on M^{op} determined as follows

$$\alpha \odot m = \alpha' \searrow m, \qquad \alpha \in \mathfrak{Q}, \, m \in M \tag{37}$$

In fact, the following relations are an immediate corollary of (M1) and (M2):

$$e \odot m = e' \searrow m = e \searrow m = m,$$

$$\alpha \odot (\beta \odot m) = \alpha' \searrow (\beta' \searrow m) = (\beta' * \alpha') \searrow m = (\alpha * \beta) \odot m.$$

Hence M^{op} is a left $\mathfrak{Q}$-module w.r.t. the left action defined by (37).

Proposition 2 *Let $\mathfrak{Q}$ be an involutive and unital quantale and $X \xrightarrow{h} Y$ be a left $\mathfrak{Q}$-module homomorphism. If $h^{\vdash}$ is the right adjoint of h and the complete lattices X^{op} and Y^{op} are provided with the respective left actions according to (37), then $Y^{op} \xrightarrow{h^{\vdash}} X^{op}$ is again a left $\mathfrak{Q}$-module homomorphism.*

Proof Let us choose $\alpha \in \mathfrak{Q}$, $x \in X$ and $y \in Y$. Then the following chain of equivalences hold:

$$x \leq \alpha \odot h^{\vdash}(y) \Leftrightarrow \alpha' \odot x \leq h^{\vdash}(y) \Leftrightarrow h(\alpha' \odot x) = \alpha' \odot h(x) \leq y.$$

Hence $x \leq \alpha \odot h^{\vdash}(y) \Leftrightarrow h(x) \leq \alpha' \searrow y \Leftrightarrow x \leq h^{\vdash}(\alpha \odot y)$ follows—i.e. $\alpha \odot h^{\vdash}(y) = h^{\vdash}(\alpha \odot y)$. □

If $\mathfrak{Q}$ is an involutive and unital quantale, then we conclude from the previous proposition that the category Mod($\mathfrak{Q}$) is self-dual.

6.1 Two-Forms in Sup

Let $\otimes$ be the tensor product and $\mathbb{1}$ be the unit object in Sup (cf. Sect. 2). A *2-form* on a complete lattice M is a join preserving map $M \otimes M \xrightarrow{\varphi} \mathbb{1}$. A 2-form on M is *symmetric* if the diagram

$$\begin{array}{ccc} M \otimes M & \xrightarrow{c_{MM}} & M \otimes M \\ & \searrow^{\varphi} & \downarrow \varphi \\ & & \mathbb{1} \end{array}$$

is commutative where c_{MM} is determined by Lemma 1.

Since Sup is monoidal closed (cf. Theorem 3), every 2-form on M can be identified with its monoidal adjoint map $M \xrightarrow{\ulcorner \varphi \urcorner} [M, \mathbb{1}]$. A symmetric 2-form φ on M is *faithful* if its monoidal adjoint is a monomorphism in Sup—this means an injective and join preserving map.

In order to give a characterization of 2-forms we recall the simple fact that for every complete lattice X a join preserving map $X \xrightarrow{h} \mathbb{1}$ can be identified with a *unique element* $z \in X$—i.e.

$$h(x) = \begin{cases} 0, & x \leq z, \\ 1, & x \nleq z. \end{cases}$$

Hence by definition of the tensor product (cf. Sect. 2) a 2-form $M \otimes M \xrightarrow{\varphi} \mathbb{1}$ can be identified with a join reversing map $M \xrightarrow{f} M$. In particular, a 2-form φ is symmetric iff the chain of equivalences

$$n \leq f(m) \quad \Leftrightarrow \quad m \otimes n \leq f \quad \Leftrightarrow \quad n \otimes m \leq f \quad \Leftrightarrow \quad m \leq f(n)$$

holds for all $m, n \in M$—this means that φ is symmetric iff the corresponding join reversing map f satisfies the following condition

$$m \leq f(f(m)), \qquad m \in M. \tag{38}$$

Since f is antitone and $\leq$ is antisymmetric, it is interesting to see that (38) also implies the subsequent relation

$$f(m) = f(f(f(m))), \quad m \in M. \tag{39}$$

Further, the monoidal adjoint map $\ulcorner\varphi\urcorner$ of a 2-form φ can be characterized by its corresponding join reversing map f as follows:

$$[\ulcorner\varphi\urcorner(m)](n) = 0 \quad \Leftrightarrow \quad \varphi(m \otimes n) = 0 \quad \Leftrightarrow \quad m \otimes n \leq f \quad \Leftrightarrow \quad n \leq f(m)$$

i.e.
$$[\ulcorner\varphi\urcorner(m)](n) = \begin{cases} 0, & n \leq f(m), \\ 1, & n \nleq f(m), \end{cases} \qquad n \in M.$$

Because of (39) a symmetric 2-form φ on M is faithful iff the corresponding join reversing map f is an involution. Hence we have established the following important result due to Resende [24].

Proposition 3 *Let M be a complete lattice. There exists a bijective map between the set of all symmetric and faithful 2-forms φ on M and the set of all order reversing involutions ^{0} on M such that the condition*

$$\varphi(m \otimes n) = 0 \quad \Leftrightarrow \quad n \leq m^0$$

holds for all $m, n \in M$.

Because of the previous proposition we gain the important understanding that the lattice-theoretic concept of a complete De Morgan algebra (cf. Example 5) can be formulated entirely in terms of categorical data provided by Sup. In this sense order reversing involutions on complete lattices are not only related to non-classical negations, but they play also an important *geometric* role expressed by their associated *orthogonality relation*:

$$m \perp n \quad \Leftrightarrow \quad n \leq m^0.$$

6.2 Involutive Left $\mathfrak{Q}$-Modules

Let $(\mathfrak{Q}, *, e, ')$ be a unital quantale with involution. We enrich left $\mathfrak{Q}$-modules by symmetric 2-forms (cf. [24]).

Definition 3 A pair (M, φ) is called an *involutive left $\mathfrak{Q}$-module* if M is left $\mathfrak{Q}$-module and φ is a symmetric 2-form on M such that the relation

$$\varphi\big((\alpha \odot x) \otimes y\big) = \varphi(x \otimes (\alpha' \odot y) \tag{40}$$

holds for all $\alpha \in \mathfrak{Q}$ and $x, y \in M$ where $\odot$ is the left action on M.

The next theorem is a refinement of Proposition 1.

Theorem 7 *Let $(M, {}^0)$ be a complete De Morgan algebra and $[M, M]$ be the involutive and unital quantale of all join preserving self-maps of M (cf. Example 5). Further, let φ be the faithful and symmetric 2-form on M corresponding to the order reversing involution $\alpha \mapsto \alpha^0$. Then there exists a bijective map between the set of all left actions $\odot$ on M with the property that (M, φ) is an involutive left $\mathfrak{Q}$-module and the set of all involutive and unital homomorphisms $\mathfrak{Q} \xrightarrow{h} [M, M]$ such that the following diagram is commutative:*

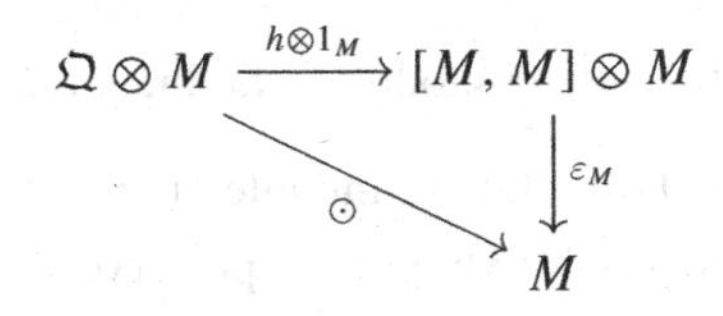

where ε_M is the join preserving map corresponding to the evaluation map ev_M.

Proof Let $\odot$ be a left action on M. Referring to the proof of Proposition 1 it is sufficient to show that $\odot$ satisfies (40) iff the monoidal adjoint map of $\odot$ is an involutive (unital) homomorphism.

(a) Let us assume that $\odot$ fulfills (40) and h is the monoidal adjoint of $\odot$. We maintain the notation of Example 5 and obtain the following chain of equivalences for all $\alpha \in \mathfrak{Q}$ and $m, n \in M$:

$$\begin{aligned} m \le [h(\alpha)^{\vdash}](n^0) &\Leftrightarrow [h(\alpha)](m) \le n^0 \\ &\Leftrightarrow \alpha \odot m \le n^0 \\ &\Leftrightarrow \varphi\big((\alpha \odot m) \otimes n\big) = 0 \\ &\Leftrightarrow \varphi\big(m \otimes (\alpha' \odot n)\big) = 0 \\ &\Leftrightarrow \alpha' \odot n \le m^0 \\ &\Leftrightarrow [h(\alpha')](n) \le m^0 \\ &\Leftrightarrow m \le \big([h(\alpha'](n)\big)^0. \end{aligned}$$

Hence $[h(\alpha)'](n) = [h(\alpha'](n)$ follows for all $n \in M$.

(b) Let us assume that $\mathfrak{Q} \xrightarrow{h} [M, M]$ is an involutive and unital homomorphism. Since the left action $\odot$ on M is determined by:

$$\alpha \odot m = [h(\alpha)](m), \quad \alpha \in \mathfrak{Q},\ m \in M, \tag{41}$$

we obtain:

$$\begin{aligned} \varphi\big((\alpha \odot m) \otimes n\big) = 0 &\Leftrightarrow \alpha \odot m \le n^0 \\ &\Leftrightarrow [h(\alpha)](m) \le n^0 \\ &\Leftrightarrow m \le [h(\alpha)^{\vdash}](n^0) \\ &\Leftrightarrow [h(\alpha)'](n) \le m^0 \\ &\Leftrightarrow [h(\alpha')](n) \le m^0 \\ &\Leftrightarrow \alpha' \odot n \le m^0 \\ &\Leftrightarrow \varphi\big(m \otimes (\alpha' \odot n)\big) = 0. \end{aligned}$$

Hence (M, φ) is an involutive left $\mathfrak{Q}$-module w.r.t. the left action defined in (41). □

Morphisms between involutive left $\mathfrak{Q}$-modules (M, φ) and (N, ψ) are left $\mathfrak{Q}$-module homomorphisms $M \xrightarrow{h} N$ which also preserve the respective symmetric 2-forms—i.e. the commutativity of the following diagram:

$$\begin{array}{ccc} M \otimes M & \xrightarrow{h \otimes h} & N \otimes N \\ & \searrow^{\varphi} & \downarrow \psi \\ & & \mathbb{1} \end{array}$$

In the case of faithful and symmetric 2-forms a left $\mathfrak{Q}$-module homomorphism $M \xrightarrow{h} N$ is a morphism iff h is *orthogonal*—i.e. h preserves and reflects the respective orthogonality relations—i.e. $m_1 \perp m_2 \Leftrightarrow h(m_1) \perp h(m_2)$ for all $m_1, m_2 \in M$.

6.3 Representations of C^*-Algebras and Involutive Left Modules

For the convenience of the reader we recall the axioms of a C^*-algebra. A Banach algebra $\mathfrak{A} = (A, +, \cdot)$ with unit e (cf. [13, 19]) is a *C^*-algebra with unit* iff A is provided with a conjugate-linear map $a \mapsto a^*$ of A into itself satisfying the following conditions:

(C1) $(a^*)^* = a$ for all $a \in A$.
(C2) $(a \cdot b)^* = b^* \cdot a^*$ for all $a, b \in A$.
(C3) $\| a^* \cdot a \| = \| a \|^2$ for all $a \in A$.

Sometimes the conjugate-linear map $a \mapsto a^*$ is called the *involution* of $\mathfrak{A}$. In this context the condition (C3) is also known as *C^*-property*.

Because of (C1), (C3) and the submultiplicativity of the norm the involution $a \mapsto a^*$ is always an isometry.

Morphisms between C^*-algebras are $*$-homomorphisms—these are algebra homomorphisms $A \xrightarrow{\pi} B$ with the property $\pi(a^*) = \pi(a)^*$ for all $a \in A$. Hence $*$-homomorphisms are algebra homomorphisms preserving the corresponding involutions. It follows from the spectral theory of self-adjoint elements of A and the C^*-property that a $*$-homomorphism π satisfies always the condition

$$\| \pi(a) \| \leq \| a \|, \qquad a \in A.$$

Hence $*$-homomorphisms are continuous.

Example 7 ([19]) Let $\mathcal{H}$ be a Hilbert space and $L(\mathcal{H})$ be the Banach algebra of all bounded and linear operators $\mathcal{H} \xrightarrow{T} \mathcal{H}$. Then $L(\mathcal{H})$ is a C^*-algebra w.r.t. to the involution given by the formation of adjoint operators—i.e.

$$\langle T(x), y \rangle = \langle x, T^*(y) \rangle, \qquad x, y \in \mathcal{H}.$$

In the next definition we summarize basic properties of representations of C^*-algebras (cf. [13, 14]).

Definition 4 (a) A *representation* of a unital C^*-algebra $\mathfrak{A} = (A, +, \cdot)$ is a pair $(\mathcal{H}, \pi)$ where $\mathcal{H}$ is a Hilbert space and π is a $*$-homomorphism from $\mathfrak{A}$ to the C^*-algebra of all bounded and linear operators on $\mathcal{H}$—i.e. $A \xrightarrow{\pi} L(\mathcal{H})$.

(b) A representation $(\pi, \mathcal{H})$ of $\mathfrak{A}$ is called *cyclic* if there exists a vector $x \in \mathcal{H}$ with $x \neq 0$ s.t. the closure of $\{\pi(a)(x) \mid a \in A\}$ coincides with $\mathcal{H}$. In this context x is termed a *cyclic vector for* π.

(c) A representation $(\pi, \mathcal{H})$ of $\mathfrak{A}$ is *irreducible* iff every non-trivial closed linear subspace U of $\mathcal{H}$ being invariant under $\pi(A)$ (i.e. $\{\pi(a)(x) \mid a \in A,\, x \in U\} \subseteq U$) coincides with $\mathcal{H}$.

The aim of the following considerations is to show that every representation of a unital C^*-algebra $\mathfrak{A}$ induces an involutive left $\mathrm{Max}(\mathfrak{A})$-module with faithful and symmetric 2-form where $\mathrm{Max}(\mathfrak{A})$ is the spectrum of $\mathfrak{A}$ (cf. Example 4). The next theorem is due to Mulvey and Pelletier [21].

Theorem 8 *Let $(\pi, \mathcal{H})$ be a representation of a unital C^*-algebra $\mathfrak{A}$. Further, let $P(\mathcal{H})$ be the complete De Morgan algebra of all closed linear subspaces of $\mathcal{H}$ provided with the orthogonal complement as order reversing involution ^{0}, and let $[P(\mathcal{H}), P(\mathcal{H})]$ be the involutive and unital quantale of all join preserving self-maps of $P(\mathcal{H})$ (cf. Example 5). Then π induces an involutive and unital homomorphism $\mathrm{Max}(\mathfrak{A}) \xrightarrow{h_\pi} [P(\mathcal{H}), P(\mathcal{H})]$ by*

$$[h_\pi(I)](U) = \text{top. closure}\big(\text{lin.hull}\{\pi(a)(x) \mid a \in I,\, x \in U\}\big) \qquad (42)$$

where $I \in \mathrm{Max}(\mathfrak{A})$ and $U \in P(\mathcal{H})$.

Because of Theorems 7 and 8 every representation $(\pi, \mathcal{H})$ of a unital C^*-algebra $\mathfrak{A}$ induces a left action $\odot_\pi$ on $P(\mathcal{H})$ in the sense of Sup determined by:

$$I \odot_\pi U = \text{top. closure}\big(\text{lin.hull}\{\pi(a)(x) \mid a \in I,\, x \in U\}\big). \qquad (43)$$

In this context the faithful and symmetric 2-form φ corresponds to the orthogonal complementation in $P(\mathcal{H})$. The pair $(P(\mathcal{H}), \odot_\pi)$ is also called the *canonical* involutive left $\mathrm{Max}(\mathfrak{A})$-module *associated with the representation* $(\pi, \mathcal{H})$ of $\mathfrak{A}$.

The next results are due to Kruml and Resende [16] showing that involutive left $\mathfrak{Q}$-modules play a significant role in the theory of operator algebras.

Theorem 9 (a) *Let $(\pi, \mathcal{H})$ be a representation of a unital C^*-algebra $\mathfrak{A} = (A, +, \cdot)$ and $(P(\mathcal{H}), \odot_\pi)$ be the* canonical *involutive left $\mathrm{Max}(\mathfrak{A})$-module associated with $(\pi, \mathcal{H})$.*

(a) A vector $x \in \mathcal{H}$ with $x \neq 0$ is cyclic for π iff $\top \odot_\pi \langle x \rangle = \top$ where $\langle x \rangle$ is the 1-dimensional subspace of $\mathcal{H}$ generated by x.
(b) The representation $(\pi, \mathcal{H})$ is irreducible iff every atom a of $P(\mathcal{H})$ is a generator of $P(\mathcal{H})$—i.e. $P(\mathcal{H}) = \{I \odot_\pi a \mid I \in \mathrm{Max}(\mathfrak{A})\}$.
(c) Two representations $(\pi_1, \mathcal{H}_1)$ and $(\pi_2, \mathcal{H}_2)$ of $\mathfrak{A}$ are equivalent (i.e. there exists a unitary transformation $\mathcal{H}_1 \xrightarrow{T} \mathcal{H}_2$ s.t. $\pi_2(a) = T \circ \pi_1(a) \circ T^*$ holds for all $a \in A$) iff the respective canonical *involutive left $\mathrm{Max}(\mathfrak{A})$-modules associated with $(\pi_1, \mathcal{H})$ and $(\pi_2, \mathcal{H})$ are isomorphic.*

It is worthwhile to note that the previous results make use of fundamental principles of the theory of C^*-algebras (see e.g. Proposition 4.5.3 in [13], Theorems 10.2.7 and 10.2.10 in [14]).

References

1. Adámek, J.: Free algebras and automata realization in the language of categories. Comment Math. Univ. Carol. **15**, 589–602 (1974)
2. Adámek, J., Trnková, V.: Automata and Algebras in Categories. Kluwer Academic Publishers, Dordrecht (1990)
3. Banaschewski, B., Nelson, E.: Tensor products and bimorphisms. Cand. Math. Bull. **19**, 385–402 (1976)
4. Birkhoff, G.: Lattice Theory, Colloquium Publications, vol. 25, 3rd edn., eighth printing. American Mathematical Society, Rhode Island (1995)
5. Denniston, J.T., Melton, A., Rodabaugh, S.E.: Enriched categories and many-valued preorders: categorical, semantical and topological perspectives. Fuzzy Sets Syst. **256**, 4–56 (2014)
6. Eklund, P., Höhle, U., Kortelainen, J.: A survey on the categorical term construction with applications. Fuzzy Sets and Syst. doi:10.1016/j.fss.2015.07.003
7. Galatos, N., Jipsen, P., Kowalski, T., Ono, H.: Residuated Lattices: An Algebraic Glimpse at Substructural Logics, Studies in Logic, vol. 151. Elsevier, Amsterdam (2007)
8. Gutiérrez García, J., Höhle, U., Kubiak, T.: Tensor products in *Sup* and their application in constructing quantales (submitted)
9. Höhle, U.: Many Valued Topology and Its Applications. Kluwer Academic Publishers, Boston (2001)
10. Höhle, U.: Categorical foundations of topology with applications to quantaloid enriched topological spaces. Fuzzy Sets Syst. **256**, 166–210 (2014)
11. Höhle, U.: Many-valued preorders I: the basis of many-valued mathematics. In: Magdalena, L., et al. (eds.) Enric Trillas: A Passion for Fuzzy Sets, Studies in Fuzziness and Soft Computing, vol. 322, pp. 125–150. Springer, Heidelberg (2015)
12. Joyal, A., Tierney, M.: An Extension of the Galois Theory of Grothendieck, Memoirs of the American Mathematical Society, vol. 51, Number 309. American Mathematical Society (1984)
13. Kadison, R.V., Ringrose, J.R.: Fundamentals of the Theory of Operator Algebras, Volume I Elementary Theory, Graduate Studies in Mathematics Volume 15. American Mathematical Society (1997)
14. Kadison, R.V., Ringrose, J.R.: Fundamentals of the Theory of Operator Algebras, Volume II Adavanced Theory, Graduate Studies in Mathematics Volume 16. American Mathematical Society (1997)
15. Kelly, G.M.: Basic Concepts of Enriched Category Theory, London Mathematical Society Lecture Notes Series 64. Cambridge University Press (1982)
16. Kruml, D., Resende, P.: On quantales that classify C^*-algebras. Cahiers Topol. Géom. Différ. Catég. **45**, 287–296 (2004)
17. Mac Lane, S.: Categories for the Working Mathematician, 2nd edn. Springer (1998)
18. Manes, E.G.: Algebraic Theories. Springer, New York (1976)
19. Meise, R., Vogt, D.: Introduction to Functional Analysis, Oxford Gruaduate Texts in Mathematics. Oxford University Press (1997)
20. Mulvey, C.J., Pelletier, J.W.: A quantisation of the calculus of relations, CMS Proceedings, vol. 13, pp. 345–360. American Mathematical Society, Providence (1992)
21. Mulvey, C.J., Pelletier, J.W.: On the quantisation of points. J. Pure Appl. Algebra **159**, 231–295 (2001)
22. Pelletier, J.W., Rosický, J.: Simple involutive quantales. J. Algebra **195**, 367–386 (1987)
23. Pu, Q., Zhang, D.: Preordered sets valued in a GL-monoid. Fuzzy Sets Syst. **187**, 1–32 (2012)
24. Resende, P.: Sup-lattice 2-forms and quantales. J. Algebra **276**, 143–167 (2004)
25. Rodabaugh, S.E.: Powerset operator foundations for poslat fuzzy theories and topologies. In: Höhle, U., Rodabaugh, S.E. (eds.) Logic, Topology, Theory, Measure, Mathematics of Fuzzy Sets, pp. 91–116. Kluwer Academic Publishers (1999)
26. Rosenthal, K.I.: Quantales and Their Applications, Pitman Research Notes in Mathematics, vol. 234. Longman Scientific Technical, Longman House, Burnt Mill, Harlow (1990)

27. Shmuely, Z.: The structure of Galois connections. Pac. J. Math. **54**, 209–225 (1974)
28. Solovyov, S.A.: Powerset operator foundation for catalg fuzzy set theories. Iran. J. Fuzzy Syst. **8**, 1–46 (2011)
29. Stubbe, I.: Categorical structures enriched in a quantaloid tensored and cotensored categories. Theory Appl. Categ. **16**, 283–306 (2006)
30. Zadeh, L.A.: Fuzzy sets. Inf. Control **8**, 338–353 (1965)
31. Zadeh, L.A.: The concept of a linguistic variable and its application to approximate reasoning I. Inf. Sci. **8**, 119–249 (1975)
32. Zadeh, L.A.: The concept of a linguistic variable and its application to approximate reasoning II. Inf. Sci. **8**, 301–357 (1975)
33. Zadeh, L.A.: The concept of a linguistic variable and its application to approximate reasoning III. Inf. Sci. **9**, 43–80 (1975)

A Geometric Approach to MV-Algebras

Daniele Mundici

Abstract Markov unrecognizability theorem puts an end to the classical program of equipping any combinatorial manifold M with a *computable* set $\mathscr{I}_M$ of invariants such that a manifold N is homeomorphic to M iff $\mathscr{I}_M = \mathscr{I}_N$. To make sense of the statement of the theorem manifolds are replaced by finite strings of symbols for triangulated rational polyhedra, and homeomorphisms are understood as rational PL-homeomorphisms. Thus, objects and arrows undergo a radical transformation—and yet with no essential loss of generality for the original recognition problem. A further restriction on the arrows arises if one views the recognizability problem from the viewpoint of algorithmic complexity theory: here one must take into account the amount of information needed to specify rational polyhedra. We are thus left with the category of rational polyhedra (objects) with integer PL-maps (arrows). A new geometry arises, where the affine group over the integers takes on the same role as the isometry group does in euclidean space. Differently from the category of rational polyhedra with rational PL-maps, a wealth of new geometric computable invariants emerges in this new category. We discuss in particular the rational measure of rational polyhedra. Its role and applicability is amplified by the duality between rational polyhedra and finitely presented MV-algebras.

1 Where Do the Łukasiewicz Axioms Come From?

Boolean logic L_2 deals with $\{0, 1\}$-observables/events. For instance, in the reduction of the COLORABILITY problem to the boolean satisfiability problem, given a graph G and a palette of k colors, the basic observable "the first vertex of G gets the third color" is coded by a variable X_{13} and every composite observable (such as "each vertex of G

D. Mundici (✉)
Department of Mathematics and Computer Science, University of Florence, Florence, Italy
e-mail: mundici@math.unifi.it

S. Saminger-Platz and R. Mesiar (eds.), *On Logical, Algebraic, and Probabilistic Aspects of Fuzzy Set Theory*, Studies in Fuzziness and Soft Computing 336,
DOI 10.1007/978-3-319-28808-6_4

gets precisely one among the k available colors") is coded by a boolean combination of the X_{ij} in such a way that the k-colorability of G amounts to the satisfiability of a suitable boolean formula ϕ_G in the basic observables X_{ij}. The faithfulness of the map $G \mapsto \phi_G$ is accounted for by saying that G is k-colorable iff ϕ_G is satisfiable. The efficiency of this map follows from its being computable in polynomial time.

Now most observables in physics, as well as most random variables in real life, are not $\{0, 1\}$-valued; measurements are not infinitely precise and their outcome can only be given by specifying a real number together with an error interval. Since physical laws are formulated in terms of relations between real-valued quantities rather than relations between intervals, errors are implicitly taken care of by assuming that observables are continuous: continuity ensures that small errors in the measurement of the basic observables have small effects on the evaluation of compound observables.

For any bounded observable $\mathscr{O}$ one may rescale the measurement unit in such a way that the result of any measurement of $\mathscr{O}$ fits into the real interval $[0, 1]$. Once a $[0, 1]$-valued logic L is chosen to deal with $[0, 1]$-valued observables as boolean logic L_2 does for $\{0, 1\}$-observables, compound observables are formalized in L by applying continuous connectives to the output (i.e., the truth-value) of basic observables: the latter are the "variables" of L.

The functional completeness of boolean logic ensures that all n-variable boolean functions are obtainable from the variables via the boolean connectives. By contrast, a brute force counting argument shows that no $[0, 1]$-valued logic L can be functionally complete, and hence one must judiciously select the most appropriate connectives for L.

If, as is often the case, L is defined in terms of a consequence relation and the L-consequence relation is formulated via Modus Ponens (MP), then inevitably L must be equipped with an "implication" operation $\Rightarrow_L : [0, 1]^2 \to [0, 1]$. By the above discussion, $\Rightarrow_L$ must be continuous. If $\Rightarrow_L$ is to be (minimally) reminiscent of boolean implication then the order of premises is irrelevant, and for any two truth-values $x, y \in [0, 1]$, $x \Rightarrow_L y$ equals 1 precisely when $x \leq y$.

Elementary as they are, these three conditions characterize the implication $x \to_{Ł_\infty} y = \min(1, 1 - x + y)$ of the Łukasiewicz infinite-valued calculus $Ł_\infty$, [19]:

Lemma 1 *For any map* $\Rightarrow : [0, 1]^2 \to [0, 1]$ *the following conditions are equivalent:*

(i) $\Rightarrow$ *is continuous,* $x \Rightarrow (y \Rightarrow z) = y \Rightarrow (x \Rightarrow z)$, *and* $(1 = x \Rightarrow y$ *iff* $x \leq y)$.

(ii) There exists precisely one increasing bijection ϕ *of* $[0, 1]$ *onto* $[0, 1]$ *such that* $x \Rightarrow y = \phi^{-1}(\phi(x) \to_{Ł_\infty} \phi(y))$ *for all* $x, y \in [0, 1]$.

Proof This is known as the Smets-Magrez Theorem [31]. Some conditions assumed in [31] are redundant (see Fodor and Roubens [16, Theorem 1.15] and Baczyński [1]. Also see [2]). Related results had been obtained earlier by Trillas and Valverde in [33, Theorem 3.4]. □

Theorem 1 *Let* $I = ([0,1], 0, 1, \Rightarrow, \neg)$ *be the unit real interval equipped with a binary operation* $\Rightarrow$ *satisfying the three conditions in Lemma 1(i), and with the derived operation* $\neg x = x \Rightarrow 0$. *Then the algebra* I *has no nontrivial congruences and satisfies the following axioms:*

(i) $1 \Rightarrow x = x$
(ii) $(x \Rightarrow y) \Rightarrow ((y \Rightarrow z) \Rightarrow (x \Rightarrow z)) = 1$
(iii) $((x \Rightarrow y) \Rightarrow y) = ((y \Rightarrow x) \Rightarrow x)$
(iv) $(\neg x \Rightarrow \neg y) \Rightarrow (y \Rightarrow x) = 1.$

Conversely, every algebra $B = ([0,1], 0, 1, \Rightarrow, \neg)$ *without nontrivial congruences and satisfying axioms (i)–(iv) is isomorphic to an algebra* $I' = ([0,1], 0, 1, \Rightarrow', \neg')$, *where* $\Rightarrow'$ *satisfies the three conditions in Lemma 1(i), and* $\neg' x = x \Rightarrow' 0$.

Proof The map ϕ of Lemma 1(ii) is an isomorphism between I and the standard Wajsberg algebra. The converse statement follows from [11, 3.5] together with the well known fact that the *standard MV-algebra* $[0,1]_{MV} = ([0,1], 0, 1, \neg, \oplus)$ (where $\neg x = 1 - x$ and $x \oplus y = \min(1, x + y)$) is not isomorphic to any of its proper subalgebras, [11, 7.2.6]. □

Equations (i)–(iv) characterize *Wajsberg algebras*—i.e., MV-algebras up to term equivalence, [11]. In this way, *MV-algebras* $(= HSP([0,1]_{MV}))$ can be introduced in the fastest possible way. Interpreted as tautologies, equations (i)–(iv) amount to the Łukasiewicz axioms [19]. Thus MV-algebras can be redefined as the algebras of the only [0, 1]-valued logic whose Modus Ponens relation is framed in terms of an implication connective satisfying the three elementary conditions of Lemma 1(i). A related approach to Łukasiewicz logic is given by the following result:

Proposition 1 ([23]) *Let* $([0,1], \vee, \wedge, *, \Rightarrow, 0, 1)$ *be a residuated lattice in which* $\vee$ *and* $\wedge$ *are the natural* max *and* min *operations. If the map* $\Rightarrow : [0,1]^2 \to [0,1]$ *is continuous then* $*$ *and* $\Rightarrow$ *are the Łukasiewicz conjunction* $x * y = \max(0, x + y - 1)$ *and implication* $x \Rightarrow y = \min(1, 1 - x + y)$.

In the language of *t-norms* (i.e., commutative associative monotone binary operations on [0, 1] having 1 as their neutral element, [20]) the above results show that among all continuous t-norms, Łukasiewicz conjunction is the only one yielding a logic with a continuous implication connective.

Our next aim in this paper is to approach MV-algebras from an entirely different viewpoint, starting from Markov's celebrated unrecognizability theorem for manifolds.

2 A New Geometry: Rational Polyhedra with Integer PL-Maps

You're browsing, let us imagine, in a music shop, and come across a box of faded pianola rolls. One of them bears an illegible title, and you unroll the first foot or two, to see if you can recognize the work from the pattern of holes in the paper. Are there four beats in the bar, or only three? Does the piece begin on the tonic, or some other note? Eventually you decide that the only way of finding out is to buy the roll, take it home, and play it on the pianola. Within seconds your ears have told you what your eyes were quite unable to make out – that you are now the proud possessor of a piano arrangement of "Colonel Bogey".

Longuet-Higgins, H. C. (1979). "Review Lecture: The Perception of Music". Proceedings of the Royal Society B: Biological Sciences 205 (1160) page 307.

A similar situation occurs in the recognition of geometrical figures P, Q. How can we effectively determine that P is a tetrahedron up to homeomorphism? How can we prove that P is *not* homeomorphic to Q? One should first note that for the statement of the problem to make sense, P and Q must be presented as finite strings of symbols. Not all presentations are equally good. The evolution of notational systems for the natural numbers shows that notations allowing more efficient computations supersede less efficient notations.

To code a combinatorial manifold P by a finite string of symbols, one usually proceeds as follows: (i) first equips P with a triangulation Δ_0, (ii) next replaces P by the underlying set of a suitable linearized counterpart Δ of Δ_0, and (iii) finally assumes that each simplex in Δ has rational vertices. In this way, P becomes a *rational polyhedron* P in euclidean space $\mathbb{R}^n$, i.e., a finite union of simplexes $S_1, \ldots, S_k \subseteq \mathbb{R}^n$ with rational vertices. P need not be convex, nor connected [32].

The original recognition problem for combinatorial manifolds P and Q has been transformed into an essentially equivalent problem for rational polyhedra, but P and Q are disfigured into finite unions of rational simplexes, (like music is disfigured into score bars) and homeomorphisms are now replaced by *rational PL-homeomorphisms*, i.e., invertible PL-maps ϕ such that every linear piece of both ϕ and its inverse has rational coefficients.

By definition, a triangulation is *rational* if so are all its simplexes. Rational polyhedra are the same as underlying sets (supports) of rational triangulations [32]. Thus they provide the following precise formulation of Markov theorem:

Theorem 2 (A.A. Markov, 1958, see [17, 30]) *No Turing-computable procedure can decide if two rational polyhedra P and Q are rationally PL-homeomorphic.*

This result puts an end to the time-honored program of equipping every combinatorial manifold with a computable set of invariants sufficient to recognize homeomorphic objects. The program was successful for curves and surfaces but fails for higher-dimensional manifolds. To investigate the computability of homeomorphism, manifolds are replaced by rational polyhedra, and homeomorphisms are replaced by rational PL-homeomorphisms. In this way—at the very least—the recognizability problem becomes recursively enumerable: some Turing machine can effectively enumerate all pairs of rationally PL-homeomorphic rational polyhedra.

Are rational PL-maps the only reasonable arrows for rational polyhedra?

As problem instances in computability theory are coded by finite strings of symbols, in algorithmic complexity theory the length of these input strings are related to the time needed to compute the output. Accordingly, in any category of rational polyhedra where space complexity is to have a role, it is natural to assume that invertible arrows between two rational polyhedra P and Q preserve the space complexity of the strings representing P and Q.

The following is a precise definition: For every point $y = (y_1, \ldots, y_n) \in \mathbb{Q}^n$ let us denote by $\mathrm{den}(y)$ the least common denominator of the coordinates of y. We say that $\mathrm{den}(y)$ is the *denominator* of y.

The vector $\tilde{y} = (\mathrm{den}(y) \cdot y_1, \ldots, \mathrm{den}(y) \cdot y_n, \mathrm{den}(y)) \in \mathbb{Z}^{n+1}$ is called the *homogeneous correspondent* of y. Given two rational polyhedra $P \subseteq [0, 1]^n$ and $Q \subseteq [0, 1]^m$, a rational PL-homeomorphism η of P onto Q is said to be a $\mathbb{Z}$-*homeomorphism* if $\mathrm{den}(x) = \mathrm{den}(\eta(x))$ for each rational point $x \in P$. Equivalently, [27], each linear piece of both η and η^{-1} has integer coefficients. (The number of linear pieces of η is always finite.)

At the end of the day we are left with a category of rational polyhedra where arrows are given by *integer PL-maps*, for short $\mathbb{Z}$-*maps*, i.e., piecewise linear maps $\zeta : P \to Q$ such that every linear piece of ζ has integer coefficients, [27]. Then $\mathbb{Z}$-homeomorphisms coincide with those invertible maps η from a rational polyhedron $P \subseteq \mathbb{R}^n$ onto a rational polyhedron $Q \subseteq \mathbb{R}^m$ such that both η and its inverse are $\mathbb{Z}$-maps. A new geometry arises, where the affine group over the integers has the same role as that of the isometry group in euclidean space, [9].

Differently from the category of rational polyhedra with *rational* PL-maps, when rational PL-maps are specialized to *integer* PL-maps a wealth of new geometric computable invariants for any rational polyhedron P emerges: the number n_d of points of denominator d lying in P, $d = 1, 2, \ldots$; the number of simplexes in the smallest regular triangulation of P (see below for the definition of regularity); the smallest n such that P is $\mathbb{Z}$-embeddable into $\mathbb{R}^n$ with preservation of denominators; the rational volume of P (to be defined later on in this paper).

These invariants make the $\mathbb{Z}$-homeomorphism of rational polyhedra more easily recognizable than rational PL-homeomorphism, just like the music we listen to is better recognizable than the music we see coded on a pianola roll. Modulo the dualities described in the next section, both finitely presented MV-algebras and unital ℓ-groups inherit these invariants—although the latter need not be immediately apparent within the purely algebraic framework.

The counterpart of Markov's unrecognizability theorem for this new category of rational polyhedra is still open: it is not known whether the $\mathbb{Z}$-homeomorphism of rational polyhedra is a decidable problem.

3 The Rational Measure of Rational Polyhedra

In this section, we introduce the $\mathbb{Z}$-homeomorphism invariant length, area, volume,... of rational polyhedra.

Following [27], for any triangulation ∇ of P we denote by $\nabla^{\max}$ the set of maximal simplexes in ∇.

For all $i = 0, 1, 2, \ldots$ we let $P^{(i)} = \bigcup\{T \in \nabla^{\max} \mid \dim(T) = i\}$, and we say that $P^{(i)}$ is the *i-dimensional part of P*. $P^{(i)}$ is a (possibly empty) polyhedron and does not depend on the triangulation ∇ of P. If $P^{(i)}$ is nonempty, then it is an i-dimensional polyhedron. The j-dimensional part of such $P^{(i)}$ is empty iff $j \neq i$.

An m-simplex $U = \operatorname{conv}(w_0, \ldots, w_m) \subseteq [0, 1]^n$ is said to be *regular* (*unimodular*, in [26]) if it is rational and the set of integer vectors $\{\tilde{w}_0, \ldots, \tilde{w}_m\}$ (the homogeneous correspondents of $w_0, \ldots, w_m$) can be extended to a basis of the free abelian group $\mathbb{Z}^{n+1}$. A simplicial complex is said to be a *regular triangulation* (of its support) if all its simplexes are regular. Regular triangulations are the affine counterparts of the regular fans of toric algebraic geometry, [14, 36] (the "nonsingular fans" of [29]).

For every regular m-simplex $T = \operatorname{conv}(v_0, \ldots, v_m) \subseteq \mathbb{R}^n$, $m = 0, \ldots, n$, we use the abbreviation $\operatorname{den}(T) = \operatorname{den}(v_0) \cdots \operatorname{den}(v_m)$. Then the *rational measure* $\lambda(T)$ of a regular k-simplex T in $\mathbb{R}^n$ is given by $\lambda(T) = (k! \operatorname{den}(T))^{-1}$.

For any rational polyhedron $P \subseteq \mathbb{R}^n$ and regular triangulation Δ of P, the *rational measure* $\lambda^{(i)}_{\Delta}(P)$ of the i-dimensional part $P^{(i)}$ of P is given by $\lambda^{(i)}_{\Delta}(P) = \sum\{\lambda(S) \mid \dim(S) = i,\ S \in \Delta^{\max}\}$, where the sum equals zero if there are no maximal i-simplexes in Δ.

The following result ensures that (i) every rational polyhedron has a regular triangulation, (ii) the rational measure is independent of the chosen triangulation, and (iii) is invariant under $\mathbb{Z}$-homeomorphisms:

Theorem 3 ([26]) *Let P be a rational polyhedron in $\mathbb{R}^n$. We then have:*

(a) P is the support of a regular triangulation.
(b) For $i = 0, 1, \ldots, n$, and arbitrary regular triangulations Δ, ∇ of a rational polyhedron $P \subseteq \mathbb{R}^n$, we have the identity $\lambda^{(i)}_{\Delta}(P) = \lambda^{(i)}_{\nabla}(P)$. Thus we can write $\lambda_i(P)$ instead of $\lambda^{(i)}_{\Delta}(P)$, and call $\lambda_i(P)$ the i-dimensional rational measure of P.
(c) $\lambda_i(P)$ is invariant under $\mathbb{Z}$-homeomorphisms.

Proof (a) [28, Lemma 2.1]. The proof relies upon toric desingularization, [14, VI, 8.5], [29, pp. 23, 31].

(b) [28, Theorem 2.3]. The proof follows from the Morelli-Włodarckzyk solution of the weak Oda conjecture, [24, 36].

(c) [28, Theorem 1.1]. The proof follows from the De Concini-Procesi theorem on elimination of points of indeterminacy in toric varieties, [29, p. 39]. □

Thus $\mathbb{Z}$-homeomorphisms preserve not only the topological properties but also the rational measure of rational polyhedra.

The rational measure has the following characterization:

Theorem 4 *(a) Let $\mathscr{L}^{(n)}$ and $\mathscr{H}^{(n)}$ respectively denote n-dimensional Lebesgue and Hausdorff measure,* [13, 15]. *Let $\mathcal{P}^{(n)}$ denote the set of all rational polyhedra in $\mathbb{R}^n$. Then for each $n = 1, 2, \ldots$ and $d = 0, 1, \ldots,$ the map $\lambda_d \colon \mathcal{P}^{(n)} \to \mathbb{R}_{\geq 0}$ has the following properties, for all $P, Q \in \mathcal{P}^{(n)}$:*

(i) (Invariance) *If $P = \gamma(Q)$ for some map γ belonging to the n-dimensional affine group over the integers, then $\lambda_d(P) = \lambda_d(Q)$.*

(ii) (Valuation) $\lambda_d(\emptyset) = 0$, $\lambda_d(P) = \lambda_d(P^{(d)})$, *and the restriction of λ_d to the set of all rational polyhedra P, Q in $\mathbb{R}^n$ having dimension at most d is a* valuation*:* $\lambda_d(P) + \lambda_d(Q) = \lambda_d(P \cup Q) + \lambda_d(P \cap Q)$.

(iii) (Conservativity) *For any $P \in \mathcal{P}^{(n)}$ let $(P, 0) = \{(x, 0) \in \mathbb{R}^{n+1} \mid x \in P\}$. Then $\lambda_d(P) = \lambda_d(P, 0)$.*

(iv) (Pyramid) *For $k = 1, \ldots, n$, if* $\mathrm{conv}(v_0, \ldots, v_k)$ *is a regular k-simplex in $\mathbb{R}^n$ with $v_0 \in \mathbb{Z}^n$ then $\lambda_k(\mathrm{conv}(v_0, \ldots, v_k)) = \lambda_{k-1}(\mathrm{conv}(v_1, \ldots, v_k))/k$.*

(v) (Normalization) *Let $j = 1, \ldots, n$. Suppose the set $B = \{w_1, \ldots, w_j\} \subseteq \mathbb{Z}^n$ is part of a basis of the free abelian group $\mathbb{Z}^n$. Let the closed parallelepiped $P_B \subseteq \mathbb{R}^n$ be defined by $P_B = \left\{x \in \mathbb{R}^n \mid x = \sum_{i=1}^{j} \gamma_i w_i,\ \ 0 \leq \gamma_i \leq 1\right\}$. Then $\lambda_j(P_B) = 1$.*

(vi) (Proportionality) *Let A be an m-dimensional rational affine subspace of $\mathbb{R}^n$ for some $m = 0, \ldots, n$. Then there is a constant $\kappa_A > 0$, only depending on A, such that $\lambda_m(Q) = \kappa_A \cdot \mathscr{H}^{(m)}(Q)$ for every rational m-simplex $Q \subseteq A$.*

***

(b) The six properties above uniquely characterize the rational measures $\lambda_o, \ldots, \lambda_n$, among all maps from $\mathcal{P}^{(n)}$ to $\mathbb{R}_{\geq 0}$, for each $n = 1, 2, \ldots$.

Proof (a) [28, 4.2].
(b) [28, 8.2]. □

4 Enter ℓ-Groups, Unital ℓ-Groups, and MV-Algebras

An ℓ-*group* is an (always abelian) group equipped with a translation invariant lattice order. Baker and Beynon proved the following duality theorem:

Theorem 5 ([3–5]) *The category of rational polyhedra with rational PL-maps is dually equivalent to finitely presented ℓ-groups with their homomorphisms.*

From the effectiveness of this duality we get the following equivalent reformulation of Markov theorem:

Theorem 6 ([17]) *The isomorphism problem for finitely presented ℓ-groups is Turing-undecidable.*

A *unital* ℓ-group is an ℓ-group with a distinguished positive archimedean element. Just as ℓ-groups originate as a modern formalization of classical euclidean magnitudes, unital ℓ-groups also take care of the (archimedean property of the) unit of measurement. While the archimedean property is undefinable in first-order logic, the following result yields an equational counterpart of unital ℓ-groups:

Theorem 7 ([25]) *There is a categorical equivalence Γ between unital ℓ-groups and MV-algebras.*

Among others, this result allows us to speak of "finitely presented" unital ℓ-groups, as the Γ-correspondents of finitely presented MV-algebras—which turn out to coincide with finitely presented unital ℓ-groups in the sense of Gabriel and Ulmer, [10, Remark 5.10], [22, Lemma 3.1].

Finitely presented MV-algebras and unital ℓ-groups have the following geometric counterpart:

Theorem 8 ([21, 27]) *The category of rational polyhedra with $\mathbb{Z}$-maps is dually equivalent to finitely presented MV-algebras with their homomorphisms. The duality sends each rational polyhedron $P \subseteq [0, 1]^n$ to the MV-algebra $\mathcal{M}(P) = \{f \upharpoonright P \mid f \in \mathcal{M}([0, 1]^n)\}$, the symbol "$\upharpoonright$" denoting restriction.*

Combining Γ with this duality we have:

Theorem 9 *The category of rational polyhedra with $\mathbb{Z}$-maps is dually equivalent to finitely presented unital ℓ-groups with their unital ℓ-homomorphisms.*

Summing up :

$$\frac{\text{rational polyhedra with rational PL-maps}}{\text{finitely presented } \ell\text{-groups}} = \frac{\text{rational polyhedra with integer PL-maps}}{\text{finitely presented unital } \ell\text{-groups}}$$

5 Applying the Rational Measure to Projective MV-Algebras

As a particular case of a general definition, an MV-algebra A is *projective* if whenever $\psi : B \to C$ is a surjective homomorphism and $\phi : A \to C$ is a homomorphism, there is a homomorphism $\theta : A \to B$ such that $\phi = \psi \circ \theta$.

Finitely generated projective MV-algebras are an interesting subclass of finitely presented MV-algebras: among others, they clarify such notions as exactness and admissibility in the proof-theory of Łukasiewicz logic, [6, Sect. 4.5].

While Baker and Beynon [3–5] showed that an ℓ-group G is finitely generated projective iff it is finitely presented, the situation is different for unital ℓ-groups and MV-algebras. As shown by the following result, in combination with Theorem 8, being finitely generated projective is a much stricter condition than being finitely presented.

Theorem 10 ([8]) *Let A be an n-generator projective MV-algebra. Then A is isomorphic to the MV-algebra $\mathcal{M}(P)$ obtained by restricting to P the functions of $\mathcal{M}([0,1]^n)$, for some set P satisfying the following conditions:*

(i) P is a rational polyhedron in $[0,1]^n$ containing a vertex of the cube $[0,1]^n$;
(ii) P is contractible;
(iii) For every regular triangulation Δ of P and maximal simplex T of Δ, the greatest common divisor of the denominators of the vertices of T is equal to 1.

Through a further excursion in algebraic topology [18, 34, 35], Cabrer [7] has recently shown that conditions (i)–(iii) are also sufficient for $\mathcal{M}(P)$ to be isomorphic to an n-generator projective MV-algebra.

Property (iii) is known as the "strong regularity" of P, equivalently, its "anchoredness", [58]. It is equivalent to asking that the affine hull of T contains an integer point of $\mathbb{R}^n$.

A folklore general result in universal algebra is to the effect that an n-generator MV-algebra A is projective iff it is isomorphic to a retract R of the free n-generator MV-algebra $\mathcal{M}([0,1]^n)$ of McNaughton functions over the unit n-cube $[0,1]^n$. Stated otherwise, there is a *retraction* (idempotent endomorphism) ρ of $\mathcal{M}([0,1]^n)$ onto $R \cong A$. Let us consider the following innocent looking problem:

Problem 1 What is the number of retractions of $\mathcal{M}([0,1]^n)$ onto R?

Note that this number is ≥ 1 precisely because A is projective. The answer is given by Theorem 11 below, whose statement is surprisingly simple–although the proof uses the rational measure of rational polyhedral in a fairly sophisticated way.

For every finitely generated projective MV-algebra C we define the *index* $\iota(C)$ as $\iota(C) = \sup\{\text{number of retractions of } \mathcal{M}([0,1]^n) \text{ onto } C'\}$, where n is the smallest number of generators of C, and C' ranges over arbitrary retracts of $\mathcal{M}([0,1]^n)$ isomorphic to C.

An easy verification shows that $\iota(\mathcal{M}([0,1]^n)) = 1$ for all $n = 1, 2, \ldots$.

For every n-generator MV-algebra B, the construction introduced in [27, Corollary 4.18], yields a canonical (Yosida) homeomorphism of the maximal spectral space μ_B onto a closed subset M of $[0,1]^n$. If $M = \mathrm{cl}(\mathrm{int}(M))$ then following Kuratowski, [12, p. 20] we unambiguously say that μ_B is a *closed domain* in $[0,1]^n$.

Theorem 11 (L.M. Cabrer, D.M.) *Let A be a finitely generated projective MV-algebra. Let n be the smallest number of generators of A. Then the index of A is finite iff the maximal spectral space of A is a closed domain in* $[0,1]^n$.

The proof uses Theorem 10, along with the properties of the rational measure of the maximal spectral space of A, (Theorem 3).

Last, but not least, another interesting application of the rational measure is in the classification of orbits of affine subspaces of $\mathbb{R}^n$ under the action of the n-dimensional affine group over the integers. This uses the orbit classification of [9] as a preliminary step.

6 Appendix: Recent Applications of MV-Algebras (a Selection)

As we have seen, MV-algebra theory heavily draws from algebraic topology and toric geometry. Conversely, the book [27] shows that MV-algebras have many applications to diverse areas of mathematics. Here is a selection of recent developments subsequent to the publication of [27]:

- Riesz spaces, [45, 48, 72]
- Differential geometry, [41, 42, 45, 69]
- Algebraic geometry, [39, 40]
- Categories, duality, sheafs, [10, 21, 37, 46, 47, 56, 65, 67]
- Semirings, tropical and idempotent mathematics, [37, 38, 49, 50, 55, 59]
- Probability, [52, 75]
- Games, [60–64, 66]
- Multisets, [46, 71]
- Semantics of Łukasiewicz logic, [68, 70]
- Proof-theory of Łukasiewicz logic, [6, 43, 58]
- Modal logic, Belief, [53, 54, 57, 64]
- Quantum structures, [51, 73, 74, 76, 77]
- Topological groups, [78]
- Discrete dynamical systems, [9]
- Interval Algebras, [44].

All these interactions between algebraic, geometric, measure-theoretic, logic-algorithmic notions are typical of mathematics. The latter is pervaded by functors that connect one part with another, and transfer information, as blood circulation does in a living body. In this paper we have just seen the action of functors on finitely presented MV-algebras and rational polyhedra.

Acknowledgments I am grateful to my friend Peter Klement, whose many papers and books [20, and references therein] taught me the importance of t-norms, and whose kind hospitality at Magdalena Bildungshaus allowed me to get in contact with a community of mathematicians—of which he has been for decades one of the focal points—involved in all aspects of fuzzy logic.

References

1. Baczyński, M.: Residual implications revisited. Notes Smets-Magrez Theorem, Fuzzy Sets Syst. **145**, 267–277 (2004)
2. Baczyński, M., Jayaram, B.: (S, N)- and R-implications: a state-of-the-art survey. Fuzzy Sets Syst. **159**, 1836–1859 (2008)
3. Baker, K.A.: Free vector lattices. Can. J. Math. **20**, 58–66 (1968)
4. Beynon, W.M.: On rational subdivisions of polyhedra with rational vertices. Can. J. Math. **29**, 238–242 (1977)
5. Beynon, W.M.: Applications of duality in the theory of finitely generated lattice-ordered abelian groups. Can. J. Math. **29**, 243–254 (1977)
6. Cabrer, L.M.: Simplicial geometry of unital lattice-ordered abelian groups. Forum Math. **27**, 1309–1344 (2015). doi:10.1515/forum-2011-0131
7. Cabrer, L.M.: Rational simplicial geometry and projective lattice-ordered abelian groups. arXiv:1405.7118v1 [math.RA] 28 May 2014
8. Cabrer, L.M., Mundici, D.: Rational polyhedra and projective lattice-ordered abelian groups with order unit. Commun. Contemp. Math. **14**(3), 1250017 (20 pages) (2012). doi:10.1142/S0219199712500174
9. Cabrer, L.M., Mundici, D.: Classifying orbits of the affine group over the integers, to appear in Ergodic Theory Dyn. Syst. doi:10.1017/etds.2015.45
10. Caramello, O., Russo, A.C.: The Morita-equivalence between MV-algebras and lattice-ordered abelian groups with strong unit. J. Algebra **422**, 752–787 (2015)
11. Cignoli, R., D'Ottaviano, I.M.L., Mundici, D.: Algebraic Foundations of Many-Valued Reasoning, Trends in Logic, vol. 7. Kluwer, Dordrecht (2000)
12. Engelking, R.: General Topology, Revised and completed edition, Sigma Series in Pure Mathematics, vol. 6. Heldermann Verlag, Berlin (1989)
13. Evans, L.C., Gariepy, R.F.: Measure Theory and Fine Properties of Functions. CRC Press, Boca Raton (1992)
14. Ewald, G.: Combinatorial Convexity and Algebraic Geometry. Springer, New York (1996)
15. Federer, H.: Geometric Measure Theory. Springer, New York (1969)
16. Fodor, J., Roubens, M.: Fuzzy Preference Modeling and Multicriteria Decision Support. Kluwer Academic Publishers, Dordrecht (1994)
17. Glass, A.M.W., Madden, J.J.: The word problem versus the isomorphism problem. J. Lond. Math. Soc. (2), **30**, 53–61 (1984)
18. Hatcher, A., Algebraic Topology. Cambridge University Press (2001)
19. Łukasiewicz, J., Tarski, A.: Untersuchungen über den Aussagenkalkül, Comptes Rendus des séances de la Société des Sciences et des Lettres de Varsovie, Classe III, 23, pp. 30–50 (1930). English translation: Investigations into the Sentential Calculus, Chapter IV. In: A. Tarski, Logic, Semantics, Metamathematics. Clarendon Press, Oxford (1956). Reprinted: Hackett, Indianapolis (1983)

20. Klement, E.P., Mesiar, R., Pap, E.: Triangular Norms. Kluwer, Dordrecht (2000)
21. Marra, V., Spada, L.: Duality, projectivity, and unification in Łukasiewicz logic and MV-algebras. Ann. Pure Appl. Logic **164**, 192–210 (2013)
22. Marra, V., Spada, L.: Two isomorphism criteria for directed colimits. arXiv:1312.0432v1, 2 Dec 2013
23. Menu, J., Pavelka, J.: A note on tensor products on the unit interval. Comment. Math. Univ. Carol. **17**, 71–83 (1976)
24. Morelli, R.: The birational geometry of toric varieties. J. Algebraic Geom. **5**, 751–782 (1996)
25. Mundici, D.: Interpretation of AF C^*-algebras in Łukasiewicz sentential calculus. J. Funct. Anal. **65**, 15–63 (1986)
26. Mundici, D.: The Haar theorem for lattice-ordered abelian groups with order-unit. Discret. Contin. Dyn. Syst. **21**, 537–549 (2008)
27. Mundici, D.: Advanced Łukasiewicz calculus and MV-algebras. Trends in Logic, vol. 35. Springer, Berlin (2011)
28. Mundici, D.: Invariant measure under the affine group over $\mathbb{Z}$, combinatorics. Probab. Comput. **23**, 248–268 (2014)
29. Oda, T.: Convex bodies and algebraic geometry. Convex Bodies and Algebraic Geometry. Springer, New York (1988)
30. Shtan'ko, M.A.: Markov's theorem and algorithmically non-recognizable combinatorial manifolds, Izvestiya RAN. Ser. Math. **68**, 207–224 (2004)
31. Smets, P., Magrez, P.: Implication in fuzzy logic. Int. J. Approx. Reason. **1**, 327–347 (1987)
32. Stallings, J.R.: Lectures on Polyhedral Topology. Tata Institute of Fundamental Research, Mumbay (1967)
33. Trillas, E., Valverde, L.: On implication and indistinguishability in the setting of fuzzy logic. In: Kacprzyk, J., Yager, R.R. (eds.) Management Decision Support Systems using Fuzzy Sets and Possibility Theory, pp. 198–212. Technical University Rhineland, Cologne (1985)
34. Whitehead, J.H.C.: On subdivisions of complexes. Math. Proc. Camb. Philos. Soc. **31**, 69–75 (1935)
35. Whitehead, J.H.C.: Simplicial spaces, nuclei and m-groups. Proc. Lond. Math. Soc. **45**, 243–327 (1939)
36. Włodarczyk, J.: Decompositions of birational toric maps in blow-ups and blow-downs. Trans. Am. Math. Soc. **349**, 373–411 (1997)

Additional Recent Literature Cited in Section 5

37. Belluce, L.P., Di Nola, A., Ferraioli, A.R.: MV-semirings and their sheaf representations. Order **30**, 165–179 (2013). doi:10.1007/s11083-011-9234-0
38. Belluce, L.P., Di Nola, A., Ferraioli, A.R.: Ideals of MV-semirings and MV-algebras. In: Litvinov, G.L., Sergeev, S.N. (eds.) Tropical and Idempotent Mathematics and Applications. Contemporary Mathematics, vol. 616, pp. 59–76 (2014)
39. Belluce, L.P., Di Nola, A., Lenzi, G.: On generalizing the Nullstellensatz for MV-algebras. J. Logic Comput. **25**, 701–707 (2015). doi:10.1093/logcom/exu042
40. Belluce, L.P., Di Nola, A., Lenzi, G.: Algebraic geometry for MV-algebras. J. Symb. Logic **79**(4), 1061–1091 (2014)
41. Busaniche, M., Mundici, D.: Bouligand-Severi tangents in MV-algebras. Revista Matemática Iberoamericana **30**(1), 191–201 (2014)
42. Cabrer, L.M.: Bouligand-Severi k-tangents and strongly semisimple MV-algebras. J. Algebra **404**, 271–283 (2014)
43. Cabrer, L.M.: Exact Unification. arXiv:1410.5583v1 [math.LO] 21 Oct 2014
44. Cabrer, L.M., Mundici, D.: Interval MV-algebras and generalizations. Int. J. Approx. Reason. **55**, 1623–1642 (2014)

45. Cabrer, L.M., Mundici, D.M.: Severi-Bouligand tangents, Frenet frames and Riesz spaces. Adv. Appl. Math. **64**, 1–20 (2015)
46. Cignoli, R., Marra, V.: Stone duality for real-valued multisets. Forum Math. **24**, 1317–1331 (2012)
47. Di Nola, A., Ferraioli, A.R., Lenzi, G.: Algebraically closed MV-algebras and their sheaf representation. Ann. Pure Appl. Logic **164**, 349–355 (2013)
48. Di Nola, A., Leustean, I.: Łukasiewicz logic and Riesz spaces. Soft Comput. **18**, 2349–2363 (2014). doi:10.1007/s00500-014-1348-z
49. Di Nola, A., Russo, C.: Semiring and semimodule issues in MV-algebras. Comm. Algebra **41**, 1017–1048 (2013)
50. Di Nola, A., Russo, C.: MV-semirings as a new perspective on mathematical fuzzy set theory: a survey. arXiv:1102.1999v4, 14 Nov 2014
51. Dvurečenskij, A.: Quantum structures versus partially ordered groups. Int. J. Theor. Phys. doi:10.1007/s10773-014-2479-9
52. Fedel, M., Keimel, K., Montagna, F., Roth, W.: Imprecise probabilities, bets and functional analytic methods in Łukasiewicz logic. Forum Math. **25**, 405–441 (2013). doi:10.1515/FORM.2011.123
53. Flaminio, T., Godo, L., Kroupa, T.: Belief functions on MV-algebras of fuzzy sets: an overview. In: Torra, V., Narukawa, Y., Sugeno, M. (eds.) Non-Additive Measures, Studies in Fuzziness and Soft Computing, vol. 310, pp. 173–200. Springer (2014)
54. Flaminio, T., Godo, L., Hosni, H.: Coherence in the aggregate: a betting method for belief functions on many-valued events. Int. J. Approx. Reason. **58**, 71–86 (2015). doi:10.1016/j.ijar.2015.01.001
55. Gavalec, M., Nemcová, Z., Sergeev, S.: Tropical linear algebra with the Łukasiewicz T-norm. Fuzzy Sets Syst. **276**, 131–148 (2015). doi:10.1016/j.fss.2014.11.008
56. Gehrke, M., van Gool, S.J., Marra, V.: Sheaf representations of MV-algebras and lattice-ordered abelian groups via duality. J. Algebra **417**, 290–332 (2014)
57. Hansoul, G., Teheux, B.: Extending Łukasiewicz logics with a modality: algebraic approach to relational semantics. Stud. Logica **101**, 505–545 (2013)
58. Jeřábek, E.E.: The complexity of admissible rules of Łukasiewicz logic. J. Logic Comput. **23**, 693–705 (2013)
59. Kala, V.: Lattice-ordered abelian groups finitely generated as semirings, to appear in the J. Commut. Algebra. arXiv:1502.01651
60. Kroupa, T.: Core of coalition games on MV-algebras. J. Logic Comput. **21**, 479–492 (2011)
61. Kroupa, T.: A generalized Möbius transform of games on MV-algebras and its application to a Cimmino-type algorithm for the core, optimization theory and related topics. Contemp. Math. **568**, 139–158 (2012)
62. Kroupa, T.: States in Łukasiewicz logic correspond to probabilities of rational polyhedra. Int. J. Approx. Reason. **53**, 435–446 (2012)
63. Kroupa, T., Majer, O.: Optimal strategic reasoning with McNaughton functions. Int. J. Approx. Reason. **55**, 1458–1468 (2014)
64. Kroupa, T., Teheux, B.: Modal extension of Łukasiewicz logic for reasoning about coalitional power. arXiv:1411.6452v1, 24 Nov 2014
65. Lawson, M.V., Scott, P.: AF inverse monoids and the structure of countable MV-algebras. arXiv:1408.1231v2, 13 Oct 2014
66. Marchioni, E., Woolridge, M.: Łukasiewicz games, In: Huhns (eds.) Proceedings of the 13th International Conference on Autonomous Agents and Multiagent Systems (AAMAS 2014), Paris, France, pp. 837–844 (2014)
67. Marra, V., Spada, L.: The dual adjunction between MV-algebras and Tychonoff spaces. Studia Logica, special issue in memoriam Leo Esakia **100**, 253–278 (2012)
68. Mundici, D.: The differential semantics of Łukasiewicz syntactic consequence, Chapter 7. In: Montagna, F. (ed.) Petr Hájek on Mathematical Fuzzy Logic, Outstanding Contributions, vol. 6, pp. 143–157. Springer International Publishing Switzerland (2015). doi:10.1007/978-3-319-06233-4

69. Mundici, D., Pedrini, A.: The Euler characteristic and valuations on MV-algebras. Math. Slovaca **64**, 563–570 (2014). doi:10.2478/s12175-014-0226-6
70. Mundici, D., Picardi, C.: Faulty sets of Boolean formulas and Łukasiewicz logic.J. Logic Comput., Adv. Access published Dec 8 (2014). doi:10.1093/logcom/exu073
71. Nganou, J.B.: Profinite MV-algebras and multisets. Order, doi:10.1007/s11083-014-9345-5
72. Pedrini, A.: The Euler characteristic of a polyhedron as a valuation on its coordinate vector lattice. arXiv:1209.3248v1, 14 Sep 2012
73. Pulmannová, S.: Representations of MV-algebras by Hilbert-space effects. Int. J. Theoret. Phys. **52**, 2163–2170 (2013)
74. Pulmannová, S., Vinceková, E.: MV-pairs and state operators. Fuzzy Sets Syst. **260**, 62–76 (2015)
75. Riečan, B.: Variation on a Poincaré theorem. Fuzzy Sets Syst. **232**, 39–45 (2013)
76. Xie, Y., Li, Y., Yang, A.: The pasting construction for effect algebras. Math. Slovaca **64**, 1051–1074 (2014). doi:10.2478/s12175-014-0258-y
77. Shang, Y., Lu, X., Lu, R.: Computing power of turing machines in the framework of unsharp quantum logic. Theoret. Comput. Sci. **598**, 2–14 (2015). doi:10.1016/j.tcs.2014.12.015
78. Weber, H.: On topological MV-algebras and topological ℓ-groups. Topology Appl. **159**, 3392–3395 (2012)

On the Equational Characterization of Continuous t-Norms

Francesc Esteva and Lluís Godo

Abstract A (continuous) t-norm is called equationally definable when the corresponding standard BL-algebra $[\mathbf{0}, \mathbf{1}]_*$ defined by $*$ and its residuum is the only (up to isomorphism) standard BL-algebra that generates the same variety $Var([\mathbf{0}, \mathbf{1}]_*)$. In this chapter we check that a continuous t-norm $*$ is equationally definable if and only if the t-norm is a finite ordinal sum of copies of the three basic continuous t-norms, i.e. Łukasiewicz, Gödel and Product t-norms.

1 Introduction

A core constituent of *fuzzy logic in narrow sense* [15], from where the discipline of Mathematical fuzzy logic has been intensively developed in the last two decades [5, 10, 11, 14], is the family of residuated many-valued logical calculi with truth values on the real unit interval [0, 1], and with min, max, a (left-continuous) t-norm $*$ and its residuum $\rightarrow_*$ as basic truth functions, interpreting respectively the lattice meet and joint connectives, a strong conjunction and its adjoint implication. These logics are also known as *t-norm based fuzzy logics*.

In this framework, Hájek introduced in [11, 12] the so-called *Basic Fuzzy logic*, BL for short, to capture the 1-tautologies common to all many-valued calculi in [0, 1] defined by a *continuous* t-norm and its residuum, as proved in [4]. Thus, BL is in fact a common sublogic of three well-known fuzzy logics: Łukasiewicz's infinitely-valued logic, Gödel's infinitely-valued logic and Product logic, corresponding to the three basic t-norms, i.e. Łukasiewicz, minimum and product t-norms.

F. Esteva · L. Godo (✉)
Artificial Intelligence Research Institute (IIIA - CSIC), Campus UAB,
08193 Bellaterra, Spain
e-mail: godo@iiia.csic.es

F. Esteva
e-mail: esesteva@iiia.csic.es

S. Saminger-Platz and R. Mesiar (eds.), *On Logical, Algebraic, and Probabilistic Aspects of Fuzzy Set Theory*, Studies in Fuzziness and Soft Computing 336,
DOI 10.1007/978-3-319-28808-6_5

The variety of BL-algebras constitutes the algebraic semantics of Hájek's BL, which is generated by the so-called *standard* BL-algebras $[\mathbf{0}, \mathbf{1}]_*$, that is, the BL-algebras defined on the real unit interval [0, 1], and that in turn are induced by continuous t-norms $*$ and their residuum $\rightarrow_*$. Some subvarieties of **BL** generated by a single standard BL-chain $[\mathbf{0}, \mathbf{1}]_*$ are well-known, in particular the subvarieties of MV algebras, Gödel algebras and Product algebras, the algebraic counterparts of Łukasiewicz, Gödel and Product logics respectively. These varieties are respectively generated by the standard algebras defined by Łukasiewicz, minimum and product t-norms, and are fully described and equationally characterized in the literature. A step further was done in [8], where all varieties $Var([\mathbf{0}, \mathbf{1}]_*)$ of BL-algebras generated by a single standard BL-chain $[\mathbf{0}, \mathbf{1}]_*$ was proved to be finitely axiomatizable.

Then the question arises of whether such an axiomatization of $Var([\mathbf{0}, \mathbf{1}]_*)$ (i.e. a set of equations) univocally characterizes $*$ itself, in the sense of whether $[\mathbf{0}, \mathbf{1}]_*$ is the only (up to isomorphism) standard BL-algebra that generates the same variety $Var([\mathbf{0}, \mathbf{1}]_*)$. When this is so, we say that $*$ is *equationally definable*.

As a rather direct consequence of results in [8], in this short note, and after introducing some needed preliminaries, we check in Sect. 3 that a continuous t-norm is equationally definable if and only if the t-norm is a finite ordinal sum of the three basic continuous t-norms, while in Sect. 4 we show how to effectively find a set of equations of $Var([\mathbf{0}, \mathbf{1}]_*)$ for an arbitrary equationally definable continuous t-norm $*$.

2 Preliminaries

We start with some elementary and well-known definitions and results about t-norms, just for the sake of the paper being self-contained. A t-norm is a binary operation on [0, 1] that is commutative, associative, non-decreasing (monotone) in both variables and that have 0 as absorbent and 1 as unity. A t-norm is continuous if it is continuous as real function of two variables. The three basic continuous t-norms are minimum (min), product (the usual product of reals, $\odot$) and Łukasiewicz (denoted $*_L$ and defined by $x *_L y = \max(0, x + y - 1)$). The greatest and smallest continuous t-norms are the minimum and the Łukasiewicz t-norms respectively, i.e., for all continuous t-norm $*$ and for all $x, y \in [0, 1]$, we have $x *_L y \leq x * y \leq \min(x, y)$.

The following are some basic results on continuous t-norms, see e.g. [13] for further details and results:

- Any continuous t-norm is an ordinal sum of (possibly infinitely-many) copies[1] of the minimum, product and Łukasiewicz t-norms.
- A t-norm $*$ is continuous if and only if it satisfies the divisibility condition: for all $x, y \in [0, 1]$ with $x > y$ there exists $z \in [0, 1]$ such that $y = x * z$.

[1] If we allow for at most a countable number of degenerated components with a single idempotent element.

- Each left-continuous t-norm $*$ uniquely defines a binary operation $\rightarrow_*$, called the residuum of $*$, that satisfies the following condition: for all $x, y, z \in [0, 1]$, $x * y \leq z$ if and only if $x \leq y \rightarrow_* z$ (*residuation or adjunction condition*).
- The residuum $\rightarrow_*$ of a left-continuous t-norm $*$ is actually defined as $x \rightarrow_* y = \max\{z \in [0, 1] : x * z \leq y\}$ (*residuated implication*).
- A left-continuous t-norm $*$ is continuous if and only if the following equation is satisfied: for all $x, y \in [0, 1]$, $x * (x \rightarrow_* y) = \min(x, y)$ (*Divisibility equation*).

On the oher hand, it is also well known that the algebraic counterpart of Hájek's BL logic [11] is given by the variety of *BL-algebras*, i.e. algebraic structures $\mathbf{A} = (A, \wedge, \vee, *, \rightarrow, 0, 1)$ satisfying:

- $(A, \wedge, \vee, 0, 1)$ is a bounded distributive lattice,
- $(A, *, 1)$ is a commutative monoid with unit 1,
- $*$ and $\rightarrow$ form and adjoint pair, i.e. they satisfy the residuation condition: for all $x, y, z \in A$, $x * y \leq z$ if and only if $x \leq y \rightarrow z$,
- Prelinearity: for all $x, y \in A$, $(x \rightarrow y) \vee (y \rightarrow x) = 1$,
- Divisibility: for all $x, y \in A$, $x * (x \rightarrow y) = x \wedge y$.

In other words, BL-algebras are a subclass of residuated lattices, namely, the class of bounded, commutative, integral residuated lattices further satisfying pre-linearity and divisibility.

A *standard BL-chain* is a BL-algebra defined over the real unit interval [0, 1]. It is easy to prove that:

- A continuous t-norm and its residuum defines a standard BL-chain,
- Each standard BL-chains is defined by a continuous t-norm and its residuum.

The last items shows that there is a bijection between continuous t-norms and standard BL-chains. From now on, we will denote by $[\mathbf{0}, \mathbf{1}]_*$ the BL-algebra $([0, 1], \min, \max, *, \rightarrow_*, 0, 1)$ defined by a continuous t-norm $*$ and its residuum.

The ordinal sum representation for continuous t-norms extends to an ordinal sum representation for standard BL-chains in the obvious way, the only new thing to consider is the definition of the residuum over the whole ordinal sum in terms of the residuum over each component. Using a similar representation for BL-chains, in [4] it was proved that the logic BL is complete with respect to the class of standard BL-chains, or in other words, that the whole variety of BL-algebras is generated by the class of standard BL-chains.

A related class of algebraic structures is that of *hoops*. In what follows we introduce some basic definitions and results about hoops and the decomposition theorem for BL-chains as ordinal sums of hoops that we will use in the next section, see [2, 3, 9] for more details.

Definition 1 A *hoop* is an algebraic structure $\mathbf{A} = (A, *, \rightarrow, 1)$ such that:

- $*$ is a binary commutative operation with unit 1, i.e. $x * y = y * x$ and $1 * x = x$ for all $x, y \in A$

- $\rightarrow$ is a binary operation satisfying:
  - for all $x \in A$, $x \rightarrow x = 1$,
  - for all $x, y, z \in A$, $(x * y) \rightarrow z = x \rightarrow (y \rightarrow z)$,
  - for all $x, y \in A$, $x * (x \rightarrow y) = y * (y \rightarrow x)$.

The associated order relation is defined by: $x \leq y$ if $x \rightarrow y = 1$.

A *basic* hoop is a hoop satisfying the following condition:

- $((x \rightarrow y) \rightarrow z) * (y \rightarrow x) \rightarrow z) \rightarrow z = 1$

A *Wajsberg* hoop is a hoop satisfying the following condition:

- for all $x, y \in A$, $(x \rightarrow y) \rightarrow y = (y \rightarrow x) \rightarrow x$.

A *cancellative* hoop is a hoop such that:

- for all $x, y, z \in A$, $x * y \leq x * z$ implies that $y \leq z$.

From this definition, one can check the following facts and properties:

(i) $\leq$ as defined above is indeed an ordering and 1 is maximal
(ii) $*$ is associative
(iii) $*$ is monotonically increasing w.r.t. $\leq$: $x \leq y$ implies $x * z \leq y * z$
(iv) $(*, \rightarrow)$ is an adjoint pair: $x \rightarrow y \leq z$ iff $x * y \leq z$
(v) $x * (x \rightarrow y) \leq y$
(vi) $1 \rightarrow x = x$

Furthermore, regarding the classes of basic, Wajsberg and cancellative hoops, the following relationship among them hold: every Wajsberg hoop is basic and each cancellative hoop is Wajsberg (hence basic as well). Note that hoops have an greatest element, but they may lack a least element. A hoop $\mathbf{A} = (A, *, \rightarrow, 1)$ is called *bounded* if $(A, \leq)$ has a least element. Then it turns out that cancellative hoops coincide with *unbounded* Wajsberg hoops, while bounded Wajsberg hoops coincide with MV-algebras.
Prominent examples of Wajsberg hoops are the following:

- **2**, defined on a set of two elements $\{a, 1\}$, that is in fact a two-element Boolean algebra.
- $Ł = ([0, 1], *_Ł, \rightarrow_Ł, 1)$, the (bounded) Wajsberg hoop defined over $[0, 1]$ by the Łukasiewicz t-norm and its residuum.
- $C = ((0, 1], \odot, \rightarrow_\odot, 1)$, the (unbounded) cancellative hoop defined over $(0, 1]$ by the product t-norm and its residuum.

A similar construction to the ordinal sums for t-norms and BL-chains can be also defined for hoops.

Definition 2 Let $(I, \leq)$ be a totally ordered set, and for all $i \in I$ let $\mathbf{A}_i = (A_i, *_i, \rightarrow_i, 1)$ be a hoop such that $A_i \cap A_j = \{1\}$ for every $j \neq i$. Then the ordinal sum of this family is the structure $\bigoplus_{i \in I} \mathbf{A_i} = (\bigcup_{i \in I} A_i, *, \rightarrow, 1)$, where the operations are defined as follows:

$$x * y := \begin{cases} x *_i y & \text{if } x, y \in A_i, \\ x & \text{if } x \in A_i \backslash \{1\}, y \in A_j, \text{ and } i < j, \\ y & \text{if } y \in A_i \backslash \{1\}, x \in A_j, \text{ and } i < j. \end{cases}$$

$$x \rightarrow y := \begin{cases} x \rightarrow_i y & \text{if } x, y \in A_i, \\ y & \text{if } x \in A_i, y \in A_j, \text{ and } i > j, \\ 1 & \text{otherwise.} \end{cases}$$

Notice that in an ordinal sum of hoops, the greatest element is common to all the hoops and to the ordinal sum as well. For instance, the product standard chain $[0, 1]_\Pi = ([0, 1], \odot, \rightarrow_\odot, 0, 1)$, viewed as a bounded hoop, can be decomposed as the ordinal sum of **2** and C, i.e. $[0, 1]_\Pi = \mathbf{2} \oplus \mathrm{C}$. Actually, in [1] the authors prove that any BL-chain, viewed as a bounded basic hoop, can be decomposed as an ordinal sum of linearly ordered Wajsberg hoops. Restricted to standard BL-chains, this result amounts to say that any standard BL-chain, as a hoop, can be decomposed as an ordinal sum of (suitably arranged) copies of the Wajsberg hoops **2**, C and Ł. In this way, besides viewing the standard product algebra as the ordinal sum of **2** plus C, we can understand the standard Gödel chain as being isomorphic to the ordinal sum of continuum many of copies of **2** (one for each element of a Gödel component), while the standard Łukasiewicz chain $[0, 1]_Ł = ([0, 1], *_Ł, \rightarrow_Ł, 0, 1)$ coincides with Ł as hoop.

As already mentioned, regarding the ordinal sums of hoops just defined, one can notice that the main difference with respect to the ordinal sum of BL-chains is that the top elements of the components are identified with the top element of the ordinal sum. Therefore, for instance, when considering the decomposition of a BL-chain as an ordinal sum of Wajsberg hoops (**2**, C or Ł in the case of standard BL-chains), the top of any component is the top of the ordinal sum, and given two consecutive components, the bottom (if it exists) of the second component is not in the first component. Notice also that the decomposition of any standard BL-chain as ordinal sum of hoops has always a first component that is either **2** (if it is an SBL-chain[2]) or Ł otherwise.

Finally recall that a set of equations determine a *variety* (or equational class) of algebraic structures. By inspecting their definition, it is clear that the classes of hoops, basic hoops and Wajsberg hoops are indeed varieties. The class of cancellative hoops turns out to be a variety as well, since the condition used in Definition 2 can be shown to be equivalent to the validity of the equation $x = y \rightarrow (y * x)$.

Thus it is interesting to know how the varieties generated by the main three prominent Wajsberg hoops, **2**, C and Ł, are related to each other. To do so we consider the following three terms:

- $e_Ł(x) = (x \rightarrow x^2) \vee ((x \rightarrow x^3) \rightarrow x^2)$
- $e_C(x) = (x \rightarrow x^2)$
- $e_{\mathbf{2}}(x) = (x \rightarrow x^3) \rightarrow x^2$

---

[2] That is, a standard BL-chain defined by an strict continuous t-norm.

where x^n stands for $x * .\overset{n}{.}. * x$. An easy computation shows that the equation $e_Ł(x) = 1$ is valid in **2** and C and not in Ł, $e_C(x) = 1$ is a valid equation in **2** and neither in C nor in Ł, and finally, the equation $e_{\mathbf{2}}(x) = 1$ is valid in C and neither in **2** nor in Ł.

Therefore, it is clear that C and Ł do not belong to variety of hoops $Var(\mathbf{2})$ generated by **2**, while **2** and Ł do not belong to the variety $Var(C)$ generated by C. On the other hand, it is easy to check that both **2** and C belong to the variety of hoops $Var(Ł)$ generated by Ł, since **2** is a subhoop of Ł and C is a subhoop of the well-known Chang algebra, which is an MV-algebra, and thus belongs to $Var(Ł)$.

Summarising, we have

$$\mathbf{2}, C \in Var(Ł), \qquad C, Ł \notin Var(\mathbf{2}), \qquad \mathbf{2}, Ł \notin Var(C),$$

and thus, the following strict inclusions among varieties hold:

$$Var(\mathbf{2}) \subset Var(Ł), \qquad Var(C) \subset Var(Ł).$$

3 Characterization of Standard BL-Chains that Are Equationally Definable

Let us denote by $[\mathbf{0}, \mathbf{1}]_*$ either the standard BL-chain, or its corresponding hoop when no confusion exists, defined over $[0, 1]$ by a continuous t-norm $*$ and its residuum $\rightarrow_*$. The goal of this section is to characterize those continuous t-norms $*$ that admit an *equational characterization* in the sense that the variety $Var([\mathbf{0}, \mathbf{1}]_*)$ is uniquely generated by $[\mathbf{0}, \mathbf{1}]_*$, that is, for any other standard BL-chain $[\mathbf{0}, \mathbf{1}]_\circ$ with $\circ$ being a t-norm non isomorphic to $*$, $Var([\mathbf{0}, \mathbf{1}]_*) \neq Var([\mathbf{0}, \mathbf{1}]_\circ)$. In such a case, we can say that the set of equations defining $Var([\mathbf{0}, \mathbf{1}]_*)$ characterize $*$.

Actually, generalizing the well-known Mostert and Shields representation theorem of continuous t-norms, Hájek showed in [12] that every standard BL-chain $[\mathbf{0}, \mathbf{1}]_*$ can be isomorphically decomposed as an ordinal sum (over a bounded ordered index set) of Gödel, Łukasiewicz and Product BL-chain components. However, as hoops, each Gödel BL-chain is isomorphic to an ordinal sum of (possibly infinite) copies of **2**, while Łukasiewicz and Product components on a closed real interval are isomorphic to Ł and $\Pi = \mathbf{2} \oplus C$ respectively. Then any standard BL-chain, as a hoop, will be isomorphic to a (possibly infinite) ordinal sum of Wajsberg hoops Ł, C and **2**.

The following definition and proposition are particular cases of more general definitions and results given in [8], and therefore here we only state them without proofs.

Definition 3 (i) We will denote by Fin the set of ordinal sums (as hoops) of finitely-many copies of Ł, 2 and C, and whose first component is either Ł or 2.

(ii) Let $\mathbf{A}$ be a standard BL-chain whose decomposition as ordinal sum of hoops is $\mathbf{A} = \mathbf{A_0} \oplus (\bigoplus_{i \in I} \mathbf{A_i})$. Then $Fin(\mathbf{A})$ is the set of all finite ordinal sums $\bigoplus_{i=0,\ldots,n} \mathbf{B_i}$ of Wajsberg hoops satisfying the following conditions:

- Each $\mathbf{B_i}$ is either $\mathbf{2}$, C or Ł,
- $\mathbf{B_0}$ is either $\mathbf{2}$ or Ł,
- There are components $\mathbf{A_0} < \mathbf{A_1} < \cdots < \mathbf{A_n}$ of $\mathbf{A}$ such that for every $i = 0, \ldots, n$: (i) if $\mathbf{B_i} =$ Ł then $\mathbf{A_i}$ is isomorphic to Ł; (ii) if $\mathbf{B_i} =$ C, then $\mathbf{A_i}$ is isomorphic either to C or to Ł; and (iii) if $\mathbf{B_i} = \mathbf{2}$, then $\mathbf{A_i}$ is isomorphic either to $\mathbf{2}$ or to Ł.

Example 1 Consider the standard BL-chain $\mathbf{A} = G \oplus Ł \oplus \Pi$. Then, for instance, $\mathbf{2} \oplus Ł$ and $\mathbf{2} \oplus \mathbf{2} \oplus Ł \oplus C$ are in $Fin(\mathbf{A})$, while neither $Ł \oplus \mathbf{B}$ for any $\mathbf{B} \in Fin$, nor $\mathbf{2} \oplus Ł \oplus Ł$ are in $Fin(\mathbf{A})$.

As shown next, the set of $Fin([\mathbf{0}, \mathbf{1}]_*)$ of BL-chains univocally determines the variety $V([\mathbf{0}, \mathbf{1}]_*)$ induced by the t-norm $*$.

Proposition 1 (c.f. Theorem 3.9 of [8]) *Let* $[\mathbf{0}, \mathbf{1}]_*$, $[0, 1]_\circ$ *be two standard BL-chains. Then* $Var([\mathbf{0}, \mathbf{1}]_*) \subseteq Var([\mathbf{0}, \mathbf{1}]_\circ)$ *if, and only if,* $Fin([\mathbf{0}, \mathbf{1}]_*) \subseteq Fin([\mathbf{0}, \mathbf{1}]_\circ)$. *Hence,* $Var([\mathbf{0}, \mathbf{1}]_*) = Var([\mathbf{0}, \mathbf{1}]_\circ)$ *if, and only if,* $Fin([\mathbf{0}, \mathbf{1}]_*) = Fin([\mathbf{0}, \mathbf{1}]_\circ)$.

Notation convention: In the following, given two continuous t-norms $*$ and $\circ$, we will write $* \equiv \circ$ to denote that they isomorphic in the usual sense of t-norms, that is, when there exists an increasing bijection $f : [0, 1] \to [0, 1]$ such that, for any $x, y \in [0, 1]$, $x \circ y = f^{-1}(f(x) * f(y))$.

The following lemma is straightforward to check.

Lemma 1 *If* $*$ *and* $\circ$ *are two continuous t-norms such that both* $[0, 1]_*$ *and* $[0, 1]_\circ$ *have a finite ordinal sum decomposition in terms of BL-components, then* $* \equiv \circ$ *if, and only if, they have the same decomposition,*

From the above proposition and lemma, the characterization of the equationally definable standard BL-chains follows.

Proposition 2 *A continuous t-norm* $*$ *admits an equational characterization if, and only if, the corresponding standard BL-chain* $[\mathbf{0}, \mathbf{1}]_*$ *can be decomposed as an ordinal sum with finitely-many copies of components* Ł, G *and* Π.

Proof First we prove that for a continuous t-norm $*$ whose decomposition as ordinal has a finite number of components, $Var([\mathbf{0}, \mathbf{1}]_\circ) = Var([\mathbf{0}, \mathbf{1}]_*)$ if and only if $\circ \equiv *$ (the components of their decomposition as ordinal sums are the same). By the previous proposition, this is equivalent to prove that if $\circ$ is a continuous t-norm such that $\circ \not\equiv *$, then $Fin(\circ) \neq Fin(*)$. We prove this claim by cases, adapting a more general proof in [8]:

- If the decomposition of $[\mathbf{0},\mathbf{1}]_\circ$ has more components than the decomposition of $[\mathbf{0},\mathbf{1}]_*$ then it is evident that there exist BL-chains in $Fin(\circ)$ that are not in $Fin(*)$. For example let $\circ$ be a continuous t-norm obtained as $Ł \oplus G$, and let $*$ be a continuous t-norm obtained as $Ł \oplus \Pi \oplus G$. Then it is clear that $2 \oplus C \in Fin([\mathbf{0},\mathbf{1}]_*)$ but $2 \oplus C \notin Fin([\mathbf{0},\mathbf{1}]_\circ)$.
- An analogous reasoning proves the statement when the decomposition of $[\mathbf{0},\mathbf{1}]_\circ$ has more components than the decomposition of $[\mathbf{0},\mathbf{1}]_*$.
- If the number of components of the decomposition $[\mathbf{0},\mathbf{1}]_*$ and $[\mathbf{0},\mathbf{1}]_\circ$ is the same, then they need to differ in some component and thus we can find BL-chains that are in $Fin(*)$ and not in $Fin(\circ)$ and viceversa. For example, let $\circ$ be the continuous t-norm obtained as $Ł \oplus Ł \oplus G$ and let $*$ be the continuous t-norm obtained as $Ł \oplus \Pi \oplus G$. Then we have that $2 \oplus C \in Fin([\mathbf{0},\mathbf{1}]_*)$ but $2 \oplus C \notin Fin([\mathbf{0},\mathbf{1}]_\circ)$, while $Ł \oplus Ł \in Fin([\mathbf{0},\mathbf{1}]_\circ)$ and $Ł \oplus Ł \notin Fin([\mathbf{0},\mathbf{1}]_*)$.

In the case the decomposition of $[\mathbf{0},\mathbf{1}]_*$ has infinitely many components, it is easy to prove that there exist infinitely-many continuous t-norms $\circ$ such that $* \not\equiv \circ$ but $Fin([\mathbf{0},\mathbf{1}]_*) = Fin([\mathbf{0},\mathbf{1}]_\circ)$. We do not formally prove the statement but we give some examples:

- If the decomposition of $[\mathbf{0},\mathbf{1}]_*$ consists of an infinite number of Łukasiewicz components Ł, then any other standard BL-chain $[\mathbf{0},\mathbf{1}]_\circ$ whose decomposition begins with an Ł component and contains infinitely many Łukasiewicz components together with (finitely or infinitely many) components Π or G, defines the same variety, namely, the full variety of BL-algebras, see [1].
- If the decomposition of $[\mathbf{0},\mathbf{1}]_*$ begins with a **2** component and contains an infinite number of Łukasiewicz components, then any other standard BL-chain $[\mathbf{0},\mathbf{1}]_\circ$ whose decomposition begins with a **2** component and contains infinitely many Łukasiewicz components together with (finitely or infinitely many) components Π or G, defines the same variety, namely, the full variety of SBL-algebras, see [1].

4 How to Find a Set of Equations of an Equationally Definable t-Norm

After identifying in the last section which t-norms are equationally definable, in this section we show how to find an effective set of equations for each of them, again relying in results from [8]. It has to be remarked that the equations actually characterise the variety generated by the standard algebra $[\mathbf{0},\mathbf{1}]_*$ for a given equationally definable t-norm $*$, and hence the equations will involve not only the operation corresponding to the t-norm but the operation corresponding to its residuum as well.

First we introduce an equation that will have a key role in axiomatizing the varieties $V([\mathbf{0},\mathbf{1}]_*)$.

Definition 4 Let **A** be a BL-chain whose decomposition as ordinal sum of Wajsberg hoops has finitely many components, i.e., $\mathbf{A} = \bigoplus_{\mathbf{i}=\mathbf{0},\mathbf{1},\ldots,\mathbf{n}} \mathbf{A_i}$. Then we will denote by e_A the following equation on $n+1$ variables,

$$\left[\left(\bigwedge_{i=0,\ldots,n-1} ((x_{i+1} \to x_i) \to x_i) * (\neg\neg x_0 \to x_0)\right) \to \left(\bigvee_{i=0,\ldots,n} x_i\right)\right] \vee \bigvee_{i=0,\ldots,n} e_i^A(x_i) = 1 \quad (e_A)$$

where $e_i^A(x) = e_Ł(x)$ if $\mathbf{A_i} = Ł$, $e_i^A(x) = e_C(x)$ if $\mathbf{A_i} = \mathrm{C}$, and $e_i^A(x) = e_2(x)$ if $\mathbf{A_i} = \mathbf{2}$.

Notation convention: for the sake of a simpler notation, from now on we will use $Fin(*)$ and $Var(*)$ to respectively denote $Fin([\mathbf{0},\mathbf{1}]_*)$ and $Var([\mathbf{0},\mathbf{1}]_*)$.

Lemma 2 *Let $*$ be a continuous t-norm whose corresponding standard BL-chain has a decomposition as ordinal sum with finitely many components Ł, Π and G, and let $\mathbf{A} \in Fin$. Then e_A is valid in all BL-chains $\mathbf{B} \in Fin(*)$ if and only if $\mathbf{A} \notin Fin(*)$.*

And from this result, we can prove the following equational characterization as a particular case of a more general result in [8, Theorem 5.2].

Proposition 3 *Let $*$ be a continuous t-norm whose corresponding standard BL-chain $[\mathbf{0},\mathbf{1}]_*$ has a decomposition as ordinal sum with finitely many components Ł, Π and G. Then,*

$Var(*)$ is axiomatized by the set of equations $AX(*) = \{e_B : \mathbf{B} \in Fin(*^\perp)\}$,

where $Fin(*^\perp) = Fin \setminus Fin(*)$.

Note that $AX(*)$ may contain an infinite number of equations. However we can do it better. Actually, one can show that one needs only a finite subset of $AX(*)$ to axiomatize $Var(*)$. Indeed, it is only necessary to keep from $Fin(*^\perp)$ only those BL-chains that are *minimal* in the following sense. Define an ordering relation in the set Fin as follows: for all $\mathbf{A}, \mathbf{B} \in Fin$, define $A \preceq B$ if $\mathbf{A} \in Var(\mathbf{B})$. And denote by $Min(*^\perp)$ the minimal elements of $Fin(*^\perp)$ with respect to the order $\preceq$. It is then clear that it is enough to consider the set of equations corresponding to the BL-chains of $Min(*^\perp)$, and moreover, it can be shown that $Min(*^\perp)$ is always finite, and hence that $Var(*)$ can be axiomatized by a finite set of equations.

Proposition 4 *Let $*$ be a continuous t-norm whose decomposition as ordinal sum of t-norms has finitely many components. Then:*

(i) The set $Min(*^\perp)$ is finite.
(ii) $Var(*)$ is axiomatized by the finite set of equations

$$AX_{min}(*) = \{e_B : \mathbf{B} \in Min(*^\perp)\}.$$

Following [8], given an arbitrary continuous t-norm $*$ and its decomposition as ordinal sum of Ł, G and Π components, an algorithmic procedure to find the set $Min(*^\perp)$ can be given. The idea to find the minimal elements of Fin which are not

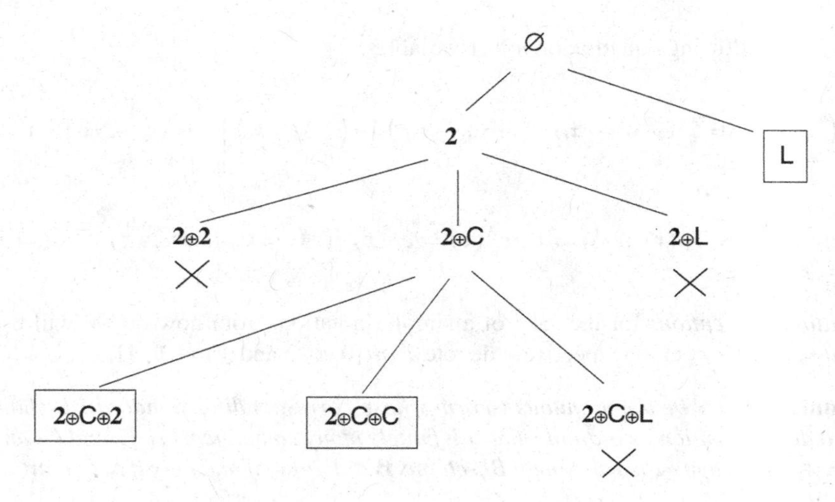

Fig. 1 Analysis for $* = G \oplus Ł$

in $Fin(*)$ is to iteratively checking ordinal sums from Fin of increasing length (1, 2, 3, etc.). At a given step i, a given current ordinal sum $\mathbf{B}$ of length i is checked whether there is another non-discarded ordinal sum $\mathbf{B}'$ of length $\leq i$ such that $\mathbf{B}' \preceq \mathbf{B}$. If so, the current ordinal sum is discarded for further analysis at step $i + 1$. Otherwise $\mathbf{B} \in Min(*^{\perp})$ only if $\mathbf{B}$ is checked to not belong to $Fin(*)$. At next step $i + 1$, only those non-discarded ordinal sums at step i are expanded with a new component, and the procedure starts over. This iterative procedure ends in a finite number of steps. We exemplify this procedure with two examples.

Example 2 Consider a continuous t-norm $*$ isomorphic to $G \oplus Ł$. The above iterative procedure, depicted in Fig. 1 as a spanning tree, yields:

$$Min(*^{\perp}) = \{Ł, \mathbf{2} \oplus C \oplus \mathbf{2}, \mathbf{2} \oplus C \oplus C\}.$$

Example 3 Consider a continuous t-norm $*$ isomorphic to $G \oplus Ł \oplus \Pi \oplus Ł$. The above iterative procedure, depicted in Fig. 2, yields:

$$Min(*^{\perp}) = \{Ł, \mathbf{2} \oplus C \oplus Ł \oplus \mathbf{2}, \mathbf{2} \oplus C \oplus Ł \oplus C\}.$$

Therefore using the result of the previous proposition, we automatically have a finite set of equations $AX_{min}(*)$ univocally characterising $*$, since the only continuous t-norm algebra (up to isomorphism) belonging to $Var(*)$ is $[\mathbf{0}, \mathbf{1}]_*$ itself.

Dedication

This short note is dedicated to Peter Klement in the occasion of his retirement. We are deeply indebted to Peter, not only for his outstanding and numerous scientific

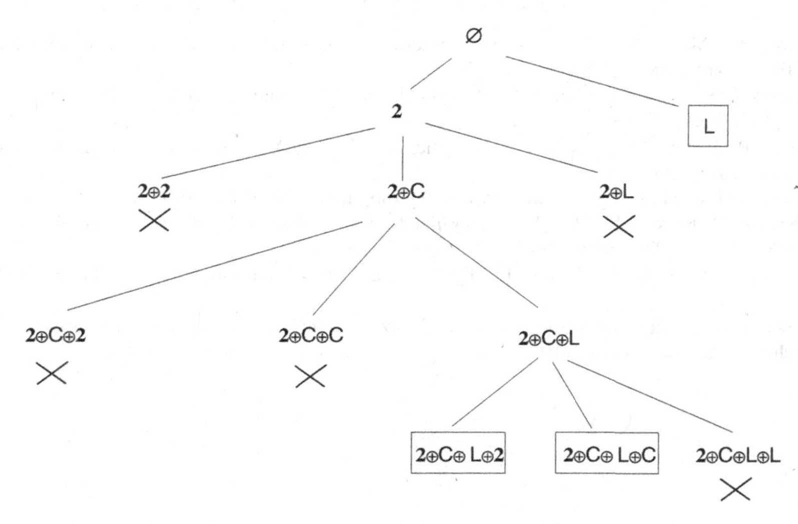

Fig. 2 Analysis for $* = G \oplus \text{Ł} \oplus \Pi \oplus \text{Ł}$

contributions to the field of fuzzy logic, but also for his incredible task of fostering the exchange of ideas and the collaboration among researchers in our community, mainly (but not only) through his Linz Seminars on Fuzzy Set Theory since 1979. Congratulations Peter!

Acknowledgments The authors have been partially supported by the Spanish MINECO project EdeTRI TIN2012-39348-C02-01.

References

1. Aglianó, P., Montagna, F.: Varieties of BL-algebras I: general properties. J. Pure Appl. Algebra **181**, 105–129 (2003)
2. Aglianó, P., Ferreirim, I.M.A., Montagna, F.: Basic hoops: an algebraic study of continuous t-norms. Stud. Logica **87**(1), 73–98 (2007)
3. Blok, W.J., Ferreirim, I.M.A.: On the structure of hoops. Algebra Univers. **43**, 233–257 (2000)
4. Cignoli, R., Esteva, F., Godo, L., Torrens, A.: Basic logic is the logic of continuous t-norms and their residua. Soft Comput. **4**, 106–112 (2000)
5. Cintula, P., Hájek, P., Noguera, C. (eds.): Handbook of Mathematical Fuzzy Logic (in 2 volumes), Studies in Logic, Mathematical Logic and Foundations, vols. 37 and 38. College Publications, London (2011)
6. Di Nola, A., Esteva, F., Garcia, P., Godo, L., Sessa, S.: Subvarieties of BL-algebras generated by single-component chains. Arch. Math. Logic **41**, 673–685 (2002)
7. Di Nola, A., Lettieri, A.: Equational characterization of all varieties of MV-algebras. J. Algebra **221**, 463–474 (1999)
8. Esteva, F., Godo, L., Montagna, F.: Equational characterization of the subvarieties of BL generated by t-norm algebras. Stud. Logica **76**(2), 161–200 (2004)

9. Ferreirim, I.M.A.: On varieties and quasivarieties of hoops and their reducts. Thesis, University of Illinois at Chicago (1992)
10. Gottwald, S.: A Traitise on Multiple-valued Logics. Studies in Logic and Computation. Research Studies Press, Baldock (2001)
11. Hájek, P.: Metamathematics of Fuzzy Logic. Trends in Logic, vol. 4. Studia Logica Library, Kluwer, Dordercht (1998)
12. Hájek, P.: Basic logic and BL-algebras. Soft Comput. **2**, 124–128 (1998)
13. Klement, P., Mesiar, R., Pap, E.: Triangular Norms. Trends in Logic, vol. 8, Studia Logica Library. Kluwer, Dordrecht (2000)
14. Novák, V., Perfilieva, I., Močkoř, J.: Mathematical Principles of Fuzzy Logic. Kluwer, Boston (1999)
15. Zadeh, L.A.: Preface. In: Marks-II, R.J. (ed.) Fuzzy Logic Technology and Applications. IEEE Technical Activities Board (1994)

The Semantics of Fuzzy Logics: Two Approaches to Finite Tomonoids

Thomas Vetterlein and Milan Petrík

Abstract Fuzzy logic generalises classical logic; in addition to the latter's truth values "false" and "true", the former allows also intermediary truth degrees. The conjunction is, accordingly, interpreted by an operation acting on a chain, making the set of truth degrees into a totally ordered monoid. We present in this chapter two different ways of investigating this type of algebras. We restrict to the finite case.

1 Introduction

The idea on which fuzzy logic is built is best understood in relationship with the canonical way in which reasoning is formalised: with classical propositional logic. The latter is the logic of "false" and "true" and propositions are evaluated in this two-element set. Among the connectives we find the logical "and", "or", and "not", interpreted in the well-known way. In addition to the two classical truth values, fuzzy logic uses intermediary degrees of truth [14]. Usually, "false" and "true" are identified with the real numbers 0 and 1, respectively; the remaining real numbers serve as further truth degrees and may express relative tendencies.

The difficulty of this approach is that there is no straightforward way to tell how the logical connectives should be interpreted. We rather have to make a decision, for instance, about the interpretation of the conjunction. Different interpretations will in general lead to different logics. As a consequence of this situation, fuzzy logic has in fact emerged as a family of many-valued logics, each of which may bring its own challenges. According to a common agreement, the binary operation on the real

T. Vetterlein (✉)
Department of Knowledge-Based Mathematical Systems,
Johannes Kepler University, Linz, Austria
e-mail: Thomas.Vetterlein@jku.at

M. Petrík
Department of Mathematics Faculty of Engineering, Czech University
of Life Sciences, Prague, Czech Republic
e-mail: petrik@cs.cas.cz

S. Saminger-Platz and R. Mesiar (eds.), *On Logical, Algebraic, and Probabilistic Aspects of Fuzzy Set Theory*, Studies in Fuzziness and Soft Computing 336,
DOI 10.1007/978-3-319-28808-6_6

unit interval taken for this purpose should be a t-norm: associative, commutative, possessing 1 as an identity, and monotone in each argument. If the set of truth values is not taken to be an uncountable set but for instance a finite chain, the operation should still fulfil the same algebraic conditions. It is natural to assume that the chain of truth degrees is a negative, commutative totally ordered monoid.

The present work is to be seen among the efforts of classifying these algebraic structures. A considerable amount of work has been done on this topic during recent years. In line with the given background, residuation has usually been additionally assumed and MTL-algebras were considered [8, 22]. Our paper [29] is devoted to MTL-algebras based on the real unit interval. For residuated lattices in general, see [3, 11]. MTL-chains fulfilling certain additional properties were considered in several works as well. For instance, MTL-chains with the weak cancellation property are the topic of [21] as well as [15]. Idempotent residuated chains are studied in [4]. The paper [16] deals with finite MTL-chains and their relationship to Abelian totally ordered groups.

The present chapter is devoted to the finite case. The tomonoids considered are assumed to be either finite, or at least to be finitely generated. We present two different approaches based on our work [24, 28], respectively. We provide an introduction to the main ideas; further details can be found in the indicated papers. We note that in [27] two further approaches to the structures under consideration are offered.

2 Totally Ordered Monoids

We investigate in this chapter the following structures [5, 9, 12, 13, 23, 30].

Definition 1 An algebra $(L; +, 0)$ is a *monoid* if (i) $+$ is an associative binary operation and (ii) 0 is an identity for $+$. A monoid $(L; +, 0)$ is called *commutative* if $+$ is commutative.

A partial order $\leq$ on a monoid L is called *compatible* if, for any $a, b, c, d \in L$, $a \leq b$ and $c \leq d$ imply $a + c \leq b + d$. A structure $(L; \leq, +, 0)$ such that $(L; +, 0)$ is a monoid and $\leq$ is a compatible total order on L is called a *totally ordered monoid*, or *tomonoid* for short.

Moreover, a tomonoid $(L; \leq, +, 0)$ is called *commutative* if so is its monoidal reduct. L is called *positive* if 0 is the bottom element. L is called *finitely generated* if L, as a monoid, is generated by finitely many elements.

For instance, let [0,1] be the real unit interval and let $\oplus \colon [0,1]^2 \to [0,1]$ be a t-conorm, that is, associative, commutative, behaving neutrally w.r.t. 0, and monotone in each argument [20]. Then $([0,1] ; \leq, \oplus, 0)$ is commutative, positive tomonoid. Similarly, let $L \subset [0,1]$ be a finite subset of [0,1] containing 0 and 1 and let $\oplus \colon L^2 \to L$ be a discrete t-conorm [7]. This is equivalent to say that $(L; \leq, \oplus, 0)$ is a finite, commutative, positive tomonoid.

We have written tomonoids in the additive way; alternatively, we may deal with the dual structures. In this case, the order is reversed and the multiplicative notation is used. In particular, the monoidal operation is then denoted by a product-like symbol and the monoidal identity by 1. The aforementioned examples would in this case become tomonoids based on a t-norm or a discrete t-norm, respectively. The choice of order and notation is not solely a matter of taste. In many-valued logics, a larger value corresponds to a higher degree of presence and hence the multiplicative notation is common. In the context of free monoids, in contrast, the additive notation is predominant. Within the present chapter, both possibilities will be made use of.

A tomonoid consisting of the monoidal identity alone is called *trivial*. We will tacitly assume throughout this paper that all tomonoids are non-trivial. A set of generators of a (non-trivial) tomonoid L will be understood to be a non-empty, finite set of elements distinct from 0 that generate L as a monoid.

Congruences of tomonoids are defined as follows; cf. [9]. Recall that a subset C of a poset is called *convex* if $a, c \in C$ and $a \leq b \leq c$ imply $b \in C$.

Definition 2 Let $(L; \leq, +, 0)$ be a tomonoid. A *tomonoid congruence* on L is a congruence $\approx$ of L as a monoid such that all $\approx$-classes are convex. On the quotient $\langle L \rangle_{\approx}$, we then denote the operation induced by $+$ again by $+$ and, for $a, b \in L$, we let $\langle a \rangle_{\approx} \leq \langle b \rangle_{\approx}$ if $a \approx b$ or $a < b$.

We immediately check that this definition is as intended.

Lemma 1 *Let $\approx$ be a tomonoid congruence on a tomonoid $(L; \leq, +, 0)$. Then the quotient $(\langle L \rangle_{\approx}; \leq, +, \langle 0 \rangle_{\approx})$ is a tomonoid again. Furthermore, if L is commutative, positive, finitely generated, then so is $\langle L \rangle_{\approx}$, respectively.*

It is difficult to classify the congruences of tomonoids. There are, however, certain special types that allow an easy description. For instance, an ideal of a commutative, positive tomonoid induces a congruence in a natural way [3]. For a discussion of this type of congruences, see, e.g., [29]. Moreover, there is an order-theoretic analogue of a Rees quotient; this type of congruences will be central in the second part of this chapter.

3 Representation of Tomonoids by Direction Cones

The first part of the present chapter is devoted to finitely generated, positive, commutative tomonoids; we will write "fg.p.c. tomonoids" for short. In particular, the finite, positive, commutative tomonoids, which correspond to the so-called discrete t-norms [7], are included in the discussion.

We investigate a particular way of representing such tomonoids. We are guided by the following ideas. First of all, any monoid can be identified with a congruence on a free monoid. Similarly, we may describe tomonoids by what we call monomial preorders. Second, the order of totally ordered Abelian groups is characterised by their cone. We introduce for tomonoids an analogous object; the so-called direction cones are certain subsets of $\mathbb{Z}^n$ that describe tomonoids and each fg.p.c. tomonoid is a quotient of a tomonoid arising in this way.

The results of this section originate from the paper [28], to which we refer for further details. A continuation of this work, in which the finite case is especially emphasised, can be found in [26].

3.1 Congruences and Monomial Preorders

Free commutative monoids play a central role in what follows. We identify the free commutative monoid over $n \geq 1$ elements with $\mathbb{N}^n$. The addition is defined pointwise and the identity is $\bar{0} = (0, \ldots, 0)$, the n-tuple consisting of zeros only. We also define $u_i = (0, \ldots, 0, 1, 0, \ldots, 0)$, "1" being at the i-th position. Clearly then, $U(\mathbb{N}^n) = \{u_1, \ldots, u_n\}$ is a set of generators of $\mathbb{N}^n$.

We endow $\mathbb{N}^n$ with the componentwise natural order. That is, for $(a_1, \ldots, a_n)$, $(b_1, \ldots, b_n) \in \mathbb{N}^n$, we put

$$(a_1, \ldots, a_n) \trianglelefteq (b_1, \ldots, b_n) \quad \text{if} \quad a_1 \leq b_1, \ldots, \; a_n \leq b_n. \tag{1}$$

Clearly, $\trianglelefteq$ is a lattice order on $(\mathbb{N}^n; +, \bar{0})$ and $\trianglelefteq$ is compatible with the addition.

Fg.p.c. tomonoids can be conveniently described on the basis of the free commutative monoid $\mathbb{N}^n$ as follows.

We call a reflexive and transitive binary relation $\preccurlyeq$ on a set A a *preorder*. We write $a \prec b$ if $a \preccurlyeq b$ but not $b \preccurlyeq a$. Any preorder $\preccurlyeq$ gives rise to an equivalence relation $\approx$, called its *symmetrisation*, where $a \approx b$ if $a \preccurlyeq b$ and $b \preccurlyeq a$. We call the equivalence class of some a w.r.t. $\approx$ a $\preccurlyeq$*-class* and we denote it by $\langle a \rangle_{\preccurlyeq}$. The preorder $\preccurlyeq$ induces on the quotient $\langle A \rangle_{\preccurlyeq}$ a partial order, which we denote by $\preccurlyeq$ again.

We call a preorder $\preccurlyeq$ *total* if $a \preccurlyeq b$ or $b \preccurlyeq a$ for any pair $a, b \in A$. Moreover, we call $\preccurlyeq$ *positive* if $0 \prec a$ for all $a \neq 0$. Finally, if $\preccurlyeq$ is defined on a monoid $(L; +, 0)$, we call $\preccurlyeq$ *compatible* if $a \preccurlyeq b$ implies $a + c \preccurlyeq b + c$.

In computational mathematics, the notion "monomial ordering" refers to compatible, positive, total orders on $\mathbb{N}^n$; see, e.g., [6]. Analogously, we call a preorder $\preccurlyeq$ on $\mathbb{N}^n$ *monomial* if $\preccurlyeq$ is compatible, positive, and total. The significance of monomial preorders becomes clear in the following proposition.

Proposition 1 *Let $\preccurlyeq$ be a monomial preorder on $(\mathbb{N}^n; +, \bar{0})$. Then its symmetrisation is a monoid congruence whose classes are convex and such that $\langle \bar{0} \rangle_{\preccurlyeq} = \{\bar{0}\}$. Moreover, $(\langle \mathbb{N}^n \rangle_{\preccurlyeq}; \preccurlyeq, +, \{\bar{0}\})$ is a fg.p.c. tomonoid.*

Conversely, let $(L; \leq, +, 0)$ be a fg.p.c. tomonoid; assume that the $n \geq 1$ elements $g_1, \ldots, g_n \in L \backslash \{0\}$ generate L. Let $\iota \colon \mathbb{N}^n \to L$ be the surjective monoid homomorphism determined by $\iota(u_i) = g_i$, $i = 1, \ldots, n$. For $a, b \in \mathbb{N}^n$ define

$$a \preccurlyeq b \quad \text{if} \quad \iota(a) \leq \iota(b). \tag{2}$$

Then $\preccurlyeq$ *is a monomial preorder of* $\mathbb{N}^n$*, and* ι *induces an isomorphism between* $(\langle\mathbb{N}^n\rangle_{\preccurlyeq}; \preccurlyeq, +, \{\bar{0}\})$ *and* $(L; \leq, +, 0)$.

Proof Let $\preccurlyeq$ be a monomial preorder on $\mathbb{N}^n$. Then, for $a, b, c, d \in \mathbb{N}^n$, $a \approx c$ and $b \approx d$ imply $a + b \approx c + d$ by the compatibility of $\preccurlyeq$; hence $\approx$ is a monoid congruence. As $\preccurlyeq$ is also positive, $\preccurlyeq$ extends $\trianglelefteq$, and it follows that the $\preccurlyeq$-classes are convex. Again by the positivity, the $\preccurlyeq$-class of $\bar{0}$ consists of $\bar{0}$ alone.

As $\preccurlyeq$ is compatible, the partial order $\preccurlyeq$ induced on $\langle\mathbb{N}^n\rangle_{\preccurlyeq}$ is compatible as well; that is, $(\langle\mathbb{N}^n\rangle_{\preccurlyeq}; \preccurlyeq, +, \langle\bar{0}\rangle_{\preccurlyeq})$ is a commutative pomonoid. Since, for any $a, b \in \mathbb{N}^n$, $a \preccurlyeq b$ or $b \preccurlyeq a$, $\langle\mathbb{N}^n\rangle_{\preccurlyeq}$ is actually a tomonoid. Moreover, since $\bar{0} \prec a$ for any $a \in \mathbb{N}^n\backslash\{\bar{0}\}$, $\langle\mathbb{N}^n\rangle_{\preccurlyeq}$ is a positive, commutative tomonoid, which is generated by the finitely many elements $\langle u_1\rangle_{\preccurlyeq}, \ldots, \langle u_n\rangle_{\preccurlyeq}$.

For the second part, assume that $(L; \leq, +, 0)$ is a fg.p.c. tomonoid and $g_1, \ldots, g_n \in L\backslash\{0\}$ generate L as a monoid. Let furthermore $\iota\colon \mathbb{N}^n \to L$ be as indicated and let $\preccurlyeq$ be defined by (2). By construction, $\preccurlyeq$ is transitive and reflexive, that is, a preorder. $\preccurlyeq$ is compatible because so is $\leq$ and ι is a monoid homomorphism. Moreover, $\preccurlyeq$ is positive because L is positive and hence $\iota(a) \leq 0$ holds only if $a = \bar{0}$. Hence $\preccurlyeq$ is a monomial preorder. Finally, for $a, b \in \mathbb{N}^n$, we have $a \approx b$ if and only if $a \preccurlyeq b$ and $b \preccurlyeq a$ if and only if $\iota(a) = \iota(b)$; hence ι induces an isomorphism as claimed.

We conclude that any monomial preorder $\preccurlyeq$ on $\mathbb{N}^n$ gives rise to a fg.p.c. tomonoid L. We call L in this case the tomonoid *represented by* $\preccurlyeq$.

Proposition 1 also states that, up to isomorphism, any fg.p.c. tomonoid L arises in this way from a monomial preorder. In other words, describing fg.p.c. tomonoids can be done by describing monomial preorders. This is what we will do in the sequel.

3.2 Tomonoids Arising from Totally Ordered Abelian Groups

The positive cones of totally ordered Abelian groups give rise to typical examples of fg.p.c. tomonoids. We will discuss these examples in some detail because they motivate our way of representing fg.p.c. tomonoids in general.

Definition 3 Let $(G; \leq, +, 0)$ be a totally ordered Abelian group and let $G^+ = \{g \in G\colon g \geq 0\}$ be its positive cone. Assume that G is generated by $g_1, \ldots, g_n \in G^+\backslash\{0\}$, where $n \geq 1$. Let L be the submonoid of G generated by $g_1, \ldots, g_n$ and let L be endowed with the total order inherited from G, with the group addition, and with the constant 0. Then we call $(L; \leq, +, 0)$ a *group cone tomonoid*.

Clearly, a group cone tomonoid is a fg.p.c. tomonoid. Note that in general we do not deal with the whole positive cone of a totally ordered Abelian group. In fact, the latter is in general not finitely generated even if the group is.

Group cone tomonoids are characterised by the following condition. We say that a fg.p.c. tomonoid L is *cancellative* if, for all $a, b, c \in L, a + c = b + c$ implies $a = b$. Note that in this case, for all $a, b, c \in L$, $a \leq b$ is equivalent to $a + c \leq b + c$.

Proposition 2 *A fg.p.c. tomonoid $(L; \leq, +, 0)$ is a group cone tomonoid if and only if it is cancellative.*

Proof The "only if" part follows from the construction of a group cone tomonoid.

To see the "if" part, let L be cancellative. Let G be the group consisting of the differences of elements of L; see, e.g., [10, Chap. II.2]. Viewing L as a subset of G, we introduce a total order on G as follows: for $a, b, c, d \in L$, we define $a - b \leq c - d$ if $a + d \leq b + c$ in L. Then $(G; \leq, +, 0)$ is a totally ordered Abelian group, and $(L; \leq, +, 0)$ is a subtomonoid of $(G^+; \leq, +, 0)$. The assertion follows.

Group cone tomonoids correspond by Proposition 1 to particular monomial preorders. We call a preorder $\preccurlyeq$ on $\mathbb{N}^n$ *cancellative* if, for any $a, b, c \in \mathbb{N}^n$, $a \preccurlyeq b$ is equivalent to $a + c \preccurlyeq b + c$.

Proposition 3 *Let the fg.p.c. tomonoid L be represented by the monomial preorder $\preccurlyeq$ on $\mathbb{N}^n$. Then L is a group cone tomonoid if and only if $\preccurlyeq$ is cancellative.*

Proof Let L be a group cone tomonoid. Then $(\langle \mathbb{N}^n \rangle_{\preccurlyeq}; \preccurlyeq, +, \{\bar{0}\})$ is cancellative by Proposition 2. Thus, for $a, b, c \in \mathbb{N}^n$, we have $a \preccurlyeq b$ iff $\langle a \rangle_{\preccurlyeq} \preccurlyeq \langle b \rangle_{\preccurlyeq}$ iff $\langle a \rangle_{\preccurlyeq} + \langle c \rangle_{\preccurlyeq} \preccurlyeq \langle b \rangle_{\preccurlyeq} + \langle c \rangle_{\preccurlyeq}$ iff $\langle a + c \rangle_{\preccurlyeq} \preccurlyeq \langle b + c \rangle_{\preccurlyeq}$ iff $a + c \preccurlyeq b + c$, that is, $\preccurlyeq$ is cancellative.

Conversely, let $\preccurlyeq$ be cancellative. Then $(\langle \mathbb{N}^n \rangle_{\preccurlyeq}; \preccurlyeq, +, \{\bar{0}\})$ is a cancellative fg.p.c. tomonoid and hence, by Proposition 2, a group cone tomonoid.

Recall next that the order of a partially ordered Abelian group $(G; \leq, +, 0)$ is uniquely determined by its positive cone G^+. In fact, for any $g, h \in G$, $g \leq h$ if and only if $h - g \in G^+$. We may also view the positive cone of a partially ordered group as the set of all differences of elements g and h such that $g \leq h$; indeed, $G^+ = \{h - g : g, h \in G \text{ such that } g \leq h\}$.

We may use the same object to describe group cone tomonoids. We denote by $(\mathbb{Z}^n; +, \bar{0})$ the free Abelian group generated by $n \geq 1$ elements. Furthermore, $\trianglelefteq$ will be the partial order on $\mathbb{Z}^n$ defined according to (1): for $a, b \in \mathbb{Z}^n$, we put $a \trianglelefteq b$ if $a + c = b$ for some $c \in \mathbb{N}^n$. Then $(\mathbb{Z}^n; \trianglelefteq, +, \bar{0})$ is a lattice-ordered group.

Definition 4 Let $\preccurlyeq$ be a cancellative monomial preorder on $\mathbb{N}^n$. Then the set

$$P_{\preccurlyeq} = \{b - a \in \mathbb{Z}^n : a, b \in \mathbb{N}^n \text{ such that } a \preccurlyeq b\}$$

is called the *positive cone* of $\preccurlyeq$.

A positive cone determines the preorder from which it is defined as in the case of groups.

Lemma 2 *Let $P \subseteq \mathbb{Z}^n$ be the positive cone of the cancellative monomial preorder $\preccurlyeq$ on $\mathbb{N}^n$. Then we have:*

(GO) *For any $a, b \in \mathbb{N}^n$, $a \preccurlyeq b$ if and only if $b - a \in P$.*

Proof By definition, $a \preccurlyeq b$ implies $b - a \in P$.

Conversely, let $b - a \in P$. Then there are $c, d \in \mathbb{N}^n$ such that $c \preccurlyeq d$ and $d - c = b - a$. It follows $a + d = b + c \preccurlyeq b + d$ and hence $a \preccurlyeq b$.

By Lemma 2, we have for any cancellative monomial preorder $\preccurlyeq$

$$\begin{aligned} P_{\preccurlyeq} &= \{z \in \mathbb{Z}^n : a \preccurlyeq b \text{ for some } a, b \in \mathbb{N}^n \text{ such that } z = b - a\} \\ &= \{z \in \mathbb{Z}^n : a \preccurlyeq b \text{ for all } a, b \in \mathbb{N}^n \text{ such that } z = b - a\}. \end{aligned} \tag{3}$$

The positive cones of partially ordered Abelian groups possess an intrinsic characterisation: they are exactly the cancellative commutative monoids such that $a + b = 0$ implies $a = b = 0$ [10]. The positive cones of cancellative monomial preorders can be described in a similar way.

Theorem 1 *A set $P \subseteq \mathbb{Z}^n$ is the positive cone of a cancellative monomial preorder on $\mathbb{N}^n$ if and only if the following conditions are fulfilled:*

(GC1) *If $z \in \mathbb{N}^n$, then $z \in P$. Moreover, if $z \in \mathbb{N}^n \setminus \{\bar{0}\}$, then $-z \notin P$.*
(GC2) *P is closed under addition.*
(GC3) *For any $z \in \mathbb{Z}^n$, at least one of $z \in P$ or $-z \in P$ holds.*

In this case, $P = P_{\preccurlyeq}$, where $\preccurlyeq$ is given by condition (GO) *above.*

Proof Let $\preccurlyeq$ be a cancellative monomial preorder on $\mathbb{N}^n$. Clearly, $0 \in P_{\preccurlyeq}$ then. Furthermore, any $z \in \mathbb{N}^n \setminus \{\bar{0}\}$ is in $P_{\preccurlyeq}$ because $\bar{0} \preccurlyeq z$ holds by the positivity of $\preccurlyeq$. Assume that also $-z \in P_{\preccurlyeq}$. Then there is a $b \in \mathbb{N}^n$ such that $b + z \preccurlyeq b$ and hence by the cancellativity $z \preccurlyeq 0$, in contradiction to the positivity of $\preccurlyeq$. (GC1) is shown.

For $a, b, c, d \in \mathbb{N}^n$, $a \preccurlyeq b$ and $c \preccurlyeq d$ implies $a + c \preccurlyeq b + c \preccurlyeq b + d$. We conclude that if $b - a, d - c \in P_{\preccurlyeq}$, also $(b - a) + (d - c) = (b + d) - (a + c) \in P_{\preccurlyeq}$. This shows (GC2).

For $a, b \in \mathbb{N}^n$, at least one of $a \preccurlyeq b$ or $b \preccurlyeq a$ holds because $\preccurlyeq$ is total. (GC3) follows as well.

Let now $P \subseteq \mathbb{Z}^n$ fulfil (GC1)–(GC3). For $a, b \in \mathbb{N}^n$, let $a \preccurlyeq b$ if $b - a \in P$. We claim that $\preccurlyeq$ is a cancellative monomial preorder. As $0 \in P$ by (GC1), $\preccurlyeq$ is reflexive. By (GC2), $\preccurlyeq$ is transitive. Hence $\preccurlyeq$ is a preorder. $\preccurlyeq$ is total by (GC3) and positive by (GC1). Finally, by construction, $a \preccurlyeq b$ is equivalent to $a + c \preccurlyeq b + c$; the compatibility and cancellativity of $\preccurlyeq$ follows.

It remains to show that P is actually the positive cone $P_{\preccurlyeq}$ of $\preccurlyeq$. By Lemma 2, we have that, for any $a, b \in \mathbb{N}^n$, $b - a \in P_{\preccurlyeq}$ if and only if $a \preccurlyeq b$. But by construction, $a \preccurlyeq b$ if and only if $b - a \in P$. Hence $P = P_{\preccurlyeq}$.

Finally, if $P \subseteq \mathbb{Z}^n$ is the positive cone of any cancellative monomial preorder $\preccurlyeq$, then $\preccurlyeq$ is by Lemma 2 uniquely determined by (GO). The last statement follows.

3.3 Direction Cones

Positive cones describe cancellative fg.p.c. tomonoids. In this section we will generalise this notion to cover a wider class of tomonoids. In this case we will not obtain a strict correlation, but we will be led to a Galois correspondence.

Let $\preccurlyeq$ be a monomial preorder on $\mathbb{N}^n$. If $\preccurlyeq$ is cancellative, then for any $a, b \in \mathbb{N}^n$ the question of whether or not $a \preccurlyeq b$ holds depends only on the difference $z = b - a$: we have $a \preccurlyeq b$ if and only if $c \preccurlyeq d$ for any other pair $c, d \in \mathbb{N}^n$ such that $z = d - c$. In fact, the positive cone $P_{\preccurlyeq}$ consists of these differences; $a \preccurlyeq b$ if and only if $b - a \in P_{\preccurlyeq}$.

In general, the question of whether or not we have $a \preccurlyeq b$ does not depend on the difference $b - a$ alone. For instance, it may be the case that $a + c \preccurlyeq b + c$ holds for some $c \in \mathbb{N}^n$ but not $a \preccurlyeq b$. However, let $z \in \mathbb{Z}^n$. Then the following lemma implies that still at least one of following possibilities applies: $a \preccurlyeq b$ for all $a, b \in \mathbb{N}^n$ such that $b - a = z$, or $b \preccurlyeq a$ for all $a, b \in \mathbb{N}^n$ such that $b - a = z$.

Lemma 3 *Let $z \in \mathbb{Z}^n$. Then there is a unique pair $a, b \in \mathbb{N}^n$ such that $z = b - a$ and, for any $c, d \in \mathbb{N}^n$ such that $z = d - c$, we have $c = a + t$ and $d = b + t$ for some $t \in \mathbb{N}^n$.*

Proof Put $a = -z \vee \bar{0}$ and $b = z \vee \bar{0}$. Then $z = b - a$. Moreover, if $c, d \in \mathbb{N}^n$ such that $d - c = z$, we have $c \trianglerighteq \bar{0}$ and $c = d - z \trianglerighteq -z$, thus $c \trianglerighteq a$; similarly, $d \trianglerighteq b$. As $b - a = d - c$, the differences $c - a$ and $d - b$ coincide and hence $c = a + t$ and $d = b + t$ for some $t \in \mathbb{N}^n$. The uniqueness of a, b follows from the $\trianglelefteq$-minimality.

Let $a, b \in \mathbb{N}^n$ be associated with $z \in \mathbb{Z}^n$ according to Lemma 3. Inspecting the proof, we see that b is simply the positive part of $z \in \mathbb{Z}^n$, and a is its (negated) negative part. Let us define

$$\begin{aligned} z^+ &= z \vee \bar{0}, \\ z^- &= -z \vee \bar{0}. \end{aligned}$$

Then we have

$$z = z^+ - z^-$$

and any other pair of elements of $\mathbb{N}^n$ whose difference is z arises from z^+ and z^- by adding a $t \in \mathbb{N}^n$.

For a compatible preorder $\preccurlyeq$ on $\mathbb{N}^n$, the obvious consequence is the following. Let $z \in \mathbb{Z}^n$. If $z^- \preccurlyeq z^+$, we conclude from Lemma 3 and the compatibility of $\preccurlyeq$ that $a \preccurlyeq b$ actually holds for *any* pair $a, b \in \mathbb{N}^n$ such that $b - a = z$. Thus, intuitively, we may view any $z \in \mathbb{Z}^n$ such that $z^- \preccurlyeq z^+$ as being "positively directed"; for, in this case we have $a \preccurlyeq a + z$ for any $a \in \mathbb{N}^n$ such that $a + z \in \mathbb{N}^n$. Our viewpoint is reflected in the following definition.

Definition 5 Let $\preccurlyeq$ be a monomial preorder on $\mathbb{N}^n$. Then the set

$$C_{\preccurlyeq} = \{z \in \mathbb{Z}^n : z^- \preccurlyeq z^+\}$$

is called the *direction cone* of $\preccurlyeq$.

By Lemma 3 we then have

$$C_{\preccurlyeq} = \{z \in \mathbb{Z}^n : a \preccurlyeq b \text{ for all } a, b \in \mathbb{N}^n \text{ such that } z = b - a\}. \quad (4)$$

The natural question is now if there is a characterisation of direction cones similar to the case of positive cones. Comparing with (3), we see that the direction cone of a cancellative monomial preorder is its positive cone. In the general case, we conclude from the positivity of $\preccurlyeq$ that condition (GC1) for positive cones applies here as well, and from the totality of $\preccurlyeq$ also condition (GC3) is immediate: for each $z \in \mathbb{Z}^n$, at least one of z or $-z$ is in $C_{\preccurlyeq}$.

However, a direction cone does not in general fulfil condition (GC2), that is, it is not necessarily closed under addition. The following notion can be used instead. We call a k-tuple $(x_1, \ldots, x_k)$, $k \geq 2$, of elements of $\mathbb{Z}^n$ *addable* if

$$(x_1 + \cdots + x_k)^- + x_1 + \cdots + x_i \trianglerighteq \bar{0} \quad (5)$$

for all $i = 0, \ldots, k$. Note that for addability the order matters.

Lemma 4 *The direction cone of a monomial preorder on $\mathbb{N}^n$ is a set $C \subseteq \mathbb{Z}^n$ fulfilling the following conditions:*

(C1) *Let $z \in \mathbb{N}^n$. Then $z \in C$ and, if $z \neq \bar{0}$, $-z \notin C$.*
(C2) *Let $(x_1, \ldots, x_k)$, $k \geq 2$, be an addable k-tuple of elements of C. Then $x_1 + \cdots + x_k \in C$.*
(C3) *Let $z \in \mathbb{Z}^n$. Then $z \in C$ or $-z \in C$.*

Proof (C1) We have $\mathbb{N}^n \subseteq C$ because $\preccurlyeq$ is positive. Assume that $-z \in C$, where $z \in \mathbb{N}^n$. Then $z = (-z)^- \preccurlyeq (-z)^+ = \bar{0}$ and the positivity of $\preccurlyeq$ implies $z = \bar{0}$.

Recall next that, by (4), $a \preccurlyeq b$ for any $a, b \in \mathbb{N}^n$ such that $b - a \in C$.

To see (C2), let $(x_1, \ldots, x_k)$ be as indicated, and put $z = x_1 + \cdots + x_k$. Then $z^-, z^- + x_1, \ldots, z^- + x_1 + \cdots + x_k \in \mathbb{N}^n$. By assumption, $x_1, \ldots, x_k \in C$; thus $z^- \preccurlyeq z^- + x_1 \preccurlyeq \ldots \preccurlyeq z^- + x_1 + \cdots + x_k = z^- + z = z^+$.

(C3) holds because $\preccurlyeq$ is total.

Our next aim is to show that conditions (C1)–(C3) characterise direction cones.

A preorder gives rise to a direction cone, which fulfils (C1)–(C3). Conversely, we can assign a preorder to a set fulfilling (C1)–(C3).

Definition 6 Let $C \subseteq \mathbb{Z}^n$ fulfil (C1)–(C3). Let $\preccurlyeq_C$ be the smallest preorder on $\mathbb{N}^n$ such that

(O) $a \preccurlyeq_C b$ for any $a, b \in \mathbb{N}^n$ such that $b - a \in C$.

Then we call $\preccurlyeq_C$ the monomial preorder *induced by* C.

In other words, for a subset C of $\mathbb{Z}^n$ fulfilling (C1)–(C3) and $a, b \in \mathbb{N}^n$, we have $a \preccurlyeq_C b$ if and only if there are $k \geq 1$ elements $z_1, \ldots, z_k \in C$ such that $a,\ a + z_1,\ a + z_1 + z_2,\ \ldots, a + z_1 + \cdots + z_k \trianglerighteq \bar{0}$ and $a + z_1 + \cdots + z_k = b$. We note that this is not the same as to say that $b - a$ is a sum of elements of C.

Lemma 5 *Let $C \subseteq \mathbb{Z}^n$ fulfil* (C1)–(C3). *Then $\preccurlyeq_C$, the monomial preorder induced by C, is in fact a monomial preorder.*

Proof By construction, $\preccurlyeq_C$ is a preorder, and by (C3), $\preccurlyeq_C$ is total. It is furthermore clear that $\preccurlyeq_C$ is compatible with the addition.

Assume next that, for some $a \in \mathbb{N}^n$, $a \preccurlyeq_C \bar{0}$ holds according to the prescription (O). Then $a = \bar{0}$ by (C1). It follows that $\bar{0} \prec_C a$ for all $a \in \mathbb{N}^n \backslash \{\bar{0}\}$, that is, $\preccurlyeq_C$ is positive. This completes the proof that $\preccurlyeq_C$ is a monomial preorder.

Theorem 2 *A set $C \subseteq \mathbb{Z}^n$ is the direction cone of a monomial preorder if and only if C fulfils* (C1)–(C3). *In this case, C is the direction cone of $\preccurlyeq_C$.*

Proof A direction cone fulfils (C1)–(C3) by Lemma 4.

Conversely, let C fulfil (C1)–(C3). Let $\preccurlyeq_C$ be the induced preorder. By Lemma 5, $\preccurlyeq_C$ is a monomial preorder.

It remains to show that $C_{\preccurlyeq_C}$, the direction cone of $\preccurlyeq_C$, coincides with C, that is, for $z \in \mathbb{Z}^n$, $z^- \preccurlyeq_C z^+$ if and only if $z \in C$. The "if" part holds by construction. For the "only if" part, assume that $z^- \preccurlyeq_C z^+ = z^- + z$. Then $z = x_1 + \cdots + x_k$ for some $x_1, \ldots, x_k \in C$ such that $z^- + x_1 + \cdots x_i \trianglerighteq \bar{0}$ for $i = 0, \ldots, k$. Then $(x_1, \ldots, x_k)$ is addable, hence $z \in C$ by (C2).

In the sequel, when speaking about direction cones without reference to a monomial preorder, we mean a subset of $\mathbb{Z}^n$ that fulfils the conditions (C1)–(C3).

A direction cone induces a preorder. As seen next, any preorder contains a preorder arising in this way.

Theorem 3 *Let $\preccurlyeq$ be a monomial preorder. Then $\preccurlyeq$ extends $\preccurlyeq_{C_\preccurlyeq}$, the monomial preorder induced by the direction cone of $\preccurlyeq$.*

Moreover, the direction cone of $\preccurlyeq_{C_\preccurlyeq}$ is $C_\preccurlyeq$ again.

Proof Let $a, b \in \mathbb{N}^n$ and assume that $a \preccurlyeq_{C_\preccurlyeq} b$ holds according to the prescription (O). Then $b - a \in C_\preccurlyeq$, that is, $z^- \preccurlyeq z^+$, where $z = b - a$. In view of Lemma 3, it follows $a \preccurlyeq b$. We conclude that $\preccurlyeq_{C_\preccurlyeq} \subseteq \preccurlyeq$.

The second part holds by Theorem 2.

We apply the shown facts to tomonoids.

Definition 7 Let $C \subseteq \mathbb{Z}^n$ be a direction cone. Then we call the tomonoid represented by $\preccurlyeq_C$ a *cone tomonoid*.

Theorem 4 *Each fg.p.c. tomonoid L is the quotient of a cone tomonoid.*

Proof This follows from Theorem 3.

3.4 A Galois Connection

We have seen that there is a mutual correspondence between monomial preorders and direction cones. This correspondence is not one-to-one, some monomial preorders are proper extensions of those that are induced by direction cones. However, we can established a Galois correspondence between the two sets.

Let us fix an $n \geq 1$. Let $\mathscr{P}$ be the set of all monomial preorders on $\mathbb{N}^n$ and let $\mathscr{C}$ be the set of all direction cones in $\mathbb{Z}^n$. We partially order the two sets by means of the set-theoretic inclusion. We then readily check that the two mappings

$$\mathscr{P} \to \mathscr{C}, \quad \preccurlyeq \mapsto C_{\preccurlyeq},$$
$$\mathscr{C} \to \mathscr{P}, \quad C \mapsto \preccurlyeq_C$$

are order-preserving. The mappings are not one-to-one; in fact, the former is surjective but not injective, and the latter is injective but not surjective. From Theorems 2 and 3 we conclude what happens when applying the mappings successively: any $\preccurlyeq \in \mathscr{P}$ is an extension of $\preccurlyeq_{C_{\preccurlyeq}}$; and any $C \in \mathscr{C}$ is equal to $C_{\preccurlyeq_C}$. Hence there is the following Galois connection between $\mathscr{P}$ and $\mathscr{C}$: for any $\preccurlyeq \in \mathscr{P}$ and $C \in \mathscr{C}$,

$$\preccurlyeq_C \subseteq \preccurlyeq \quad \text{if and only if} \quad C \subseteq C_{\preccurlyeq}.$$

3.5 Example

We conclude by presenting an example illustrating the results of this section. Let L be the 9-element fg.p.c. tomonoid specified as follows. Let L be generated by its two elements a and b and assume that

$$0 \; < \; a \; < \; b \; < \; 2a \; < \; a+b \; < \; 2b \; < \; 3a \; <$$
$$2a+b \; = \; a+2b \; = \; 4a \; < \; 2a+2b \; = \; 3a+b \; = \; 5a \; = \; 3b$$

and that the last indicated element is the top element. In accordance with Proposition 1, let $\iota \colon \mathbb{N}^2 \to L$ be the surjective monoid homomorphism such that $\iota((1,0)) = a$ and $\iota((0,1)) = b$, and endow $\mathbb{N}^2$ with the preorder $\preccurlyeq$ according to (2). Then we have

$$(0,0) \prec (1,0) \prec (0,1) \prec (2,0) \prec (1,1) \prec (0,2) \prec$$
$$(3,0) \prec (2,1) \approx (1,2) \approx (4,0) \prec (m,n),$$

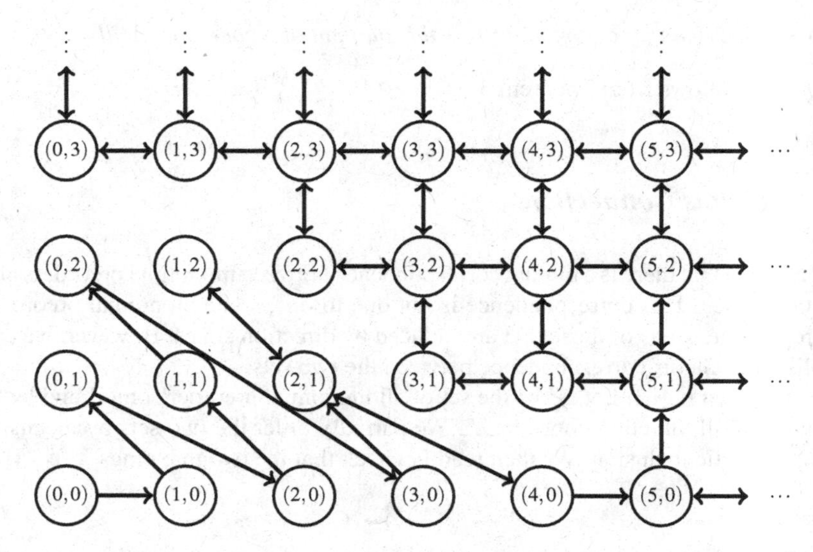

Fig. 1 The example tomonoid L. The *simple arrows* indicate the immediate-successor relation w.r.t. $\preccurlyeq$; the *double arrows* indicate $\preccurlyeq$-equivalence

where (m, n) is any of the remaining elements of $\mathbb{N}^2$. A graphical representation of $(L; \leq, +, 0)$ can be found in Fig. 1.

According to Definition 5, the direction cone is

$$\begin{aligned} C_{\preccurlyeq} &= \{(p, q) \in \mathbb{Z}^2 \colon (-p \vee 0, -q \vee 0) \preccurlyeq (p \vee 0, q \vee 0)\} \\ &= \{(p, q) \in \mathbb{Z}^2 \colon p, q \geq 0\} \cup \\ &\quad \{(-2, 2), (-1, 1), (-1, 2), (2, -1), (3, -2), (3, -1), (4, -2), (4, -1)\} \cup \\ &\quad \{(p, q) \in \mathbb{Z}^2 \colon p \leq 0 \text{ and } q \geq 3\} \cup \\ &\quad \{(p, q) \in \mathbb{Z}^2 \colon p \geq 5 \text{ and } q \leq 0\}. \end{aligned}$$

This set is depicted in Fig. 2.

Finally, we calculate $\preccurlyeq_{C_{\preccurlyeq}}$, the preorder representing a cone tomonoid whose quotient is L. The preorder $\preccurlyeq_{C_{\preccurlyeq}}$ can most easily be read off directly from Fig. 1. Namely, we collect the order relations that hold between elements of the form $(m, 0)$ and $(0, n)$, where $m, n \geq 1$; then we translate and concatenate them. The result is depicted in Fig. 3. From $\preccurlyeq_{C_{\preccurlyeq}}$, we get L by requiring the elements $(2, 1)$, $(1, 2)$, and $(4, 0)$ of $\mathbb{N}^2$ to be equivalent.

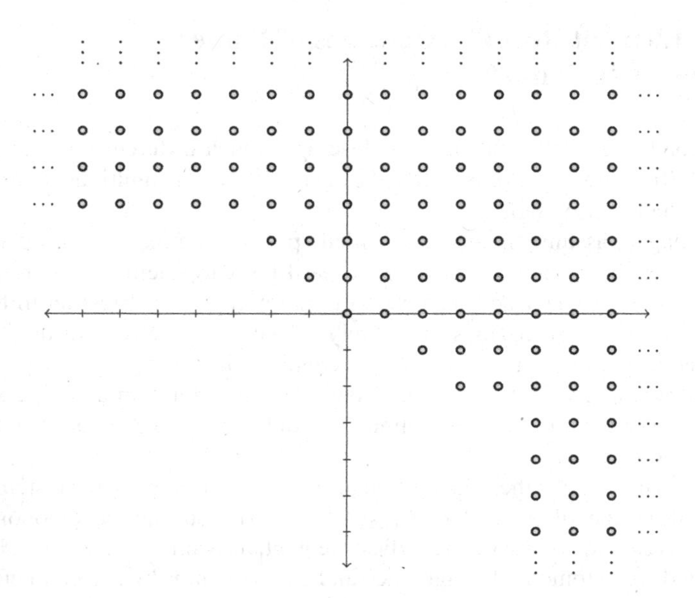

Fig. 2 The direction cone $C_{\preccurlyeq}$ of the monomial preorder $\preccurlyeq$ representing L. Each element of $C_{\preccurlyeq}$ is depicted as a circle in the $\mathbb{Z}^2$ plane

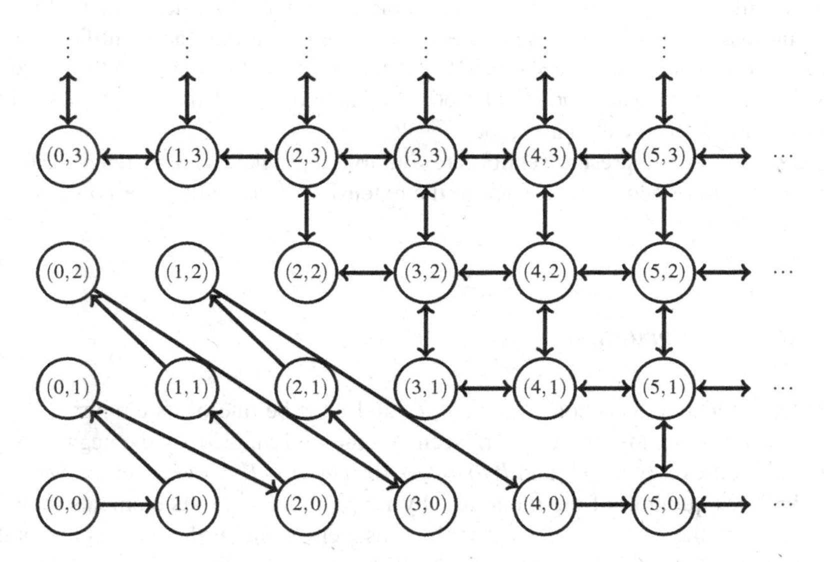

Fig. 3 The cone tomonoid represented by $\preccurlyeq_{C_{\preccurlyeq}}$, whose quotient is L

4 One-Element Rees Coextensions of Finite Negative Tomonoids

In the second part of this chapter, we develop a much different point of view on tomonoids. To begin with, we switch to the dual order and the multiplicative notation, as is common in fuzzy logic.

We will again assume the property called "positive" in the previous part. In the present context, however, positivity means that 1 is the top element; accordingly, we will refer to this property as "negative". Furthermore, we will restrict to the finite case. Finally, our considerations do not rely on the commutativity of the monoidal product and hence we will not assume this condition here.

We shall write "f.n." for "finite, negative". That is, a f.n. tomonoid is a structure $(L; \leq, \odot, 1)$ such that $(L; \odot, 1)$ is a finite monoid and $\leq$ is a compatible total order whose top element is 1.

Our aim is to describe the construction of f.n. tomonoids in a step-by-step manner. The main idea is the following. Let $(L; \leq, \odot, 1)$ be a non-trivial f.n. tomonoid and let 0 and α be its smallest and second smallest element, respectively. Then the identification of 0 and α is a tomonoid congruence and the quotient is by one element smaller than L. Continuing in the same way, we get a sequence of tomonoids that ends with the trivial one. It seems then natural to ask how to generate such a sequence in the reversed order. That is, given an f.n. tomonoid L, how can we determine all those f.n. tomonoids $\bar{L}$ that are by one element larger and such that the identification of their smallest two elements leads back to L? This is in fact the question that we will answer. We will provide a practical method of determining from L systematically all tomonoids $\bar{L}$ of the indicated type.

The results of the present section are due to [24], where further details can be found. For a more general approach to the extension of partially ordered monoids, see, e.g., [18, 19].

4.1 Rees Congruences

Consider a negative tomonoid $(L; \leq, \odot, 1)$ and let q be one of its elements. Then $I_q = \{a \in L : a \leq q\}$ is an ideal of L, seen as a monoid. Indeed, by the negativity of L, $a \leq q$ implies $a \odot b \leq q$ and $b \odot a \leq q$ for any $b \in L$. Consequently, we may form the Rees quotient of the monoid L by I_q; see, e.g., [17]. Its elements may be identified with the elements that are not in I_q as well as one further element, usually denoted by 0. Obviously, this monoid congruence has only convex classes and hence it is a tomonoid congruence; cf., e.g., [9].

Definition 8 Let $(L; \leq, \odot, 1)$ be a f.n. tomonoid and let $q \in L$. For $a, b \in L$, let $a \approx_q b$ if $a = b$ or $a, b \leq q$. Then we call $\approx_q$ the *Rees congruence* by q. We denote the quotient by L/q and call it the *Rees quotient* of L by q.

Moreover, we call L a *Rees coextension* of L/q. We call L a *one-element* Rees coextension, or simply a *one-element coextension*, if L is non-trivial and q is the atom of L.

For a finite chain L, we will denote by 0 the bottom element and we write

$$L^* = L \setminus \{0\}.$$

If L has at least two elements, we furthermore call the second smallest element of L the *atom* of L. We will use in the sequel the symbol α to denote the atom.

Given a non-trivial f.n. tomonoid L, then its Rees quotient L/α by its atom α arises from L by the identification of the smallest two elements. L is in this case a one-element coextension of L/α. Our aim is to determine all one-element coextensions of a f.n. tomonoid. We will then obviously be in the position to construct, starting from the trivial tomonoid, successively all f.n. tomonoids.

4.2 Tomonoid Partitions

A binary operation $\odot$ on a set A gives rise to a partition of $A \times A$: the blocks of the partition are the subsets of all those pairs that are mapped by $\odot$ to the same value. The blocks are commonly referred to as the *level sets* of $\odot$. This partition, together with the assignment that associates with each block the respective element of A, specifies $\odot$ uniquely.

The representation of binary operations based on level sets was first applied to the theory of tomonoids in [25]. We note that it comes along with the possibility of representing tomonoids within two dimensions only.

Definition 9 Let $(L; \leq, \odot, 1)$ be a tomonoid. We define, for any $(a, b), (c, d) \in L^2$,

$$(a, b) \sim (c, d) \quad \text{if} \quad a \odot b = c \odot d.$$

We call $\sim$ the *level equivalence* of L.

Based on the level equivalence of a tomonoid L, we will endow the set L^2 with a first-order structure as follows.

Definition 10 Let $\leq$ be a total order on a set L and let $1 \in L$. We denote the componentwise order on L^2 by $\trianglelefteq$, that is, we put

$$(a, b) \trianglelefteq (c, d) \quad \text{if } a \leq c \text{ and } b \leq d$$

for $a, b, c, d \in L$. Moreover, let $\sim$ be an equivalence relation on L^2 such that the following conditions hold:

(P1) For any $a, b, c, d, e, f \in L$, if $(1, e) \sim (a, b) \trianglelefteq (c, d) \sim (1, f)$, then $e \leq f$.

(P2) For any $(a, b) \in L^2$, there is exactly one $c \in L$ such that $(a, b) \sim (1, c) \sim (c, 1)$.

(P3) For any $a, b, c, d, e \in L$, $(a, b) \sim (d, 1)$ and $(b, c) \sim (1, e)$ imply $(d, c) \sim (a, e)$.

We then call the structure $(L^2; \trianglelefteq, \sim, (1,1))$ a *tomonoid partition.*

Proposition 4 *Let $(L; \leq, \odot, 1)$ be a tomonoid and let $\sim$ be the level equivalence of L. Then $(L^2; \trianglelefteq, \sim, (1,1))$ is a tomonoid partition.*

Proof Let $a, b, c, d \in L$. By the compatibility of $\leq$ with $\odot$, we have that $(a, b) \trianglelefteq (c, d)$ implies $a \odot b \leq c \odot d$. (P1) follows. Moreover, as 1 is the monoidal identity, we have that $(a, b) \sim (c, 1)$ iff $(a, b) \sim (1, c)$ iff $a \odot b = c$. Hence also (P2) holds. Finally, (P3) is implied by the associativity of $\odot$.

By Proposition 4, each tomonoid L gives rise to a tomonoid partition; we will speak about the tomonoid partition *associated with L*.

We next see that there is a converse of Proposition 4. We will use the following simplified notation. When L is a chain and $1 \in L$, we will identify the elements of the form $(1, c) \in L^2$, where $c \in L$, with c. It will be clear from the context if c denotes an element of L or of L^2. For instance, if $\sim$ is an equivalence relation on L^2, then $(a, b) \sim c$ means $(a, b) \sim (1, c)$. Similarly, the $\sim$-class of some $c \in L$ is meant to be the $\sim$-class containing $(1, c)$.

Proposition 5 *Let $(L^2; \trianglelefteq, \sim, (1,1))$ be a tomonoid partition. Let $\leq$ be the underlying total order of L. Moreover, for any $a, b \in L$, let*

$$a \odot b \;=\; \textit{the unique } c \textit{ such that } (a, b) \sim c. \tag{6}$$

Then $(L; \leq, \odot, 1)$ is the unique tomonoid such that $(L^2; \trianglelefteq, \sim, (1,1))$ is its associated tomonoid partition.

Proof By assumption, L is totally ordered and $\trianglelefteq$ is the induced componentwise order on L^2. Evidently, $\trianglelefteq$ determines the total order $\leq$ on L uniquely. It is furthermore clear from (P2) that $\odot$ can be defined by (6).

For $a \in L$, we have $1 \odot a = a$ by construction and $a \odot 1 = 1 \odot a$ by (P2). Furthermore, (P2) and (P3) imply the associativity of $\odot$. Thus $(L; \odot, 1)$ is a monoid. Let $a \leq b$. Then $(a, c) \trianglelefteq (b, c)$, and we conclude from (P1) that $a \odot c \leq b \odot c$. Similarly, we see that $c \odot a \leq c \odot b$. Thus $\leq$ is compatible with $\odot$ and $(L; \leq, \odot, 1)$ is a tomonoid. It is clear that $\sim$ is the level equivalence of L and we conclude that $(L^2; \trianglelefteq, \sim, (1,1))$ is its associated tomonoid partition.

0	t	u	v	w	x	y	z	1	
0	t	u	v	w	x	y	z	1	1
0	0	t	u	u	v	w	x	z	z
0	0	0	t	t	u	u	v	y	y
0	0	0	t	t	u	u	v	x	x
0	0	0	0	0	t	t	u	w	w
0	0	0	0	0	t	t	u	v	v
0	0	0	0	0	0	0	t	u	u
0	0	0	0	0	0	0	0	t	t
0	0	0	0	0	0	0	0	0	0

Fig. 4 A tomonoid partition associated with an eight-element negative tomonoid L. Rows and columns of the array correspond to the elements of L, thus each square in the array corresponds to a pair $(a, b) \in L^2$, where a is the row index and b is the column index. In order to represent $\sim$, we have indicated in each square (a, b) the product of a and b in L; two squares are $\sim$-equivalent iff they contain the same symbol. For instance, the $\sim$-class of u comprises even elements and the $\sim$-class of 1 just one

Let $(L^2; \trianglelefteq, \sim, (1,1))$ be associated to another tomonoid $(L'; \leq', \odot', 1')$. Then, by the way in which a tomonoid partition is constructed from a tomonoid, $L' = L$, $\leq' = \leq$, and $1' = 1$. Furthermore, if for some $a, b, c \in L$ we have $a \odot' b = c$, then $(a, b) \sim (1, c)$ and hence $a \odot b = c$. We conclude $\odot' = \odot$.

By Propositions 4 and 5, tomonoids and tomonoid partitions are in a one-to-one correspondence. We will present our results in the sequel mostly with reference to the latter, that is, with reference to tomonoid partitions.

Let us next devote some remarks to the geometric interpretation of the conditions (P1)–(P3) in Definition 10. Let L be a tomonoid. Then L is a chain and hence L^2 can be viewed as a square array. For elements $(a, b), (c, d) \in L^2$, $(a, b) \trianglelefteq (c, d)$ means that (a, b) is left underneath (c, d). Moreover, for negative tomonoids, 1 is the top element; in this case, $(1, 1)$ is located in the upper right corner of L^2. See Fig. 4 for an illustration.

In order to interpret (P1)–(P3), let us view the level equivalence of L as a partition of L^2. Condition (P2) has probably the most straightforward meaning. By (P2), each block contains exactly one element of the form $(1, c)$, $c \in L$. That is, we may index the blocks by the elements of the line indexed by 1. Furthermore, $(c, 1)$ and $(1, c)$ are for each $c \in L$ in the same block and hence a similar statement holds also for the column indexed by 1.

By the identification of the blocks with the line indexed by 1, the blocks are totally ordered. Condition (P1) says that the componentwise order on L^2 is in accordance with this total order. Namely, when moving from any element of a block to the right or upwards, we arrive at a block indexed by a larger element.

Condition (P3), which accounts for the associativity, possesses an appealing geometric interpretation as well. An illustration is given in Fig. 5. Here, we assume that 1 is the top element of L. Within the square array representing L^2, consider two

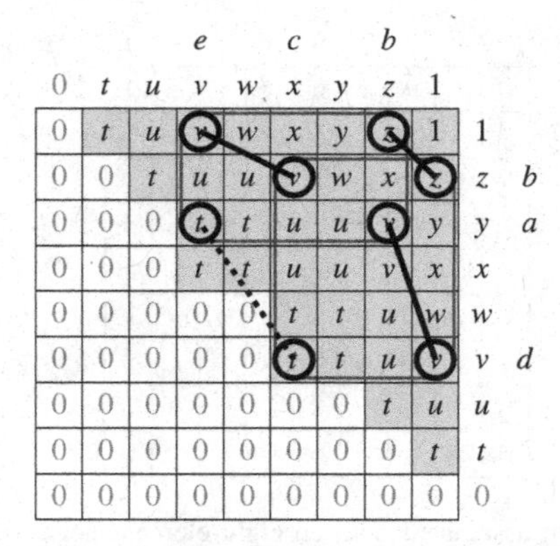

Fig. 5 The "Reidemeister" condition (P3). A (connected or broken) *bold line* between two elements of the array indicates level equivalence. By (P3), the equivalences of the pairs connected by a *solid line* imply the equivalence of the pair connected by a *broken line*

rectangles such that one hits the upper edge and the other one hits the right edge. Assume that the upper left, upper right, and lower right vertices of these rectangles are in the same blocks, respectively. By (P3), then also the remaining pair, consisting of the lower left vertices, is in the same block. A related property is known from the field of web geometry and called the "Reidemeister condition" [1, 2].

We conclude the subsection with a characterisation of those tomonoid partitions in which we are actually interested: the finite, negative ones. The slightly optimised characterisation will be useful in subsequent proofs.

Proposition 6 *Let* $(L; \leq)$ *be a finite and at least two-element chain with the top element* 1. *Let* 0 *be the bottom element of* L. *Then* $(L^2; \trianglelefteq, \sim, (1,1))$ *is a tomonoid partition if and only if* (P1), (P2), *and the following condition hold:*

(P3') *For any* $a, b, c, d, e \in L \setminus \{0, 1\}$, $(a, b) \sim d$ *and* $(b, c) \sim e$ *imply* $(d, c) \sim (a, e)$.

In this case, $(L^2; \trianglelefteq, \sim, (1,1))$ *is finite and negative.*

Proof The "only if" part is clear by definition.

To see the "if" part, let $(L^2; \trianglelefteq, \sim, (1,1))$ fulfil (P1), (P2), and (P3'). We next show that the negativity criterion of Lemma 6(i) holds:

($\star$) $(a, b) \sim (1, c)$ implies $c \leq a$ and $c \leq b$.

Indeed, in this case $(c, 1) \sim (1, c) \sim (a, b) \trianglelefteq (a, 1)$ by (P2) and the fact that 1 is the top element. Hence, by (P1), $c \leq a$. Similarly, we see that $c \leq b$.

It remains to prove (P3). Let $a, b, c, d, e \in L$ be such that $(a, b) \sim d$ and $(b, c) \sim e$. We have to show $(d, c) \sim (a, e)$ if one of the five elements equals 0 or 1. We consider certain cases only, the remaining ones are seen similarly.

Let $a = 1$. Then $(1, b) \sim (1, d)$, hence $b = d$ by (P2), and it follows $(d, c) = (b, c) \sim (1, e) = (a, e)$.

Let $d = 1$. Then $(a, b) \sim (1, 1)$, and by $(\star)$, we conclude $a = b = 1$. From $(b, c) \sim e$ it follows $e = c$. Hence $(d, c) = (a, e)$.

Note next that, for any $f \in L$, $(f, 0) \sim 0$. This follows again from $(\star)$.

Let $a = 0$. Then $(a, b) = (0, b) \sim 0$ and hence $d = 0$. Hence $(d, c) = (0, c) \sim 0 \sim (0, e) = (a, e)$.

Let $d = 0$. Then $(d, c) = (0, c) \sim 0$. From $(b, c) \sim e$, it follows by $(\star)$ that $e \leq b$. Hence $(1, 0) \sim (0, 0) \trianglelefteq (a, e) \trianglelefteq (a, b) \sim (1, 0)$ and we conclude from (P1) that $(a, e) \sim 0$. In particular, $(a, e) \sim (d, c)$.

4.3 Properties and Constructions for Tomonoid Partitions

We have seen that tomonoids and tomonoid partitions are in a one-to-one correspondence. Consequently, we can apply properties, constructions, etc. defined for tomonoids to tomonoid partitions as well. We establish in this subsection a few of such correspondences.

For convenience, we will apply to tomonoid partitions the same notions as to tomonoids. For instance, a tomonoid partition will be called negative if the corresponding tomonoid is negative.

Lemma 6 *Let* $(L^2; \trianglelefteq, \sim, (1,1))$ *be a tomonoid partition.*

(i) *The following statements are pairwise equivalent:*

- L^2 *is negative.*
- $(1, 1)$ *is the top element of* L^2.
- *The* $\sim$*-class of any* $c \in L$ *is contained in* $\{(a, b) \in L^2 \colon a, b \geq c\}$.

(ii) *The following statements are equivalent:*

- L^2 *is commutative.*
- $(a, b) \sim (b, a)$ *for any* $a, b \in L$.

A further property considered in the sequel is Archimedeanicity. In what follows, we write a^n for the n-fold product $a \odot \cdots \odot a$.

Definition 11 We call a negative tomonoid *Archimedean* if, for any $a \leq b < 1$, there is an $n \geq 1$ such that $b^n \leq a$.

Note that negative tomonoids with at most two elements are trivially Archimedean. Archimedean f.n. tomonoid partitions are characterised as follows.

Lemma 7 *Let* $(L^2; \trianglelefteq, \sim, (1,1))$ *be a f.n. tomonoid partition. The following statements are pairwise equivalent:*

- L^2 *is Archimedean.*
- $(b, a) \nsim (1, a)$ *for any* $a \in L^\star$ *and* $b < 1$.
- $(a, b) \nsim (a, 1)$ *for any* $a \in L^\star$ *and* $b < 1$.

Proof Let $(L; \leq, \odot, 1)$ be the corresponding f.n. tomonoid and let 0 be the bottom element of L. W.l.o.g., we can assume $0 \neq 1$. We show that (i) and (ii) are equivalent. The equivalence of (i) and (iii) is seen similarly.

Assume that (ii) holds. By the negativity of L, we have $b \odot a < a$ for all $a \neq 0$ and $b < 1$. Let $a < 1$. Then, for any $n \geq 1$, either $a^{n+1} < a^n$ or $a^n = 0$. As L is finite, the latter possibility applies for a sufficiently large n. It follows that L is Archimedean.

Assume that (ii) does not hold. Let $a \neq 0$ and $b < 1$ such that $b \odot a = a$. As L is negative, we then have $a \leq b$ and it follows $b^n \geq b^{n-1} \odot a = a > 0$ for any $n \geq 2$. Hence L cannot be Archimedean.

We next see how Rees quotients are formed in our framework.

Proposition 7 *Let $(L^2; \trianglelefteq, \sim, (1,1))$ be a negative tomonoid partition and let $q \in L$. Let $L_q = \{a \in L : a > q\} \dot{\cup} \{0\}$, where 0 is a new element, and endow L_q with the total order extending the total order on $\{a \in L : a > q\}$ such that 0 is the bottom element. Then, for each $c \in {L_q}^\star$, the $\sim$-class of c is contained in $({L_q}^\star)^2$. Let $\sim_q$ be the equivalence relation on ${L_q}^2$ whose classes are the $\sim$-classes of each $c \in {L_q}^\star$ as well as the subset of ${L_q}^2$ containing the remaining elements. Then $({L_q}^2; \trianglelefteq, \sim_q, (1,1))$ is the Rees quotient of L^2 by q.*

Proof Let $(L; \leq, \odot, 1)$ be the corresponding negative tomonoid. Let $\odot_q$ be the binary operation on L_q such that $(L_q; \leq, \odot_q, 1)$ is (under the obvious identifications) the Rees quotient of L by q. Let $({L_q}^2; \trianglelefteq, \sim'_q, (1,1))$ be the associated tomonoid partition.

Let $a, b, c \in L$ such that $c > q$ and $(a, b) \sim c$. Then $a, b \geq c$ by Lemma 6(i) and consequently $a, b > q$. We conclude that the $\sim$-class of each $c \in {L_q}^\star$ is contained in $({L_q}^\star)^2$.

We have to show $\sim'_q = \sim_q$. Let $a, b, c \in L_q$ such that $c \neq 0$. Then $(a, b) \sim'_q c$ iff $a \odot_q b = c$ iff $a \odot b = c$ iff $(a, b) \sim c$. Hence the $\sim'_q$-class of each $c \in {L_q}^\star$ coincides with the $\sim$-class of c. There is only one further $\sim'_q$-class, the $\sim'_q$-class of 0, which consequently consists of all elements of ${L_q}^2$ not belonging to the $\sim$-class of any $c \in {L_q}^\star$.

We may interpret Proposition 7 once again geometrically. Let L^2 be a finite negative tomonoid partition and let q be an element of the underlying tomonoid L. Then the Rees quotient by q arises from the partition on L^2 by removing all columns and rows indexed by elements $\leq q$ and by adding instead a single new column from left and a single new row from below. Moreover, all elements that originally belonged to a class of some $a \leq q$ are joined into a single class, which is the class of the new zero. In contrast, the classes of elements strictly larger than q remain unchanged.

Figure 6 shows the chain obtained from a eight-element tomonoid by applying this procedure repeatedly to the respective atom.

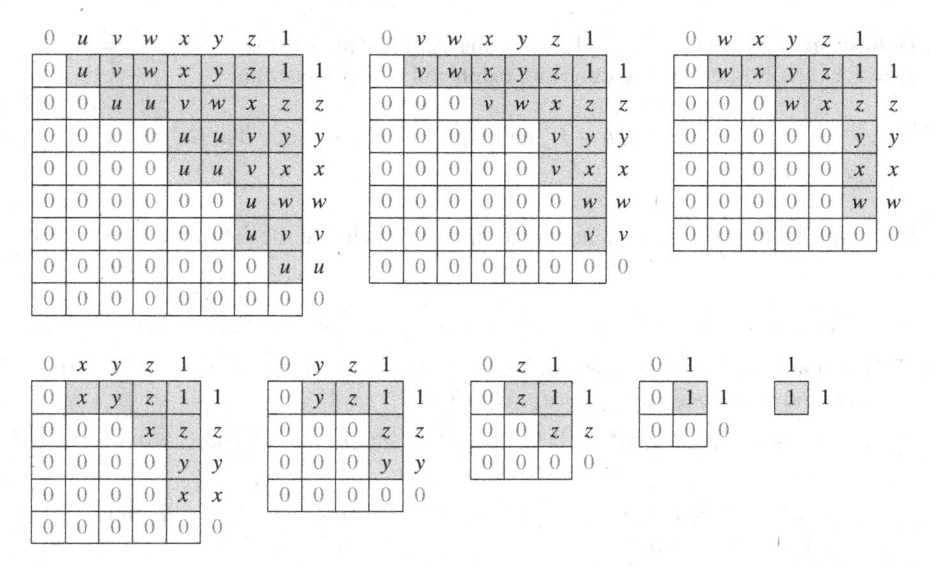

Fig. 6 Starting from the eight-element tomonoid shown in Fig. 4, the successive formation of Rees quotients by the atom leads eventually to the trivial tomonoid

4.4 One-Element Coextensions

Based on the level-set representation, we will in this subsection provide a systematic description of all one-element coextensions of a finite, negative tomonoid. We will restrict to the Archimedean case; for the general case we refer to [24]. That is, we will determine the coextensions of Archimedean f.n. tomonoids that are Archimedean again.

We will proceed, roughly, as follows. We start from a tomonoid partition, seen as a partitioned square array; cf. Fig. 4. We enlarge the sides of this square by one element, doubling the lowest row and left-most column. We determine the equivalence relation $\bar{\sim}$ that makes the enlarged square into a tomonoid partition in two steps. We first determine an intermediate equivalence relation $\dot{\sim}$, called the ramification. $\dot{\sim}$ has a universal property: the level equivalence of any Archimedean one-element coextension extends $\dot{\sim}$. Second, we choose the final equivalence relation $\bar{\sim}$, merging certain $\dot{\sim}$-classes such that the part of the square containing the classes of the new tomonoid's bottom element and atom is divided up into exactly two $\bar{\sim}$-classes.

For a chain $(L; \leq)$, let us define $\bar{L} = L^\star \dot{\cup} \{0, \alpha\}$, where $0, \alpha$ are new elements, and let us endow $\bar{L}$ with the total order extending the total order on $L^\star$ such that $0 < \alpha < a$ for all $a \in L^\star$. We call $(\bar{L}; \leq)$ the *zero doubling extension* of L.

Furthermore, let $(L; \leq, \odot, 1)$ be a f.n. tomonoid. We will assume that any one-element coextension of L is of the form $(\bar{L}; \leq, \bar{\odot}, 1)$. In particular, the intersection of L and $\bar{L}$ is exactly $L^\star$ and $a \mathbin{\bar{\odot}} b = a \odot b$ whenever $a, b, a \odot b \in L^\star$.

Definition 12 Let $(L^2; \trianglelefteq, \sim, (1,1))$ be an Archimedean f.n. tomonoid partition. Let $\bar{L} = L^\star \dot{\cup} \{0, \alpha\}$ be the zero doubling extension of L. We define

$$\begin{aligned} \mathscr{P} &= \{(a,b) \in \bar{L}^2 \colon a, b \in L^* \text{ and there is a } c \in L^\star \text{ such that } (a,b) \sim c\}, \\ \mathscr{Q} &= \bar{L}^2 \setminus \mathscr{P}. \end{aligned} \tag{7}$$

Let $\dot{\sim}$ be the smallest equivalence relation on $\bar{L}^2$ such that the following conditions hold:

(E1) For any $(a,b), (c,d) \in \mathscr{P}$ such that $(a,b) \sim (c,d)$, we have $(a,b) \dot{\sim} (c,d)$.

(E2) For any $(a,b), (b,c) \in \mathscr{P}$ and $d, e \in L^\star$ such that $(d,c), (a,e) \in \mathscr{Q}$, $(a,b) \sim d$, and $(b,c) \sim e$, we have $(d,c) \dot{\sim} (a,e)$.

(E3) For any $a, b, c, e \in L^\star$ such that $(a,b) \in \mathscr{Q}$, $(b,c) \sim e$, and $c < 1$, we have $(a,e) \dot{\sim} 0$.

Moreover, for any $a, b, c, d \in L^\star$ such that $(b,c) \in \mathscr{Q}$, $(a,b) \sim d$, and $a < 1$, we have $(d,c) \dot{\sim} 0$.

(E4) We have $(0,1) \dot{\sim} (1,0) \dot{\sim} (\alpha, b) \dot{\sim} (b, \alpha)$ for any $b < 1$, and $(\alpha, 1) \dot{\sim} (1, \alpha)$. Moreover, for any $(a,b), (c,d) \in \mathscr{Q}$ such that $(a,b) \trianglelefteq (c,d) \dot{\sim} 0$, we have $(a,b) \dot{\sim} 0$.

Then we call the structure $(\bar{L}^2; \trianglelefteq, \dot{\sim}, (1,1))$ the *ramification* of $(L^2; \trianglelefteq, \sim, (1,1))$.

A few remarks might help to clarify the meaning of Definition 12. Let the tomonoid partition $(L^2; \trianglelefteq, \sim, (1,1))$ be given. The subset $\mathscr{P}$ of $\bar{L}^2$ consists of all pairs $(a,b) \in L^{\star 2}$ whose product in L is not the bottom element. That is, $\mathscr{P}$ is the union of the $\sim$-classes of all $c \in L^\star$ and this union lies in $L^{\star 2}$. We note that $\mathscr{P}$ is an upwards closed subset of $\bar{L}^2$ and, consequently, its complement $\mathscr{Q}$ is a downward closed subset of $\bar{L}^2$.

The intermediate equivalence relation $\dot{\sim}$ is determined by successive application of conditions (E1)–(E4). We observe that $\dot{\sim}$-equivalences involving elements of $\mathscr{P}$ are required by condition (E1) only. In fact, all the $\sim$-classes contained in $\mathscr{P}$ are $\dot{\sim}$-classes as well.

The $\dot{\sim}$-classes contained in $\mathscr{Q}$ are determined by conditions (E2)–(E4). In fact, each prescription contained in (E2) and (E3) is of the form that certain $\sim$-equivalences imply that a certain pair of elements of $\mathscr{Q}$ is $\dot{\sim}$-equivalent. Finally, (E4) prescribes that the $\dot{\sim}$-class of 0 is downward closed. We remark that $\mathscr{Q}$ contains the $\dot{\sim}$-class of the bottom element 0, the $\dot{\sim}$-class of the atom α, and possibly further $\dot{\sim}$-classes, which contain neither $(1,c)$ nor $(c,1)$ for any $c \in \bar{L}$.

In the sequel, for two equivalence relations $\sim_1$ and $\sim_2$ on a set A, we say that $\sim_1$ is *coarser* than $\sim_2$ if $\sim_2 \subseteq \sim_1$. In other words, $\sim_1$ coarser than $\sim_2$ if and only if each $\sim_1$-class is a union of $\sim_2$-classes.

Lemma 8 *Let $(L^2; \trianglelefteq, \sim, (1,1))$ be an Archimedean f.n. tomonoid partition and let $(\bar{L}^2; \trianglelefteq, \bar{\sim}, (1,1))$ be an Archimedean one-element coextension of L^2. Furthermore,* *let* $(\bar{L}^2; \trianglelefteq, \dot{\sim}, (1,1))$ *be the ramification of* L^2. *Then* $\bar{\sim}$ *is coarser than* $\dot{\sim}$ *and the following holds: the* $\bar{\sim}$*-class of each* $c \in L^\star$ *coincides with the* $\dot{\sim}$*-class of* c, *the* $\bar{\sim}$*-class of* 0 *is downward closed, and each* $\bar{\sim}$*-class contains exactly one element of the form* $(1, c)$ *for some* $c \in \bar{L}$.

Proof Let $(L; \leq, \odot, 1)$ and $(\bar{L}; \leq, \bar{\odot}, 1)$, where $\bar{L} = L^\star \dot{\cup} \{0, \alpha\}$, be the two tomonoids in question.

As noted above, condition (E1) requires $\dot{\sim}$-equivalences only between elements of $\mathscr{P}$ and the remaining conditions require $\dot{\sim}$-equivalences only between elements of $\mathscr{Q}$. Furthermore, $\mathscr{P}$ is the union of the $\sim$-classes of all $c \in L^\star$. By (E1), these $\sim$-classes are also $\dot{\sim}$-classes. Moreover, by Proposition 7, each $\sim$-class of a $c \in L^\star$ is a $\bar{\sim}$-class. We conclude that the $\bar{\sim}$-class of each $c \in L^\star$ coincides with the $\dot{\sim}$-class of c and $\mathscr{P}$ is the union of these subsets.

We next check that any two elements that are $\dot{\sim}$-equivalent according to one of the conditions (E2)–(E4) are also $\bar{\sim}$-equivalent. Since $\dot{\sim}$ is, by assumption, the smallest equivalence relation with the indicated properties, it will then follow that $\dot{\sim} \subseteq \bar{\sim}$.

Ad (E2): Let $(a, b), (b, c) \in \mathscr{P}$, $d, e \in L^\star$, $(a, b) \sim d$, and $(b, c) \sim e$. Then $a, b, c \in L^\star$, hence $a \mathbin{\bar{\odot}} b = a \odot b = d$ and $b \mathbin{\bar{\odot}} c = b \odot c = e$. Consequently, $d \mathbin{\bar{\odot}} c = (a \mathbin{\bar{\odot}} b) \mathbin{\bar{\odot}} c = a \mathbin{\bar{\odot}} (b \mathbin{\bar{\odot}} c) = a \mathbin{\bar{\odot}} e$, that is $(d, c) \mathrel{\bar{\sim}} (a, e)$.

Ad (E3): Let $a, b, c, e \in L^\star$, $(a, b) \in \mathscr{Q}$, $(b, c) \sim e$, and $c < 1$. Then $a \mathbin{\bar{\odot}} b \leq \alpha$ and hence $a \mathbin{\bar{\odot}} e = a \mathbin{\bar{\odot}} (b \mathbin{\bar{\odot}} c) = (a \mathbin{\bar{\odot}} b) \mathbin{\bar{\odot}} c \leq \alpha \mathbin{\bar{\odot}} c$. As L is assumed to be Archimedean, α is the atom of $\bar{L}$, and $c < 1$, we conclude $\alpha \mathbin{\bar{\odot}} c = 0$. Hence $(a, e) \mathrel{\bar{\sim}} 0$. Similarly, we argue for the second part of (E3).

Ad (E4): As L is Archimedean, we have, for any $b < 1$, $0 \mathbin{\bar{\odot}} 1 = 1 \mathbin{\bar{\odot}} 0 = \alpha \mathbin{\bar{\odot}} b = b \mathbin{\bar{\odot}} \alpha = 0$ by Lemma 7 and hence $(0, 1) \mathrel{\bar{\sim}} (1, 0) \mathrel{\bar{\sim}} (\alpha, b) \mathrel{\bar{\sim}} (b, \alpha)$. Furthermore, we have $(\alpha, 1) \mathrel{\bar{\sim}} (1, \alpha)$. Finally, let $(a, b), (c, d) \in \mathscr{Q}$ and assume $(a, b) \trianglelefteq (c, d) \mathrel{\bar{\sim}} 0$. Then $a \mathbin{\bar{\odot}} b \leq c \mathbin{\bar{\odot}} d = 0$ and thus $(a, b) \mathrel{\bar{\sim}} 0$ as well.

It is finally clear that the $\bar{\sim}$-class of 0 is downward closed. The last statement holds by condition (P2) of a tomonoid partition.

The following theorem is the main result of this section.

Theorem 5 *Let* $(L^2; \trianglelefteq, \sim, (1,1))$ *be an Archimedean f.n. tomonoid partition and let* $(\bar{L}^2; \trianglelefteq, \dot{\sim}, (1,1))$ *be the ramification of* L^2. *Let* $\bar{\sim}$ *be an equivalence relation on* L^2 *that is coarser than* $\dot{\sim}$ *and such that the following holds: the* $\bar{\sim}$*-class of each* $c \in L^\star$ *coincides with the* $\dot{\sim}$*-class of* c, *the* $\bar{\sim}$*-class of* 0 *is downward closed, and each* $\bar{\sim}$*-class contains exactly one element of the form* $(1, c)$ *for some* $c \in \bar{L}$. *Then* $(\bar{L}^2; \trianglelefteq, \bar{\sim}, (1,1))$ *is an Archimedean one-element coextension of* L^2.

Moreover, all Archimedean one-element coextensions of L^2 *arise in this way.*

Proof $\mathscr{P}$, defined by (7), is the union of the $\sim$-classes of all $c \in L^\star$. As we have seen in the proof of Lemma 8, these subsets of $\mathscr{P}$ are also $\dot{\sim}$-classes. Recall also that $\mathscr{P}$ is upwards closed and $\mathscr{Q} = \bar{L}^2 \setminus \mathscr{P}$ is downward closed.

By (E4), we have $(1, 0) \mathrel{\dot{\sim}} (0, 1)$ and $(1, \alpha) \mathrel{\dot{\sim}} (\alpha, 1)$. We claim that $(1, 0) \mathrel{\dot{\nsim}} (1, \alpha)$. Indeed, (E1), (E2), and (E3) involve only elements (a, b) such that $a, b \in L^\star$. Hence, none of these prescriptions involves the elements $(1, \alpha)$ or $(\alpha, 1)$. Moreover, by (E4), the elements $(a, 0)$ and $(0, a)$ for any a as well as (a, α) and (α, a) for any $a \neq 1$ belong to the $\dot{\sim}$-class of $(1, 0)$. Again, $(1, \alpha)$ and $(\alpha, 1)$ are not concerned. Finally, the $\dot{\sim}$-class of $(1, 0)$ is a downward closed set. Also this prescription has no effect on $(1, \alpha)$ or $(\alpha, 1)$ because there is no element in $\mathscr{Q}$ that is larger than $(1, \alpha)$ or $(\alpha, 1)$. We conclude that $\{(1, \alpha), (\alpha, 1)\}$ is an own $\dot{\sim}$-class and our claim is shown.

Let now $\bar{\sim} \supseteq \dot{\sim}$ be as indicated. Note that, by what we have seen so far, at least one such equivalence relation exists. In accordance with Proposition 6, we will verify (P1), (P2), and (P3').

We have shown that $(1, c) \mathrel{\bar{\sim}} (c, 1)$ for all $c \in \bar{L}$. By construction, $\bar{\sim}$ fulfils (P2). Furthermore, the $\bar{\sim}$-class of 0 is downward closed and $\mathscr{Q}$, which is the union of the $\bar{\sim}$-classes of 0 and α, is downward closed as well. We conclude that (P1) holds for $\bar{\sim}$.

It remains to show that $\bar{\sim}$ fulfils (P3'). Let $a, b, c, d, e \in L \setminus \{0, 1\}$ such that $(a, b) \mathrel{\bar{\sim}} d$ and $(b, c) \mathrel{\bar{\sim}} e$. We distinguish the following cases.

Case 1. Let $d, e \in L^\star$. Then $(a, b) \sim d$ and $(b, c) \sim e$. As $\sim$ fulfils (P3), we have $(d, c) \sim (a, e)$. In particular, it follows that $(d, c) \in \mathscr{P}$ iff $(a, e) \in \mathscr{P}$. If (d, c) and (a, e) are both in $\mathscr{P}$, we have $(d, c) \mathrel{\bar{\sim}} (a, e)$ because the $\dot{\sim}$-classes contained in $\mathscr{P}$ are $\bar{\sim}$-classes as well. If (d, c) and (a, e) are both in $\mathscr{Q}$, we have $(d, c) \mathrel{\dot{\sim}} (a, e)$ by (E2) and consequently also $(d, c) \mathrel{\bar{\sim}} (a, e)$, because $\bar{\sim}$ extends $\dot{\sim}$.

Case 2. Let $d = \alpha$ and $e \in L^\star$. Then $(d, c) \mathrel{\dot{\sim}} 0$ by (E4). Furthermore, we have $a \in L^\star$ by (E4), $b, c \in L^\star$ because $(b, c) \in \mathscr{P}$, $(a, b) \in \mathscr{Q}$, and $(b, c) \sim e$. It follows $(a, e) \mathrel{\dot{\sim}} 0$ by (E3). Consequently, $(d, c) \mathrel{\bar{\sim}} 0 \mathrel{\bar{\sim}} (a, e)$.

Case 3. Let $d \in L^\star$ and $e = \alpha$. We argue similarly to Case 2.

Case 4. Let $d = e = \alpha$. Then $(d, c) \mathrel{\dot{\sim}} (a, e) \mathrel{\dot{\sim}} 0$ by (E4) and consequently also $(d, c) \mathrel{\bar{\sim}} (a, e)$.

By Proposition 6, $(\bar{L}^2; \trianglelefteq, \bar{\sim}, (1,1))$ is a f.n. tomonoid partition, which is moreover Archimedean by (E4) and Lemma 7. It is finally clear from Proposition 7 that the Rees quotient of $\bar{L}^2$ by the atom α is L^2.

The final statement follows from Lemma 8.

Let us summarise our construction and add some remarks. In order to determine the one-element coextensions of a f.n. tomonoid L, we start from its associated tomonoid partition $(L^2; \trianglelefteq, \sim, (1,1))$. We first determine its ramification $(\bar{L}^2; \trianglelefteq, \dot{\sim}, (1,1))$ according to Definition 12. This is done by means of the conditions (E1)–(E4); note that these prescriptions are largely independent, it is not necessary to apply them in a recursive way. To obtain, second, a coextension of the desired type, the set $\mathscr{Z} = \langle(1, 0)\rangle_{\bar{\sim}}$, i.e. the $\bar{\sim}$-class of the bottom element, is chosen according to Theorem 5. This is done as simple as follows: $\mathscr{Z}$ is a union of $\dot{\sim}$-classes contained in $\mathscr{Q}$ including $\langle(1, 0)\rangle_{\dot{\sim}}$ but excluding the $\dot{\sim}$-class $\{(1, \alpha), (\alpha, 1)\}$, and $\mathscr{Z}$ is downward closed. Thus, to determine a specific one-element coextension, all we have to do is to select an arbitrary set of $\dot{\sim}$-classes different from $\{(\alpha, 1), (1, \alpha)\}$ and $\mathscr{Z}$ will then be the smallest downward closed set containing them.

Note that one possible choice is $\mathscr{Z} = \mathscr{Q} \setminus \{(\alpha, 1), (1, \alpha)\}$. This means that the explained procedure always leads to a result, that is, every Archimedean, finite, negative tomonoid has at least one Archimedean one-element coextension.

Also in the general case, it is interesting that the explained procedure never requires revisions. At no place decisions are required that lead to an impossible situation, we may always proceed to end up with a coextension as desired.

Acknowledgments The support of the first author by the Austrian Science Fund (FWF): project I 1923-N25 (New perspectives on residuated posets) and the support of the second author by the Czech Science Foundation under Project 15-07724Y are gratefully acknowledged.

References

1. Aczél, J.: Quasigroups, nets and nomograms. Adv. Math. **1**, 383–450 (1965)
2. Blaschke, W., Bol, G.: Geometrie der Gewebe, topologische Fragen der Differentialgeometrie (in German). Springer, Berlin (1939)
3. Blount, K., Tsinakis, C.: The structure of residuated lattices. Int. J. Algebra Comput. **13**, 437–461 (2003)
4. Chen, W., Zhao, X.: The structure of idempotent residuated chains. Czech. Math. J. **59**, 453–479 (2009)
5. Clifford, A.H., Preston, G.B.: The Algebraic Theory of Semigroups, vol. 1. American Mathematical Society, Providence (1961)
6. Cox, D., Little, J., O'Shea, D.: Ideals, varieties, and algorithms. An introduction to Computational Algebraic Geometry and Commutative Algebra, 3rd edn. Springer, New York (2007)
7. De Baets, B., Mesiar, R.: Discrete triangular norms. In: Rodabaugh, S.E., et al. (Eds.), Topological and Algebraic Structures in Fuzzy Sets. A Handbook of Recent Developments in the Mathematics of Fuzzy Sets, pp. 389–400. Kluwer Academic Publishers, Dordrecht (2003)
8. Esteva, F., Godo, L.: Monoidal t-norm based logic: towards a logic for left-continuous t-norms. Fuzzy Sets Syst. **124**, 271–288 (2001)
9. Evans, K., Konikoff, M., Madden, J.J., Mathis, R., Whipple, G.: Totally ordered commutative monoids. Semigroup Forum **62**, 249–278 (2001)
10. Fuchs, L.: Partially Ordered Algebraic Systems. Pergamon Press, Oxford (1963)
11. Galatos, N., Jipsen, P., Kowalski, T., Ono, H.: Residuated lattices. An Algebraic Glimpse at Substructural Logics. Elsevier, Amsterdam (2007)
12. Grillet, P.A.: Semigroups. An Introduction to the Structure Theory. Marcel Dekker, New York (1995)
13. Grillet, P.A.: Commutative Semigroups. Kluwer Academic Publishers, Dordrecht (2001)
14. Hájek, P.: Metamathematics of Fuzzy Logic. Kluwer Academic Publisher, Dordrecht (1998)
15. Horčík, R.: Structure of commutative cancellative integral residuated lattices on (0, 1]. Algebra Univers. **57**, 303–332 (2007)
16. Horčík, R.: On the structure of finite integral commutative residuated chains. J. Log. Comput. **21**, 717–728 (2011)
17. Howie, J.M.: An Introduction to Semigroup Theory. Academic Press, London (1976)
18. Hulin, A.J.: Extensions of ordered semigroups. Czech. Math. J. **26**, 1–12 (1976)
19. Kehayopulu, N., Tsingelis, M.: Ideal extensions of ordered semigroups. Commun. Algebra **31**, 4939–4969 (2003)
20. Klement, E.P., Mesiar, R., Pap, E.: Triangular Norms. Kluwer Academic Publishers, Dordrecht (2000)
21. Montagna, F., Noguera, C., Horčík, R.: On weakly cancellative fuzzy logics. J. Log. Comput. **16**, 423–450 (2006)
22. Noguera, C., Esteva, F., Godo, L.: Generalized continuous and left-continuous t-norms arising from algebraic semantics for fuzzy logics. Inf. Sci. **180**, 1354–1372 (2010)
23. Petrich, M.: Introduction to Semigroups. Charles E. Merrill Publishing Company, Columbus (1973)

24. Petrík, M., Vetterlein, Th.: Rees coextensions of finite, negative tomonoids. J. Log. Comput. (to appear)
25. Petrík, M., Sarkoci, P.: Associativity of triangular norms characterized by the geometry of their level sets. Fuzzy Sets Syst. **202**, 100–109 (2012)
26. Vetterlein, Th.: A Representation of Finite, Positive, Commutative Tomonoids. www.flll.jku.at/sites/default/files/u24/Direction-f-cones.pdf
27. Vetterlein, Th.: Algebraic semantics: the structure of chains. In: Cintula, P., Fermüller, C., Noguera, C. (eds.) Handbook of Mathematical Fuzzy Logic, vol. 3 (to appear)
28. Vetterlein, Th.: On positive commutative tomonoids. Algebra Univ. (to appear)
29. Vetterlein, Th: Totally ordered monoids based on triangular norms. Commun. Algebra **43**, 2643–2679 (2015)
30. Ya, E.: Gabovich, Fully ordered semigroups and their applications. Russ. Math. Surv. **31**, 147–216 (1976)

Structure of Uninorms with Continuous Diagonal Functions

Andrea Mesiarová-Zemánková

Abstract The structure of uninorms which are continuous on some special parts of the unit square is discussed. After a summary of partial results achieved in the characterization of uninorms with continuous underlying t-norm and t-conorm in the past years, a full characterization of these uninorms is described. Representation theorems based on the set of discontinuity points of such a uninorm and the ordinal sum construction for semigroups are presented. Further generalizations yield uninorms with continuous diagonal functions. Several results related to uninorms with continuous diagonals are investigated. Further generalizations are also discussed.

1 Introduction

Uninorms (originally called uni-norms, see [49]) are functions that on the one hand generalize both t-norms and t-conorms, and on the other hand allow bipolar behaviour [50]. Moreover, uninorms linearly transformed to the interval $[-1, 1]$ are just bipolar t-conorms (see [31]). A binary function is a uninorm if it is commutative, associative, non-decreasing in each variable and has a neutral element $e \in [0, 1]$. Therefore the class of uninorms covers also the class of t-norms (for which $e = 1$) and the class of t-conorms (for which $e = 0$). To distinguish uninorms which are not t-norms or t-conorms, in later works authors assume that a uninorm has a neutral element $e \in]0, 1[$. Such uninorms are also called proper.

Due to the associativity the n-ary form of any uninorm is uniquely given and thus it can be extended to an aggregation function working on $\bigcup_{n \in \mathbb{N}} [0, 1]^n$.

In this chapter we focus on uninorms with continuous diagonals. At first we focus on uninorms with continuous underlying functions (for the definition of underlying functions see Sect. 2). Afterwards we will search for conditions under which uninorms with continuous diagonals have continuous underlying functions. The chapter is organized as follows. In Sect. 2 we will give a historical overview of results related

A. Mesiarová-Zemánková (✉)
Mathematical Institute, Slovak Academy of Sciences, Bratislava, Slovakia
e-mail: zemankova@mat.savba.sk

S. Saminger-Platz and R. Mesiar (eds.), *On Logical, Algebraic, and Probabilistic Aspects of Fuzzy Set Theory*, Studies in Fuzziness and Soft Computing 336,
DOI 10.1007/978-3-319-28808-6_7

to the characterization of uninorms with continuous underlying functions. In Sect. 3 we will show a full characterization of uninorms with continuous underlying functions. The historical overview of results related to t-norms with continuous diagonals can be found in Sect. 4, where we will discuss uninorms with continuous diagonals and some further generalizations. We describe our conclusions and further perspectives in Sect. 5.

2 Historical Overview

First results on the structure of uninorms are due to Yager and Rybalov [49] and Fodor et al. [12]. As first observation we can introduce the fact that for a uninorm with the neutral element $e \in [0, 1]$ there is $U(a, 0) = 0$ for all $a \leq e$, and $U(a, 1) = 1$ for all $a \geq e$. This observation was then extended in [12] where it was shown that for each uninorm U the restriction of U to $[0, e]^2$ is a t-norm on $[0, e]^2$, i.e., a linear transformation of some t-norm T_U on $[0, 1]^2$ and the restriction of U to $[e, 1]^2$ is a t-conorm on $[e, 1]^2$, i.e., a linear transformation of some t-conorm S_U on $[0, 1]^2$. We will call T_U (S_U) an underlying t-norm (t-conorm) of U, and together T_U and S_U will be called the underlying functions of U. If $e = 1$ ($e = 0$) then the underlying t-conorm (t-norm) is not defined and the underlying t-norm (t-conorm) is the uninorm itself. Moreover,

$$\min(x, y) \leq U(x, y) \leq \max(x, y)$$

for all $(x, y) \in [0, e] \times [e, 1] \cup [e, 1] \times [0, e]$. Thus each uninorm has a conjunctive behaviour if all inputs are below the neutral element, a disjunctive behaviour if all inputs are above the neutral element, and an averaging behaviour on the reminder.

For every uninorm the value $a = U(0, 1)$ is the annihilator of U and it holds $a \in \{0, 1\}$ (see [12]). If $U(0, 1) = 0$ the uninorm U is called conjunctive (andlike) and if $U(0, 1) = 1$ the uninorm U is called disjunctive (orlike). Moreover, if $U: [0, 1]^2 \longrightarrow [0, 1]$ is a continuous uninorm then $e = 1$ or $e = 0$, i.e., U is either a continuous t-norm, or a continuous t-conorm (see [22]). This means that there is no uninorm with a neutral element $e \in]0, 1[$ continuous on the whole unit square. In fact for every uninorm either the function $u_1: [0, 1] \longrightarrow [0, 1]$ or the function $u_0: [0, 1] \longrightarrow [0, 1]$ is not continuous, where $u_1(x) = U(1, x)$ and $u_0(x) = U(0, x)$ for $x \in [0, 1]$. This follows from the fact shown in [12], that if u_1 and u_0 are both continuous except at $x = e$ then $U(0, 1) = 0$ implies $U(x, 1) = x$ for all $x \in [0, e[$, and $U(0, 1) = 1$ implies $U(x, 0) = x$ for all $x \in]e, 1]$ This yields the following theorem.

Theorem 1 ([12]) *Suppose that* $U: [0, 1]^2 \longrightarrow [0, 1]$ *is a uninorm with neutral element* $e \in]0, 1[$ *and both functions* $x \mapsto U(x, 1)$ *and* $x \mapsto U(x, 0)$ $(x \in [0, 1])$ *are continuous except at the point* $x = e$. *Then* U *is given by one of the following forms. If* $U(0, 1) = 0$ *then*

$$U_{\min}(x, y) = \begin{cases} e \cdot T(\frac{x}{e}, \frac{y}{e}), & \textit{if } (x, y) \in [0, e]^2, \\ e + (1 - e) \cdot S(\frac{x-e}{1-e}, \frac{y-e}{1-e}), & \textit{if } (x, y) \in [e, 1]^2, \\ \min(x, y), & \textit{otherwise}, \end{cases}$$

and if $U(0, 1) = 1$ *then*

$$U_{\max}(x, y) = \begin{cases} e \cdot T(\frac{x}{e}, \frac{y}{e}), & \textit{if } (x, y) \in [0, e]^2, \\ e + (1 - e) \cdot S(\frac{x-e}{1-e}, \frac{y-e}{1-e}), & \textit{if } (x, y) \in [e, 1]^2, \\ \max(x, y), & \textit{otherwise}. \end{cases}$$

In both formulas T *is a t-norm and* S *is a t-conorm.*

On the other hand, from [22] we know that if $T: [0, 1]^2 \longrightarrow [0, 1]$ is a t-norm and $S: [0, 1]^2 \longrightarrow [0, 1]$ is a t-conorm, then for any $e \in [0, 1]$ the two functions $U_{\min}, U_{\max}: [0, 1]^2 \longrightarrow [0, 1]$ given by

$$U_{\min}(x, y) = \begin{cases} e \cdot T(\frac{x}{e}, \frac{y}{e}) & \text{if } (x, y) \in [0, e]^2, \\ e + (1 - e) \cdot S(\frac{x-e}{1-e}, \frac{y-e}{1-e}) & \text{if } (x, y) \in [e, 1]^2, \\ \min(x, y) & \text{otherwise} \end{cases}$$

and

$$U_{\max}(x, y) = \begin{cases} e \cdot T(\frac{x}{e}, \frac{y}{e}) & \text{if } (x, y) \in [0, e]^2, \\ e + (1 - e) \cdot S(\frac{x-e}{1-e}, \frac{y-e}{1-e}) & \text{if } (x, y) \in [e, 1]^2, \\ \max(x, y) & \text{otherwise} \end{cases}$$

are uninorms. We will denote the set of all uninorms of the first type by $\mathcal{U}_{\min}$ and of the second type by $\mathcal{U}_{\max}$.

In the literature we can find several aggregation functions that generalize uninorms. In [26] Mas et al. introduced left and right uninorms, where the commutativity of uninorms was relaxed and a left uninorm possesses a left neutral element and a right uninorm possesses a right neutral element. A uninorm without commutativity was called a pseudo-uninorm in [48]. By removing the associativity and the commutativity from the axioms of uninorms Liu introduced in [23] the concept of semi-uninorms (on a complete lattice) and Su et al. [46] introduced the concept of left and right semi-uninorms (on a complete lattice). Uninorms on a finite totally ordered set were studied in [25]. By replacing the neutral element of a uninorm by an n-neutral element we obtain n-uninorms that were studied in [2].

It is evident that the more we generalize the uninorm functions (on $[0, 1]^2$), the more complicated their structure will be. Thus in the clarification of the structure of uninorms we should take an opposite approach and start from some special subclasses of uninorms with less complicated structure.

From the observations made above we see that the structure of t-norms and t-conorms plays a major role in the investigation of the structure of uninorms. Since t-norms and t-conorms are dual to each other, also their structure is similar. Let us focus for a while on the class of t-norms. Although the class of left-continuous t-norms was not yet fully characterized, and several peculiar examples of t-norms belong to this class, the class of continuous t-norms (t-conorms) has a simple characterization. Each continuous t-norm is an ordinal sum of continuous Archimedean t-norms, and each continuous Archimedean t-norm is generated by a continuous additive generator. Note that a t-norm T (t-conorm S) is called Archimedean if for all $x, y \in]0, 1[$ there exists an $n \in \mathbb{N}$ such that $x_T^{(n)} < y$, ($x_S^{(n)} > y$) where

$$x_T^{(n)} = \underbrace{T(x, T(x, \ldots))}_{n\text{-times}}, \qquad x_S^{(n)} = \underbrace{S(x, S(x, \ldots))}_{n\text{-times}}$$

(see [19]). A continuous t-norm (t-conorm) is Archimedean if and only if it has only trivial idempotent elements 0 and 1. A continuous Archimedean t-norm T (t-conorm S) is either strict, i.e., strictly increasing on $]0, 1]^2$ (on $[0, 1[^2$), or nilpotent, i.e., there exists $(x, y) \in [0, 1]^2$ such that $T(x, y) = 0$ ($S(x, y) = 1$).

The simplicity of the structure of continuous t-norms and t-conorms encouraged several authors to believe that the structure of uninorms with continuous underlying functions could inherit a similar easy characterization. As we will see later, the structure of uninorms with continuous underlying functions is similar to the structure of continuous t-norms in a number of aspects, however, there are some peculiarities that are different. Anyhow, in both cases we rely on two main construction methods: the ordinal sum construction and the construction based on an additive generator.

Recall an original definition of an ordinal sum of semigroups by Clifford [5].

Theorem 2 *Let $A \neq \emptyset$ be a totally ordered set and $(G_\alpha)_{\alpha \in A}$ with $G_\alpha = (X_\alpha, *_\alpha)$ be a family of semigroups. Assume that for all $\alpha, \beta \in A$ with $\alpha < \beta$ the sets X_α and X_β are either disjoint or that $X_\alpha \cap X_\beta = \{x_{\alpha,\beta}\}$, where $x_{\alpha,\beta}$ is both the neutral element of G_α and the annihilator of G_β and where for each $\gamma \in A$ with $\alpha < \gamma < \beta$ we have $X_\gamma = \{x_{\alpha,\beta}\}$. Put $X = \bigcup_{\alpha \in A} X_\alpha$ and define the binary operation $*$ on X by*

$$x * y = \begin{cases} x *_\alpha y & \text{if } (x, y) \in X_\alpha \times X_\alpha, \\ x & \text{if } (x, y) \in X_\alpha \times X_\beta \text{ and } \alpha < \beta, \\ y & \text{if } (x, y) \in X_\alpha \times X_\beta \text{ and } \alpha > \beta. \end{cases}$$

Then $G = (X, *)$ is a semigroup. The semigroup G is commutative if and only if for each $\alpha \in A$ the semigroup G_α is commutative.

Each continuous t-norm has a closed set of idempotent points I (see [19]) and thus $[0, 1] \setminus I$ is equal to a union of open disjoint subintervals $]a_k, b_k[$ of $[0, 1]$ which define supports for summands in the ordinal sum representation, i.e., the corresponding semigroup is defined on $[a_k, b_k[$. If a t-norm is isomorphic to a strict t-norm on $[a_k, b_k]^2$ we can further divide this summand into an ordinal sum of semigroups defined on $\{a_k\}$ and on $]a_k, b_k[$, however, if T is on $[a_k, b_k]^2$ isomorphic

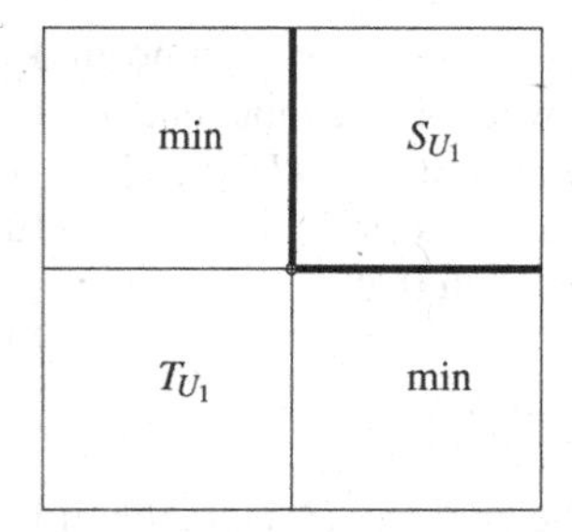

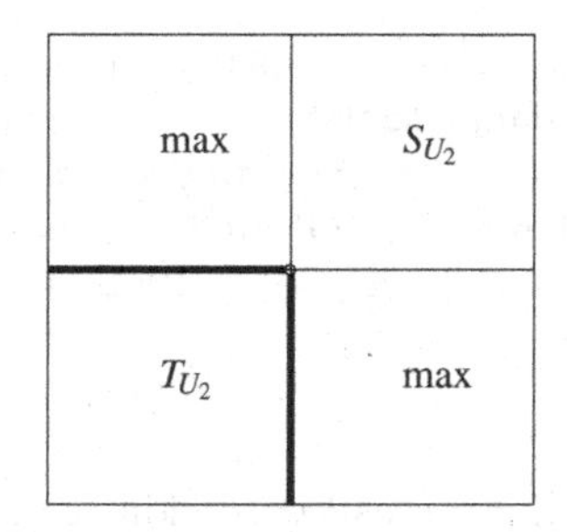

Fig. 1 The uninorm U_1 (*left*) and the uninorm U_2 (*right*) from Remark 1. The *bold lines* denote the points of discontinuity of U_1 and U_2

to a nilpotent t-norm this is not possible. In the case of t-norms the distinction between $\{a_k\}$ with $]a_k, b_k[$ and $[a_k, b_k[$ plays no role, however, as we will see, in the case of uninorms this will be a crucial observation.

Remark 1 Assume $U_1 \in \mathcal{U}_{\min}$ and $U_2 \in \mathcal{U}_{\max}$ with respective neutral elements $e_1, e_2 \in]0, 1[$. Then U_1 and U_2 are ordinal sums of semigroups in the sense of Clifford. In the first case we have three semigroups $G_a = ([0, e_1[, T_{U_1})$, $G_b = (]e_1, 1], S_{U_1})$, $G_c = (\{e_1\}, \min)$, $A_1 = \{a, b, c\}$ and the order on the set A_1 is given by $a < b < c$. In the second case we have again three semigroups $G_a = ([0, e_2[, T_{U_2})$, $G_b = (]e_2, 1], S_{U_2})$, $G_c = (\{e_2\}, \max)$, $A_2 = \{a, b, c\}$ and the order on the set A_2 is given by $b < a < c$. We can see the uninorms U_1 and U_2 in Fig. 1.

In [12] we can find another example of uninorms that are related to ordinal sum of semigroups, namely pseudo-continuous uninorms. A uninorm U with neutral element e is called pseudo-continuous on $[0, 1]^2$ if and only if U is continuous on the set $[0, 1]^2 \setminus \{(x, y) \mid x = e \text{ or } y = e\}$, i.e., U is continuous on the unit square except the segments $[(e, 0), (e, 1)]$ and $[(0, e), (1, e)]$. For a pseudo-continuous conjunctive uninorm U with neutral element $e \in]0, 1[$ we have $U \in \mathcal{U}_{\min}$, where T_U and S_U are continuous in the open unit square and thus they are either continuous, or obtained from an ordinal sum of a continuous t-norms (t-conorms) and a t-subnorm (t-superconorm). More details on uninorms such that T_U and S_U are continuous in the open unit square can be found in Sect. 4.2. Similarly, for a pseudo-continuous disjunctive uninorm U with neutral element $e \in]0, 1[$ we have $U \in \mathcal{U}_{\max}$ and T_U and S_U are continuous in the open unit square.

So far we know that each continuous t-norm is equal to an ordinal sum of Archimedean t-norms. Furthermore, each Archimedean t-norm possesses a continuous additive generator.

Proposition 1 *Let* $t\colon [0, 1] \longrightarrow [0, \infty]$ *be a continuous strictly decreasing function such that* $t(1) = 0$. *Then the binary operation* $T\colon [0, 1]^2 \longrightarrow [0, 1]$ *given by*

$$T(x, y) = t^{-1}(\min(t(0), t(x) + t(y)))$$

is a continuous t-norm. The function t is called an additive generator *of* T.

Extending the concept of additive generators also to the class of uninorms Fodor et al. defined representable uninorms in [12] (which were before independently studied as associative compensatory operators in [18]). A uninorm $U\colon [0,1]^2 \longrightarrow [0,1]$ is called *representable* if there exists a continuous, strictly increasing function $h\colon [0,1] \longrightarrow [-\infty,\infty]$, $h(0) = -\infty$, $h(1) = \infty$ such that

$$U(x, y) = h^{-1}(h(x) + h(y)).$$

The function h is called an additive generator of the uninorm U. Note that if we relax the strict monotonicity of h then the neutral element of the generated function will be lost. If we relax conditions $h(0) = -\infty$, $h(1) = \infty$ then if $h(0) = 0$, $h(1) > 0$ we obtain a t-conorm, if $h(1) = 0$, $h(0) < 0$ we obtain a t-norm, if $h(0) > 0$, $h(1) > 0$ we obtain a t-subnorm (see Sect. 4.2) and if $h(0) < 0$, $h(1) < 0$ we obtain a t-superconorm. In the case that $h(0) < 0$, $h(1) > 0$ the associativity will be lost. Further, a uninorm $U\colon [0,1]^2 \longrightarrow [0,1]$ is representable if and only if U is continuous everywhere on the unit square except for the two points $(0,1)$ and $(1,0)$ ([1, 6, 11, 31, 42]), i.e., U is almost continuous. This implies that if U is almost continuous then it is strictly increasing on the open unit square, i.e., T_U and S_U are a strict t-norm and a strict t-conorm. On the other hand, from [14] it follows that if T and S are a strict t-norm and a strict t-conorm then there exists a representable uninorm U such that $T_U = T$ and $S_U = S$.

In our summary we have already mentioned almost continuous uninorms, i.e., uninorms continuous on $[0,1] \setminus \{(0,1),(1,0)\}$, pseudo-continuous uninorms, i.e., uninorm continuous on $[0,1]^2 \setminus \{(x,y) \mid x = e \text{ or } y = e\}$, and now let us focus on uninorms continuous on $]0,1[^2$. In [14] we can find several interesting results about this class of uninorms. First, if U is a uninorm with the neutral element $e \in]0,1[$ and there exists a $u \in [0,e[$ such that $U(x,y) = x$ for all $x \in]u,e[$ and $y \in]e,1[$, then U is not continuous in $]0,1[^2$. Further, these uninorms have a very clear structure.

Theorem 3 ([14]) *Assume that $U\colon [0,1]^2 \longrightarrow [0,1]$ is a uninorm with the neutral element $e \in]0,1[$ and U is continuous on $]0,1[^2$. Then U can be represented as* (i) *or* (ii) :

(i)

$$U(x,y) = \begin{cases} e \cdot T_U(\frac{x}{e}, \frac{y}{e}), & \textit{if } x, y \in [0,u], \\ h^{-1}(h(x)+h(y)), & \textit{if } x, y \in]u,1[, \\ x, & \textit{if } x \in [0,u],\ y \in]u,1[, \\ x, & \textit{if } x \in [0,\lambda[,\ y = 1, \\ 1, & \textit{if } x \in]\lambda,1],\ y = 1, \\ x \textit{ or } 1, & \textit{if } x = \lambda,\ y = 1, \end{cases}$$

where, $u \in [0,e[$, $\lambda \in [0,u]$, $U(\lambda,\lambda) = \lambda$, function $h\colon [u,1] \longrightarrow [-\infty,\infty]$ is continuous, strictly increasing, and $h(u) = -\infty$, $h(e) = 0$, $h(1) = \infty$. Note that the values of U on the remaining parts of $[0,1]^2$ are defined in such a way that U is commutative.

(ii)

$$U(x,y) = \begin{cases} e + (1-e)\cdot S_U(\frac{x-e}{1-e}, \frac{y-e}{1-e}), & \textit{if } x, y \in [v, 1], \\ r^{-1}(r(x) + r(y)), & \textit{if } x, y \in]0, v[, \\ x, & \textit{if } x \in [v, 1], y \in]0, v[, \\ x, & \textit{if } x \in]\omega, 1], y = 0, \\ 0, & \textit{if } x \in [0, \omega[, y = 0, \\ x \textit{ or } 0, & \textit{if } x = \omega, y = 0, \end{cases}$$

where, $v \in]e, 1]$, $\omega \in [v, 1]$, $U(\omega, \omega) = \omega$, *function* $r\colon [0, v] \longrightarrow [-\infty, \infty]$ *is continuous, strictly increasing, and* $r(0) = -\infty$, $r(e) = 0$, $r(v) = \infty$.

Remark 2 Assume a uninorm with the neutral element $e \in]0, 1[$, continuous on $]0, 1[^2$, which has the form (i) with $\lambda > 0$. This uninorm is an ordinal sum of semigroups in the sense of Clifford. Here we have five semigroups $G_a = ([0, \lambda[, U)$, $G_b = (\{\lambda\}, U)$, $G_c = (]\lambda, u[, U)$, $G_d = ([u, 1[, U)$, and $G_f = (\{1\}, U)$, with $A_1 = \{a, b, c, d, f\}$ and the order on the set A_1 is given by $a < b < f < c < d$ if $U(1, \lambda) = \lambda$ and by $a < f < b < c < d$ if $U(1, \lambda) = 1$. Similarly, if U has the form (ii) with $\omega < 1$ then it is an ordinal sum of semigroups in the sense of Clifford. Here we have five semigroups $G_a = (\{0\}, U)$, $G_b = (]0, v], U)$, $G_c = (]v, \omega[, U)$, $G_d = (\{\omega\}, U)$, and $G_f = (]\omega, 1], U)$, with $A_2 = \{a, b, c, d, f\}$ and the order on the set A_2 is given by $f < d < a < c < b$ if $U(0, \omega) = \omega$ and by $f < a < d < c < b$ if $U(0, \omega) = 0$.

Another important subclass of uninorms are idempotent uninorms, i.e., uninorms where $U(x, x) = x$ for all $x \in [0, 1]$. In the case of t-norms and t-conorms there is only one idempotent t-norm—the minimum, and only one idempotent t-conorm—the maximum. Therefore idempotent uninorms are uniquely given and continuous on $[0, e]^2 \cup [e, 1]^2$. Idempotent uninorms were studied in several papers and for our purposes we will recall results of [3, 24, 41] (for more information see the references in these papers). From [3] we see that every idempotent uninorm U is internal, i.e., $U(x, y) \in \{x, y\}$ holds for all $(x, y) \in [0, 1]^2$. Further, idempotent uninorms that are left-continuous, or right-continuous were characterized in [3]. The complete characterization of idempotent uninorms from [24] was later corrected in [41]. In the following a non-increasing function $g\colon [0, 1] \longrightarrow [0, 1]$ is called *Id-symmetrical* if its completed graph F_g is Id-symmetrical, i.e., $(x, y) \in F_g$ if and only if $(y, x) \in F_g$. Note that a completed graph was defined in [41] as follows: let $g\colon [0, 1] \longrightarrow [0, 1]$ be any decreasing function and let G be the graph of g, that is

$$G = \{(x, g(x)) \mid x \in [0, 1]\};$$

for any point of discontinuity s of g, let s^- and s^+ be the corresponding lateral limits. Then, we define the completed graph of g, denoted by F_g, as the set obtained from G by adding the vertical segments in any discontinuity point s, from s^- to s^+.

Theorem 4 *Consider* $e \in]0, 1[$. *The following items are equivalent:*

(i) *U is an idempotent uninorm with neutral element e.*
(ii) *There exists a decreasing, Id-symmetrical function* $g\colon [0, 1] \longrightarrow [0, 1]$ *with fixed point e such that U is for all* $(x, y) \in [0, 1]^2$ *given by*

$$U(x, y) = \begin{cases} \min(x, y) & \textit{if } y < g(x) \textit{ or } y = g(x), x < g(g(x)), \\ \max(x, y) & \textit{if } y > g(x) \textit{ or } y = g(x), x > g(g(x)), \\ x \textit{ or } y & \textit{if } y = g(x), x = g(g(x)), \end{cases}$$

being commutative on the set of points $(x, g(x))$ *such that* $x = g(g(x))$.

For more details we recommend [41]. Idempotent uninorms on finite ordinal scales were studied in [4].

A uninorm U which is internal on $A(e) = [0, e] \times [e, 1] \cup [e, 1] \times [0, e]$, i.e., $U(x, y) \in \{x, y\}$ for all $(x, y) \in A(e)$ is called locally internal on $A(e)$. For uninorms locally internal on $A(e)$ we have a result from [8](see also [7]) which shows that if $U\colon [0, 1]^2 \longrightarrow [0, 1]$ is a uninorm locally internal on $A(e)$ with neutral element $e \in]0, 1[$ then there exists a non-increasing function $g\colon [0, 1] \longrightarrow [0, 1]$ with fixed point e, such that $\inf\{y \in [0, 1] \mid g(y) = g(x)\} \leq g(g(x)) \leq \sup\{y \in [0, 1] \mid g(y) = g(x)\}$ for all $x \in [0, 1]$, $g(x) = 0$ for all $x > g(0)$, $g(x) = 1$ for all $x < g(1)$, and for all $(x, y) \in A(e)$ there is

$$U(x, y) = \begin{cases} \min(x, y) & \textit{if } y < g(x), \text{ or } y = g(x), x < g(g(x)), \\ \max(x, y) & \textit{if } y > g(x), \text{ or } x = g(x), x > g(g(x)), \\ x \text{ or } y & \textit{if } y = g(x) \text{ and } x = g(g(x)). \end{cases}$$

Further, $U|_{[0,e]^2}$ is an ordinal sum of t-norm summands defined on intervals $[a_i, b_i]$, $i \in A_1$ such that $]a_i, b_i[\subset [0, e] \setminus \{g(x) \mid x \in [e, 1]\}$ for all $i \in A_1$, and $U|_{[e,1]^2}$ is an ordinal sum of t-conorm summands defined on intervals $[c_i, d_i]$, $i \in A_2$ such that $]c_i, d_i[\subset [e, 1] \setminus \{g(x) \mid x \in [0, e]\}$ for all $i \in A_2$.

On the other hand, satisfaction of all these conditions is not enough to ensure that U is associative. Both necessary and sufficient conditions for a uninorm to be locally internal on $A(e)$ can be found in [9].

Finally, we recall two additional important classes of uninorms: uninorms with underlying t-norm and t-conorm given as ordinal sums and Archimedean uninorms, i.e., uninorms where both the underlying t-norm as well as the underlying t-conorm are Archimedean. From [10] we can obtain the following result: let U be a uninorm, and $a, b, c, d \in [0, 1]$, $a \leq b \leq e \leq c \leq d$ be such that b is the neutral element of $U|_{[a,b]^2}$ and c is the neutral element of $U|_{[c,d]^2}$; then the set $([a, b] \cup [c, d])^2$ is closed under U. Several other results from [10] were extended in [34], where it was also shown that if the underlying t-norm and t-conorm of a uninorm U are continuous then for all idempotent points $a, b, c, d \in [0, 1]$, $a \leq b \leq e \leq c \leq d$, of a uninorm U the set $([a, b[\cup\{U(b, c)\}\cup]c, d])^2$ is closed under U. This shows that in the investigation of uninorms with continuous underlying t-norm and t-conorm given as ordinal sums, the class of Archimedean uninorms with continuous underlying functions plays an indispensable role. Further, for a uninorm with continuous underlying functions for every idempotent point $a \in [0, 1]$ of U we have $U(x, y) \in \{x, y\}$ for all $(x, y) \in \{a\} \times [0, 1] \cup [0, 1] \times \{a\}$. Thus if either $T_U = \min$, or $S_U = \max$ then U is locally internal on $A(e)$.

Let us now focus on the class of Archimedean uninorms with continuous underlying functions. Recall that a continuous Archimedean t-norm (t-conorm) is either strict or nilpotent. First let us focus on the case when both T_U and S_U are strict. The main result from [11] (compare also [40]) says that for a uninorm $U: [0, 1]^2 \longrightarrow [0, 1]$ with the neutral element $e \in]0, 1[$ such that both T_U and S_U are strict one of the following three statements hold:

(i) $U \in \mathcal{U}_{\min}$,
(ii) $U \in \mathcal{U}_{\max}$,
(iii) U is representable.

Here, if we assume any $(x_0, y_0) \in]0, e[\times]e, 1[\cup]e, 1[\times]0, e[$, then if $U(x_0, y_0) = \min(x_0, y_0)$ we have $U \in \mathcal{U}_{\min}$, if $U(x_0, y_0) = \max(x_0, y_0)$ then we have $U \in \mathcal{U}_{\max}$, and if $\min(x_0, y_0) < U(x_0, y_0) < \max(x_0, y_0)$ then U is representable. This result was later corrected in [21] where it was observed that if $U(x_0, y_0) = \min(x_0, y_0)$ $(U(x_0, y_0) = \max(x_0, y_0))$ for some $(x_0, y_0) \in]0, e[\times]e, 1[\cup,]e, 1[\times]0, e[$ the uninorm U does not necessarily belong to $\mathcal{U}_{\min}$ $(\mathcal{U}_{\max})$ since it can differ on the boundary of the unit square. More precisely, the following result was shown.

Theorem 5 ([21]) *Let $U: [0, 1]^2 \longrightarrow [0, 1]$ be a uninorm with the neutral element $e \in]0, 1[$ such that both T_U and S_U are strict then one of the following seven statements holds:*

(i) $U \in \mathcal{U}_{\min}$,
(ii)

$$U(x, y) = \begin{cases} e \cdot T_U(\frac{x}{e}, \frac{y}{e}) & \textit{if } (x, y) \in [0, e]^2, \\ e + (1 - e) \cdot S_U(\frac{x-e}{1-e}, \frac{y-e}{1-e}) & \textit{if } (x, y) \in [e, 1]^2, \\ 1 & \textit{if } x = 1 \textit{ or } y = 1, \\ \min(x, y) & \textit{otherwise,} \end{cases}$$

(iii)

$$U(x, y) = \begin{cases} e \cdot T_U(\frac{x}{e}, \frac{y}{e}) & \textit{if } (x, y) \in [0, e]^2, \\ e + (1 - e) \cdot S_U(\frac{x-e}{1-e}, \frac{y-e}{1-e}) & \textit{if } (x, y) \in [e, 1]^2, \\ 1 & \textit{if } x = 1,\ y > 0 \textit{ or } y = 1,\ x > 0, \\ \min(x, y) & \textit{otherwise,} \end{cases}$$

(iv) $U \in \mathcal{U}_{\max}$,

(v)

$$U(x,y)=\begin{cases} e\cdot T_U(\frac{x}{e},\frac{y}{e}) & \textit{if } (x,y)\in[0,e]^2,\\ e+(1-e)\cdot S_U(\frac{x-e}{1-e},\frac{y-e}{1-e}) & \textit{if } (x,y)\in[e,1]^2,\\ 0 & \textit{if } x=0 \textit{ or } y=0,\\ \max(x,y) & \textit{otherwise,}\end{cases}$$

(vi)

$$U(x,y)=\begin{cases} e\cdot T_U(\frac{x}{e},\frac{y}{e}) & \textit{if } (x,y)\in[0,e]^2,\\ e+(1-e)\cdot S_U(\frac{x-e}{1-e},\frac{y-e}{1-e}) & \textit{if } (x,y)\in[e,1]^2,\\ 0 & \textit{if } x=0,\ y<1 \textit{ or } y=0,\ x<1,\\ \max(x,y) & \textit{otherwise,}\end{cases}$$

(vii) U is representable.

As we mentioned before, if $U\in\mathcal{U}_{\min}$, or $U\in\mathcal{U}_{\max}$, then U is an ordinal sum in the sense of Clifford. Similarly, six of the previous seven forms of a uninorm with strict T_U and S_U are ordinal sums in the sense of Clifford. Since both T_U and S_U are strict we can divide a semigroup acting on $[0,e[$ into two semigroups acting on $\{0\}$ and $]0,e[$ and similarly we can divide semigroup acting on $]e,1]$ into two semigroups acting on and $]e,1[$ and $\{1\}$. By changing the order of these semigroups in the ordinal sum construction we can obtain all six cases mentioned above, except the case when U is representable (see also [35]).

Now assume that both T_U and S_U are nilpotent. In such a case we know that the semigroup acting on $[0,e[$ ($]e,1]$) cannot be further divided. In [21] we can find the following result.

Theorem 6 ([21]) *Let $U\colon[0,1]\longrightarrow[0,1]^2$ be a uninorm with the neutral element $e\in]0,1[$ such that both T_U and S_U are nilpotent. Then either one of the following two statements holds:*

(i) $U\in\mathcal{U}_{\min}$,
(ii) $U\in\mathcal{U}_{\max}$.

If we focus on a general Archimedean uninorm with continuous underlying operations, from [21, 34, 35] we can see that if T_U is strict and S_U is nilpotent (T_U is nilpotent and S_U is strict) then U can have a form (i), (iv), or (v) ((i), (ii), or (iv)) from Theorem 5.

Now we have presented all important results that form a basis for the characterization of uninorms with continuous underlying t-norm and t-conorm. In the following section we will show this characterization.

3 Characterization of Uninorms with Continuous Underlying Functions

As a first step towards the characterization of uninorms with continuous underlying t-norm and t-conorm we should recall the ordinal sum construction for uninorms which was introduced in [32]. For any $0 \leq a \leq b < c \leq d \leq 1$, $v \in [b, c]$, and a uninorm U with the neutral element $e \in [0, 1]$ we will use a transformation $f\colon [0, 1] \longrightarrow [a, b[\cup \{v\} \cup]c, d]$ given by

$$f(x) = \begin{cases} (b-a) \cdot \frac{x}{e} + a & \text{if } x \in [0, e[, \\ v & \text{if } x = e, \\ d - \frac{(1-x)(d-c)}{(1-e)} & \text{otherwise.} \end{cases} \tag{1}$$

Then f is linear on $[0, e[$ and on $]e, 1]$ and thus it is a piece-wise linear isomorphism of $[0, 1]$ to $([a, b[\cup \{v\} \cup]c, d])$ and if $U\colon [0, 1] \longrightarrow [0, 1]^2$ is a uninorm then a binary function $U_v^{a,b,c,d}\colon ([a, b[\cup \{v\} \cup]c, d])^2 \longrightarrow ([a, b[\cup \{v\} \cup]c, d])$ given by

$$U_v^{a,b,c,d}(x, y) = f(U(f^{-1}(x), f^{-1}(y))) \tag{2}$$

is a uninorm on $([a, b[\cup \{v\} \cup]c, d])^2$. Note that we assume that $a = b$ $(c = d)$ if and only if $e = 0$ $(e = 1)$. The function f is piece-wise linear, however, more generally we can use any increasing isomorphic transformation.

If U_1 and U_2 are uninorms, with respective neutral elements e_1, e_2, then for $0 \leq a < b < c < d \leq 1$ we have

$$(U_1)_v^{a,b,c,d}(x, y) = (U_2)_v^{a,b,c,d}(x, y)$$

if and only if $U_1(x, y) = \phi^{-1}(U_2(\phi(x), \phi(y)))$, where $\phi\colon [0, 1] \longrightarrow [0, 1]$ is a strictly increasing isomorphism with $\phi(e_1) = e_2$ which is linear on $[0, e_1]$ and on $[e_1, 1]$. Similar result can be obtain for the case when $a = b$ $(c = d)$, however, then only the corresponding parts of uninorms are isomorphic.

Now we have the following result

Proposition 2 ([32]) *Assume $e \in [0, 1]$. Let K be an index set which is finite or countably infinite and let $(]a_k, b_k[)_{k\in K}$ be a disjoint system of open subintervals (which can be also empty) of $[0, e]$, such that $\bigcup_{k\in K}[a_k, b_k] = [0, e]$. Similarly, let $(]c_k, d_k[)_{k\in K}$ be a disjoint system of open subintervals (which can be also empty) of $[e, 1]$, such that $\bigcup_{k\in K}[c_k, d_k] = [e, 1]$. Let further these two systems be anti-comonotone, i.e., $b_k \leq a_i$ if and only if $c_k \geq d_i$ for all $i, k \in K$. Assume a family of uninorms $(U_k)_{k\in K}$ on $[0, 1]^2$ such that if both $]a_k, b_k[$ and $]c_k, d_k[$ are non-empty then U_k is a proper uninorm, otherwise if $]a_k, b_k[$ is non-empty then U_k is a t-norm and if $]c_k, d_k[$ is non-empty then U_k is a t-conorm. If both $]a_k, b_k[$ and $]c_k, d_k[$ are empty then $a_k = b_k = a_{k_1} = b_{k_1}$ and $c_k = d_k = c_{k_1} = d_{k_1}$ does not hold for any $k_1 \in K, k \neq k_1$, and here only the value $U_k(0, 1)$ is interesting. Denote $K_* = \{k \in K \mid]a_k, b_k[\neq \emptyset\}$*

and $K^* = \{k \in K \mid]c_k, d_k[\neq \emptyset\}$. *Further, let* $B = \{b_k \mid k \in K\} \setminus \{a_k \mid k \in K_*\}$ *and* $C = \{c_k \mid k \in K\} \setminus \{d_k \mid k \in K^*\}$. *We define a function* $n\colon B \longrightarrow B \cup C$ *given for all* $b_k \in B$ *by*

$$n(b_k) = \begin{cases} b_k & \text{if } U_k(1,0) = 0, \\ c_k & \text{else.} \end{cases}$$

Let the ordinal sum $U^e = (\langle a_k, b_k, c_k, d_k, U_k\rangle \mid k \in K)^e$ *be given by*

$$U^e(x,y) = \begin{cases} y & \text{if } x = e, \\ x & \text{if } y = e, \\ (U_k)_{v_k}^{a_k,b_k,c_k,d_k} & \text{if } (x,y) \in ([a_k, b_k[\cup]c_k, d_k])^2, \\ x & \text{if } y \in [b_k, c_k], x \in [a_k, d_k] \setminus [b_k, c_k], \\ y & \text{if } x \in [b_k, c_k], y \in [a_k, d_k] \setminus [b_k, c_k], \\ \min(x,y) & \text{if } (x,y) \in [b_k, c_k]^2 \setminus (]b_k, c_k[^2 \cup \{(b_k, c_k), (c_k, b_k)\}), \\ & \text{where } b_k \in B, c_k \in C, x + y < c_k + b_k, \\ \max(x,y) & \text{if } (x,y) \in [b_k, c_k]^2 \setminus (]b_k, c_k[^2 \cup \{(b_k, c_k), (c_k, b_k)\}), \\ & \text{where } b_k \in B, c_k \in C, x + y > c_k + b_k, \\ n(b_k) & \text{if } (x,y) = (b_k, c_k) or (x,y) = (c_k, b_k), b_k \in B, c_k \in C, \\ \min(x,y) & \text{if } (x,y) \in \{b_k\} \times [b_k, c_k] \cup [b_k, c_k] \times \{b_k\} \\ & \text{and } b_k \in B, c_k \notin C, \\ \max(x,y) & \text{if } (x,y) \in \{c_k\} \times [b_k, c_k] \cup [b_k, c_k] \times \{c_k\} \\ & \text{and } b_k \notin B, c_k \in C, \end{cases}$$

where $v_k = c_k$ ($v_k = b_k$) *if there exists an* $i \in K$ *such that* $b_k = a_i$ *and* U_i *is disjunctive (conjunctive) and* $v_k = n(b_k)$ *if* $b_k \in B, c_k \in C$, $v_k = b_k$ *if* $b_k \in B, c_k \notin C$, $v_k = c_k$ *if* $b_k \notin B, c_k \in C$, *and* $(U_k)_{v_k}^{a_k,b_k,c_k,d_k}$ *is given by the formula* (2). *Then* U^e *is a uninorm.*

In the case of t-norms (t-conorms) ordinal sum can be defined also on such intervals that $\bigcup_{k\in K}[a_k, b_k] \neq [0,1]$ ($\bigcup_{k\in K}[c_k, d_k] \neq [0,1]$). In such a case the remaining parts of the unit square are simply filled in by the min (max). In the case of uninorms, if $\bigcup_{k\in K}[a_k, b_k] \neq [0,e]$ ($\bigcup_{k\in K}[c_k, d_k] \neq [e,1]$) the remaining parts of the unit square should be filled in by internal uninorms, which are, however, not unique and therefore we have to specify them.

If for a summand $\langle a_k, b_k, c_k, d_k, U_k\rangle$ for some $k \in K$ we have $a_k = b_k$ and $c_k = d_k$ we say that this summand is empty. If for a summand $\langle a_k, b_k, c_k, d_k, U_k\rangle$ for some $k \in K$ we have $a_k \neq b_k$ and $c_k \neq d_k$ we say that this summand is complete. We say that an ordinal sum $U^e = (\langle a_k, b_k, c_k, d_k, U_k\rangle \mid k \in K)^e$ is complete when all its summands are complete.

If all summands used in this ordinal sum construction have continuous underlying t-norm and t-conorm also U^e will have continuous underlying t-norm and t-conorm. Thus this ordinal sum construction can be used to construct uninorms with continuous underlying t-norm and t-conorm. From [33] we know that a uninorm U on $[0, 1]^2$ is a complete ordinal sum of representable uninorms if and only if there exists a continuous strictly decreasing function $r\colon [0, 1] \longrightarrow [0, 1]$ with $r(0) = 1$, $r(e) = e$ and $r(1) = 0$ such that U is continuous on $[0, 1] \setminus \{(x, r(x)) \mid x \in [0, 1]\}$ and U has countably many idempotent points.

On the other hand, not every uninorm with continuous underlying t-norm and t-conorm can be obtained as an ordinal sum of uninorms with continuous underlying t-norm and t-conorm. Recall Theorem 5: here all cases except when U is representable are ordinal sums in the sense of Clifford, however, only cases (i) and (iv) are ordinal sums of uninorms. Let us see where this difference is hidden. For a detailed proof we recommend [35].

We see that the ordinal sum of uninorms is based on the intervals of the form $[a, b[\cup]c, d]$. However, if the corresponding t-norm (t-conorm) that acts on $[a, b]^2$ ($[c, d]^2$) is strict we can divide the semigroup acting on $[a, b[$ ($]c, d]$) into semigroups acting on $\{a\}$ and $]a, b[$ ($]c, d[$ and $\{d\}$). By changing the order of these semigroups in the ordinal sum construction (in the sense of Clifford) we can then obtain uninorms that cannot be obtained as an ordinal sum of uninorms, although they are Archimedean. Thus in the case of Archimedean uninorms the problem is always on the border of the unit square. In the case of t-norms (t-conorms) on the border of the unit square we always have $T = \min$ ($S = \max$), however, in the case of uninorms with continuous underlying functions it is a mixture of min and max.

Recall that for a uninorm with continuous underlying t-norm and t-conorm U is internal on $\{q\} \times [0, 1]$, where $q \in [0, 1]$ is an idempotent point of U. After some computations we can see that each uninorm with continuous underlying t-norm and t-conorm is equal to an ordinal sum of uninorms with continuous underlying t-norm and t-conorm on the set $([0, 1] \setminus I_U)^2$, where I_U is the set of idempotent points of U. Thus the difference can appear only in points $(x, y) \in [0, 1]^2$ such that at least one of x and y is idempotent.

Summarizing, we need a more general construction than the ordinal sum construction that would allow semigroups defined on singletons. However, in order to keep monotonicity, the order of these singletons cannot be changed arbitrarily, but it depends on the other summands in the ordinal sum construction. As a first condition, in [35] it was shown that for every idempotent point q of U the point of change where $U(x, q) = \min(x, q)$ changes to $U(x, q) = \max(x, q)$, i.e., such a $p \in [0, 1]$ that $U(x, q) = \min(x, q)$ for all $x < p$ and $U(x, q) = \max(x, q)$ for all $x > p$, is unique and it is an idempotent point of U. Assume now that U with continuous underlying t-norm and t-conorm is not equal to an ordinal sum of uninorms. This means that for an ordinal sum of uninorms V such that $U = V$ on $([0, 1] \setminus I_U)^2$ there exists an idempotent point $q \in [0, 1]$ of U such that $U(q, x) \neq V(q, x)$ for some $x \in [0, 1]$. Assume that $q > e$ (the other case is analogical). Then both restrictions of U and V on $[0, q]^2$ are uninorms on $[0, q]^2$. Let us transform these restrictions to $[0, 1]^2$ linearly. We obtain two uninorms U^* and V^* that are not equal on the boundary of the unit square and V^* is an ordinal sum of uninorms. Now we will recall two examples of such ordinal sums of uninorms.

Recall Theorem 3 case (i). Here U is an ordinal sum of uninorms only if $U(1, x) = x$ for all $x \in [0, u]$. In such a case U is an ordinal sum of a representable uninorm, which corresponds to a complete summand, and several non-complete summands that together represent the t-norm on $[0, u]^2$. Otherwise we have a separate semigroup $(\{1\}, U)$ and values $U(1, x)$ for $x \in [0, 1]$ determine the order on the set A. We should stress that we can separate $\{1\}$ from $[u, 1]$ since U is representable on $[u, 1]^2$ and thus U is a strict t-conorm on $[e, 1]^2$. Thus we see that if we have an ordinal sum V^* of a representable uninorm and several non-complete t-norm summands we can obtain different uninorms that coincides with V^* on $]0, 1[^2$ by shifting the point of change, where $V^*(x, 1) = 1$ changes to $V^*(x, 1) = x$, through all idempotent points contained in $[0, u]$.

On the other hand, assume an ordinal sum V^* of a uninorm on $[a, b]$ for some $0 < a \leq e \leq b < 1$ and a representable uninorm on $[0, a[\cup]b, 1]$. Then since $V^*(x, 1) = 1$ for all $x \in]0, a[$ and $V^*(x, 0) = 0$ for all $x \in]b, 1[$ and $V^*(0, 1)$ is determined by the corresponding representable uninorm, we see that due to the monotonicity the values $V^*(x, 1)$ and $V^*(x, 0)$ are uniquely determined for all $x \in [0, 1]$. Thus we cannot obtain any different uninorm by redefining the values on the border of the unit square. From these two examples we see that the difference can occur only if a non-complete summand appears. These observations can be summarized into the construction method called the *extended ordinal sum of uninorms* (compare [35]). First note that a continuous t-norm $T\colon [0, 1]^2 \longrightarrow [0, 1]$ (t-conorm $S\colon [0, 1]^2 \longrightarrow [0, 1]$) is called *c-strict* if $T(x, y) \in]0, 1[$ ($S(x, y) \in]0, 1[$) for all $(x, y) \in]0, 1[^2$. In the other case T (S) will be called *c-nilpotent*.

Proposition 3 *Let $U^e\colon [0, 1]^2 \longrightarrow [0, 1]$ be a uninorm such that there is $U^e = (\langle a_k, b_k, c_k, d_k, U_k\rangle \mid k \in K)^e$, where all conditions of Proposition 2 are satisfied. Denote $G = \{b_k \mid k \in K, a_k = b_k \neq e, U^e(b_k, c_k) = b_k\}$, $H = \{c_k \mid k \in K, c_k = d_k \neq e, U^e(b_k, c_k) = c_k\}$, and for $x \in G$ denote $G_x = \{k \in K \mid b_k = x\}$, for $x \in H$ denote $H_x = \{k \in K \mid c_k = x\}$. Let G_x^{**} be the closure of the set $\{c_k \mid k \in G_x\}$ and denote $G_x^* = G_x^{**} \setminus \{d_i\}_{i \in K}$, and let H_x^{**} be the closure of the set $\{b_k \mid k \in H_x\}$ and denote $H_x^* = H_x^{**} \setminus \{a_i\}_{i \in K}$. Further, for $k \in G_x$, $x \in G$ denote*

$$F_k = \begin{cases} \{\{c_k\}, [c_k, d_k[, [c_k, d_k]\} & \text{if } S_{U_k} \text{is c-strict,} \\ \{\{c_k\}, [c_k, d_k]\} & \text{if } S_{U_k} \text{is c-nilpotent,} \end{cases}$$

for $c \in G_x^*$ denote

$$F_c^* = \begin{cases} \{[0, c]\} & \text{if } c = \inf\{c_k \mid k \in G_x\}, \\ \{[0, c[, [0, c]\} & \text{else,} \end{cases}$$

and for $k \in H_x$, $x \in H$ denote

$$J_k = \begin{cases} \{\emptyset, \{a_k\}, [a_k, b_k[\} & \text{if } T_{U_k} \text{is c-strict,} \\ \{\emptyset, [a_k, b_k[\} & \text{if } T_{U_k} \text{is c-nilpotent,} \end{cases}$$

and for $c \in H_x^*$ *denote*

$$J_b^* = \begin{cases} \{[0, b[\} & \text{if } b = \sup\{b_k \mid k \in H_x\}, \\ \{[0, b[, [0, b]\} & \text{else.} \end{cases}$$

With the convention $S \cup \{S_1, S_2\} = \{S \cup S_1, S \cup S_2\}$ *let*

$$g\colon G \longrightarrow \bigcup_{x \in G} (\bigcup_{k \in G_x} ([0, c_k[\cup F_k) \cup \bigcup_{c \in G_x^*} F_c^*)$$

be a function such that

$$g(x) \in \bigcup_{k \in G_x} ([0, c_k[\cup F_k) \cup \bigcup_{c \in G_x^*} F_c^*$$

and let

$$h\colon H \longrightarrow \bigcup_{x \in H} (\bigcup_{k \in H_x} ([0, a_k[\cup J_k) \cup \bigcup_{b \in H_x^*} J_b^*)$$

be a function such that

$$h(x) \in \bigcup_{k \in H_x} ([0, a_k[\cup J_k) \cup \bigcup_{b \in H_x^*} J_b^*,$$

where for all $x \in G$, $y \in H$ *there is* $y \in g(x)$ *if and only if* $x \in h(y)$.

Then the binary function $V^e\colon [0, 1]^2 \longrightarrow [0, 1]$ *given by*

$$V^e(x, y) = \begin{cases} U^e(x, y) & \text{if } (x, y) \in ([0, 1] \setminus (G \cup H))^2, \\ \min(x, y) & \text{if } x \in G, y \in g(x), \text{ or } y \in G, x \in g(y), \\ \max(x, y) & \text{if } x \in G, y \notin g(x), \text{ or } y \in G, x \notin g(y), \\ \min(x, y) & \text{if } x \in H, y \in h(x), \text{ or } y \in H, x \in h(y), \\ \max(x, y) & \text{if } x \in H, y \notin h(x), \text{ or } y \in H, x \notin h(y) \end{cases}$$

is a uninorm, which will be called an extended ordinal sum of uninorms. We write $V^e = (\langle a_k, b_k, c_k, d_k, U_k \rangle \mid k \in K)^e$.

Example 1 Assume an ordinal sum uninorm $U^e\colon [0, 1]^2 \longrightarrow [0, 1]$ such that $U^e = (\langle 0, e, e, e, T \rangle, \langle 0, 0, e, b, C_1 \rangle, \langle 0, 0, b, 1, C_2 \rangle)^e$, for some $b, e \in [0, 1]$, $0 < e < b < 1$ (see Fig. 2) and a t-norm T and t-conorms C_1, C_2. Assume that C_1 is c-strict and C_2 is c-nilpotent. Since T is a t-norm we have $U^e(0, e) = 0$ and thus we have $G = \{0\}$, $H = \emptyset$ for sets G, H from the previous proposition. If we denote the summands respectively as 1, 2, 3 for $K = \{1, 2, 3\}$ we get $G_0 = \{2, 3\}$. Further, $G_x^{**} = \{e, b\}$, $G_x^* = \emptyset$,

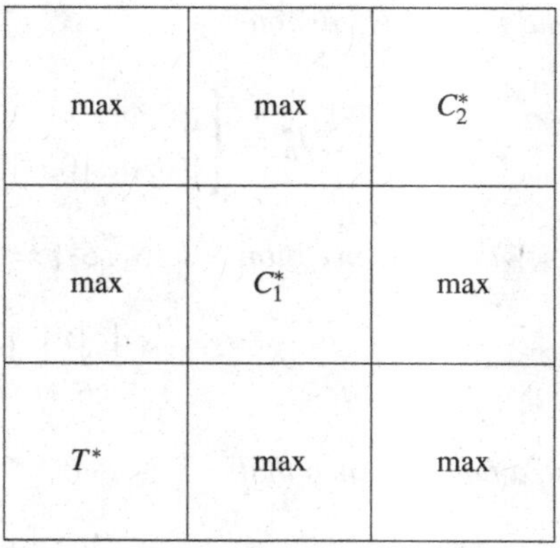

Fig. 2 The ordinal sum uninorm U^e from Example 1

max	max	C_2^*
max	C_1^*	max
T^*	max	max

$$F_2 = \{\{e\}, [e, b[, [e, b]\},$$

$$F_3 = \{\{b\}, [b, 1]\}.$$

The function g is defined only in one point 0 and its range is the set

$$\{[0, e], [0, b[, [0, b], [0, 1]\}.$$

Thus if $V^e = (\langle 0, e, e, e, T\rangle, \langle 0, 0, e, b, C_1\rangle, \langle 0, 0, b, 1, C_2\rangle)^e$, is an extended ordinal sum we have $V^e = U^e$ if $g(0) = [0, e]$. Further we can define three other different extended ordinal sums by respectively selecting a different value/interval for $g(0)$. It is evident that V^e and U^e may differ only on $\{0\} \times [0, 1] \cup [0, 1] \times \{0\}$. A sketch of a more complicated example can be seen on Fig. 3.

If all summands in the extended ordinal sum of uninorms are uninorms with continuous underlying functions also the extended ordinal sum will be a uninorm with continuous underlying functions. Now we can present an opposite result which will complete the characterization of uninorms with continuous underlying functions via the extended ordinal sum construction.

Theorem 7 *Let* $U: [0, 1]^2 \longrightarrow [0, 1]$ *be a uninorm with continuous underlying t-norm and t-conorm, with the neutral element* $e \in [0, 1]$. *Then* $U = V^e$, *where* $V^e = (\langle a_k, b_k, c_k, d_k, U_k\rangle \mid k \in K)^e$ *is an extended ordinal sum of uninorms for some systems* $(]a_k, b_k[)_{k\in K}$ *and* $(]c_k, d_k[)_{k\in K}$ *satisfying all conditions of Proposition 3, where for all* $k \in K$ *the uninorm* U_k *is either internal (including the minimum t-norm and the maximum t-conorm), or representable (including continuous Archimedean t-norms and t-conorms).*

The proof of this theorem follows from [35, Proposition12].

From the above result we see that each uninorm with continuous underlying functions can be decomposed via the extended ordinal sum construction into internal and Archimedean uninorms. Since these uninorms were already characterized (see

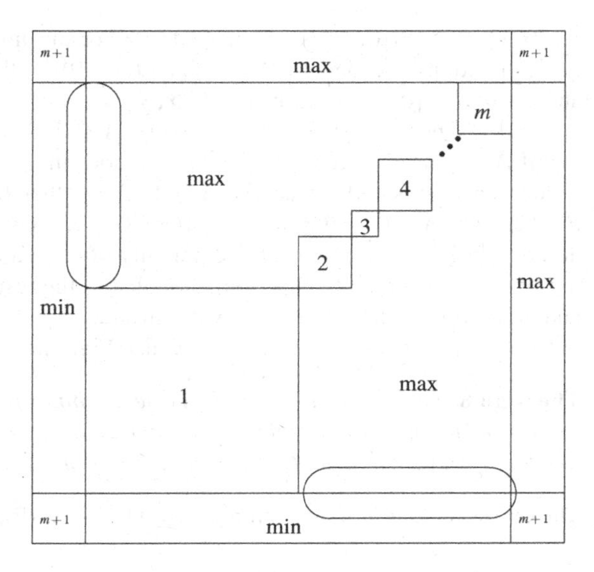

Fig. 3 Sketch of a uninorm which is an ordinal sum with $m + 1$ summands. The summands 1 and $m + 1$ are complete, the others are non-complete. The rounded area (the line in the center) designates the place where the ordinal sum construction and the extended ordinal sum construction can differ

the previous section) we have now a detailed knowledge about structure of uninorms with continuous underlying functions.

The major role in the decomposition of a uninorm with continuous underlying functions to an extended ordinal sum of uninorms plays its characterizing multi-function, which is yet another possibility how to characterize uninorms with continuous underlying t-norm and t-conorm. This characterizing multi-function in fact covers the set of points of discontinuity of such a uninorm U. We will now show results from [34] which characterize the set of points of discontinuity of a uninorm with continuous underlying functions.

The first interesting result shows that a uninorm with continuous underlying functions is either left-continuous or right continuous (or continuous) in each point from $[0, 1]^2$. Next we need the definition of a multi-function.

Definition 1 A mapping $p \colon X \longrightarrow \mathcal{P}(Y)$ is called a *multi-function* if for every $x \in X$ it assigns a subset of Y, i.e., $p(x) \subseteq Y$. A multi-function p is called

(i) *non-increasing* if for all $x_1, x_2 \in X$, $x_1 < x_2$ there is $p(x_1) \geq p(x_2)$, i.e., for all $y_1 \in p(x_1)$ and all $y_2 \in p(x_2)$ we have $y_1 \geq y_2$ and thus $\text{Card}\,(p(x_1) \cap p(x_2)) \leq 1$,
(ii) *symmetric* if $y \in p(x)$ if and only if $x \in p(y)$.

The graph of a multi-function p will be denoted by $G(p)$, i.e., $(x, y) \in G(p)$ if and only if $y \in p(x)$.

A symmetric multi-function $p \colon [0, 1] \longrightarrow \mathcal{P}([0, 1])$ is surjective, i.e., for all $y \in Y$ there exists an $x \in X$ such that $y \in p(x)$, if and only if we have $p(x) \neq \emptyset$ for all $x \in X$. The graph of a symmetric, surjective, non-increasing multi-function

$p\colon [0,1] \longrightarrow \mathscr{P}([0,1])$ is a connected line. For any uninorm with continuous underlying functions we denote $A = \inf\{x \mid U(x,0) > 0\}$, $B = \sup\{x \mid U(x,1) < 1\}$ and let $a, d \in [0,1]$ be such that $U(x,y) = e$ for some $y \in [0,1]$ if and only if $x \in]a,d[$. Then either $A = 1, B \neq 0$, or $A \neq 1, B = 0$, or $A = 1, B = 0$. Note that if $A = 1, B \neq 0$, then U is non-continuous in $(B,1)$, if $A \neq 1, B = 0$, then U is non-continuous in $(0,A)$, and if $A = 1, B = 0$ then U is non-continuous in $(0,1)$. Further, we have $0 \leq B \leq a \leq e \leq d \leq A \leq 1$. Now we can introduce the result describing the set of points of discontinuity of a uninorm with continuous underlying functions via a multi-function. Note that in the results on idempotent uninorms and uninorms locally internal on $A(e)$ introduced in the previous section we can see examples of such a multi-function (see also Fig. 1).

Theorem 8 *Let $U\colon [0,1]^2 \longrightarrow [0,1]$ be a uninorm with continuous underlying t-norm and t-conorm. Then there exists a symmetric, surjective, non-increasing multi-function r on $[0,1]^2$ such that U is continuous on $[0,1]^2 \setminus R$, where $R = G(r)$.*

The corresponding multi-function $r\colon [0,1] \longrightarrow \mathscr{P}([0,1])$ is given by

$$r(x) = \begin{cases} \{1\} & \text{if } x \in]0,B[, \\ \{0\} & \text{if } x \in]A,1[, \\ [0,B] & \text{if } x = 1, \\ [A,1] & \text{if } x = 0, \\ \{y \mid U(x,y) = e\} & \text{if } x \in]a,d[, \\ \{y \mid (x,y) \in R^*\} & \text{otherwise,} \end{cases}$$

where $R^* = \{(x,y) \in [0,1]^2 \mid U \textit{ is non-continuous in } (x,y)\}$.

Note that U need not to be non-continuous in all points of R. In fact, U is continuous in all points from $\{x\} \times [0,1]$ for all $x \in [0,B[\cup]a,d[\cup]A,1]$. The symmetric non-increasing multi-function from the previous theorem need not to be unique. The differences can appear on $]a,d[$. However, if we require additionally that $U(x,y) = e$ implies $(x,y) \in G(r)$ for all $(x,y) \in [0,1]^2$, such a multi-function is uniquely given and such a multi-function is called the characterizing multi-function of a uninorm U with continuous underlying functions.

In the following example we will show that the existence of the symmetric, surjective, non-increasing characterizing multi-function for a uninorm need not imply that the uninorm has continuous underlying functions.

Example 2 Let $U\colon [0,1]^2 \longrightarrow [0,1]$ be given by

$$U(x,y) = \begin{cases} 0 & \text{if } \max(x,y) < e, \\ x & \text{if } y = e, \\ y & \text{if } x = e, \\ \max(x,y) & \text{otherwise.} \end{cases}$$

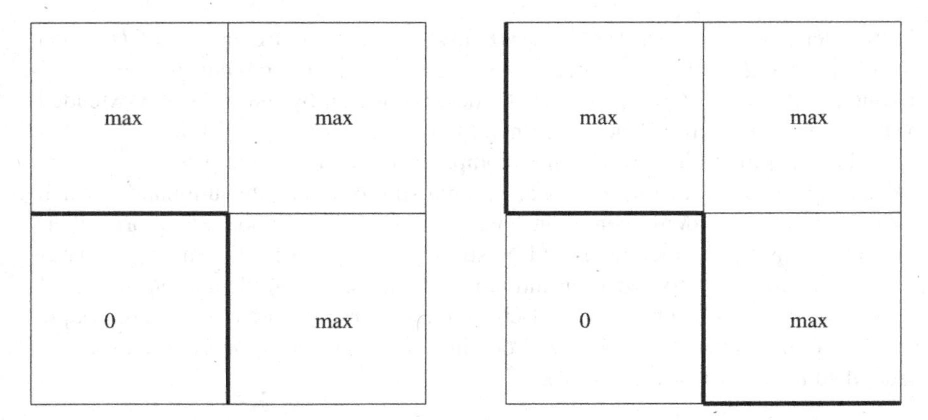

Fig. 4 The uninorm U from Example 2. The bold lines denote the points of discontinuity of U (*left*) and the characterizing multi-function r of U (*right*)

Then $U \in \mathscr{U}_{\max}$ is a uninorm, where the underlying t-norm is the drastic product and the underlying t-conorm is the maximum. This uninorm is non-continuous in points from $\{e\} \times [0, e] \cup [0, e] \times \{e\}$. Thus the corresponding multi-function is given by (see Fig. 4)

$$r(x) = \begin{cases} [e, 1] & \text{if } x = 0, \\ e & \text{if } x \in]0, e[, \\ [0, e] & \text{if } x = e, \\ 0 & \text{otherwise.} \end{cases}$$

Since $U(x, y) = e$ implies $x = y = e$ we see that U is continuous on $[0, 1]^2 \setminus R$, where $R = G(r)$ and r is a symmetric, surjective, non-increasing multi-function such that $U(x, y) = e$ implies $(x, y) \in R$. However, the drastic product t-norm is not continuous and thus U does not have continuous underlying functions.

Although the existence of the symmetric, surjective, non-increasing characterizing multi-function for a uninorm does not mean that the underlying functions are continuous, from the above example we can observe that in points from $\{e\} \times]0, e[$ the uninorm U is neither left- nor right-continuous. Therefore we get the following.

Theorem 9 ([34]) *Let $U \colon [0, 1]^2 \longrightarrow [0, 1]$ be a uninorm which is continuous on $[0, 1]^2 \setminus R$, where $R = G(r)$ and r is a symmetric, surjective, non-increasing multi-function such that $U(x, y) = e$ implies $(x, y) \in R$. Then U has continuous underlying functions if and only if in each point $(x, y) \in [0, 1]^2$ the uninorm U is either left-continuous or right-continuous (or continuous).*

Assume a uninorm U with continuous underlying functions and its characterizing multi-function. The graph of this characterizing multi-function can be decomposed into maximal segments which are either horizontal, vertical, or strictly decreasing, and border points of these segments are always idempotent elements of U. Then each horizonal and vertical segment corresponds to a non-complete summand, i.e., a continuous t-norm or t-conorm, in the decomposition by means of the extended ordinal sum construction. These t-norms (t-conorms) need not be Archimedean, but could be decomposable to further non-complete summands. Moreover, each strictly decreasing segment corresponds to an ordinal sum of complete summands, which are representable or idempotent uninorms (in the case of idempotent uninorms their characterizing multi-function should be strictly decreasing in this situation). Thus by combination of decomposition into components based on idempotent points of U and on maximal segments of its characterizing multi-function we can couple corresponding idempotent points and obtain a full decomposition via the extended ordinal sum construction (see [35]).

4 Uninorms with Continuous Diagonals and Further Generalizations

In the previous section we have shown a full characterization of uninorms with continuous underlying t-norm and t-conorm. In this section we will try to answer the question how much we can weaken the conditions in order to get the same characterization as in the previous sections. In other words, we will study uninorms continuous on some parts of the unit interval and see under which conditions we will obtain a uninorm with continuous underlying t-norm and t-conorm. In the first part of this section we will focus on uninorms continuous on the diagonal and in the second part we will discuss further generalizations.

4.1 Uninorms Continuous on Diagonal

Recall that for a binary operation $O\colon [0,1]^2 \longrightarrow [0,1]$ its corresponding diagonal function $d_O\colon [0,1] \longrightarrow [0,1]$ is given by $d_O(x) = O(x,x)$ for all $x \in [0,1]$. As we have seen above, in the structure of uninorms a major role is played by t-norms and t-conorms, and this holds especially when we discuss areas around the diagonal. Therefore in the investigation of uninorms with continuous diagonals we have to first focus on t-norms (t-conorms) with continuous diagonals. As t-norms and t-conorms are dual all results on t-norms with continuous diagonals can be immediately obtained also for t-conorms.

The first step in the study of t-norms with continuous diagonals was an open problem from [43]: whether a t-norm with continuous diagonal function have to be continuous. It is obvious that each continuous t-norm has a continuous diagonal and the construction of all continuous t-norms with the given continuous diagonal was described in [47] (see also [17, 27]). For the opposite problem a negative answer was given by Krause [20] (see also [19, 29, 44]). The Krause t-norm has a fractal-like structure, it is Archimedean and continuous on the diagonal, however, it is not right-continuous and it is not left-continuous on the border of the unit square.

Thus in general a t-norm with a continuous diagonal need not be continuous. However, there are several results showing under which conditions such a t-norm is continuous. We will start with Archimedean t-norms. In [37] continuous extensions of t-norms known on $[a, b]^2$, for $0 \leq a \leq b \leq 1$, to the whole unit square were investigated. From this paper we can obtain the following results:

(i) Let T be a t-norm continuous on $[a, 1]^2$, such that $T(x, x) < x$ for all $x \in [a, 1[$. Then T is on $[a, 1]^2$ conditionally cancellative, i.e., $T(x, y) = T(x, z) > T(a, a)$ implies $y = z$ for all $x, y, z \in [a, 1]$.
(ii) Let T be an Archimedean t-norm continuous and cancellative on $[a, 1]^2$, such that T has a continuous diagonal. Then T is continuous.
(iii) Let T be an Archimedean t-norm continuous and conditionally cancellative, but not cancellative, on $[a, 1]^2$. Then T is continuous.

Thus we see that for an Archimedean t-norm the continuity on $[a, 1]^2$ for arbitrary $0 < a < 1$ and continuity of the diagonal implies continuity on the whole unit square. Moreover, such a t-norm is uniquely determined by its values on $[a, 1]^2$ and its diagonal function. Note that if we relax the Archimedean property then an ordinal sum of the Krause t-norm on $[0, a]$ for any $0 < a < 1$ will yield a non-continuous t-norm with continuous diagonal which is continuous on $[a, 1]^2$. Further, in the following example we will see that the continuity of the diagonal is not implied by the Archimedean property and continuity on $[a, 1]^2$.

Example 3 ([38]) Assume a t-norm $T_*\colon [0, 1]^2 \longrightarrow [0, 1]$ given by

$$T_*(x, y) = \begin{cases} x \cdot y & \text{if } x \cdot y \geq \frac{1}{4}, \\ \min(x, y) & \text{if } \max(x, y) = 1, \min(x, y) < \frac{1}{4}, \\ 0 & \text{otherwise.} \end{cases}$$

Then T_* is an Archimedean t-norm which is continuous on $[\frac{1}{2}, 1]^2$. However, T_* is not continuous on the diagonal. Note that T_* is the weakest t-norm that coincide with the product t-norm on $[\frac{1}{2}, 1]^2$.

We can summarize the above results in the following Corollary.

Corollary 1 *Let $U\colon [0, 1]^2 \longrightarrow [0, 1]$ be an Archimedean uninorm which is continuous on the diagonal. Then U has continuous underlying functions if and only if it is continuous on $[a, e]^2$ and $[e, b]^2$ for some $a, b \in [0, 1]$ with $0 \leq a < e < b \leq 1$.*

From the above result we see that, for example, if for a uninorm U the T_U is the Krause t-norm and S_U is the t-conorm dual to the Krause t-norm then there are no such $a, b \in [0, 1]$, $0 \leq a < e < b \leq 1$, that U it is continuous on $[a, e]^2$ and $[e, b]^2$.

In general, we have the following result for both t-norms and t-conorms.

Corollary 2 ([36]) *A t-norm $T\colon [0,1]^2 \longrightarrow [0,1]$ (t-conorm $S\colon [0,1]^2 \longrightarrow [0,1]$) is continuous if and only if it is continuous on the diagonal and for each idempotent point $x \in]0,1]$ ($x \in [0,1[$) there exists a $\delta_x > 0$ such that T (S) is continuous in (x, y) for all $y \in [x - \delta_x, x]$ ($y \in [x, x + \delta_x]$).*

From this corollary we can conclude that a uninorm $U\colon [0,1]^2 \longrightarrow [0,1]$ has continuous underlying functions if and only if it is continuous on the diagonal and for each idempotent point $x \in]0,1[$ there exists a $\delta_x > 0$ such that if $x \le e$ then U is continuous in (x, y) for all $y \in [x - \delta_x, x]$, and if $y \ge e$ then U is continuous in (x, y) for all $y \in [x, x + \delta_x]$. For more details we recommend [36]. If we again take a uninorm U such that T_U is the Krause t-norm and S_U is the t-conorm dual to the Krause t-norm then there is no such $\delta > 0$ that U is continuous on $\{e\} \times [e - \delta, e + \delta]$.

4.2 Further Generalizations

In this subsection we will first focus on the case when we have no information about the continuity of the diagonal of the given uninorm. In this case let us recall a result from [13] which shows that if a cancellative Archimedean t-norm is (left-)continuous in the point $(1, 1)$ then it is isomorphic with the product t-norm, i.e., it is continuous and strict. This means that if a uninorm U is cancellative on $]0, e]^2 \cup [e, 1[^2$ and continuous in (e, e) then U has continuous underlying functions and thus it has a structure described in the previous section.

Moreover, in [36] it was shown that an Archimedean t-norm which is border-continuous (i.e., each point from the boundary of the unit square is the point of continuity of T) is also continuous. This means that an Archimedean uninorm which is left-continuous in all points from $[0, e] \times \{e\}$ and right-continuous in all points from $[e, 1] \times \{e\}$ has continuous underlying functions.

From these two results we see that the Archimedean property can 'spread' the continuity from some part of the unit square to the whole unit square. For both results, the Archimedean property of the corresponding uninorm is crucial. In the first case recall an example of a non-Archimedean, left-continuous, cancellative t-norm ([28, 45]) which is given by $T(x, y) = z = \sum_{i\in\mathbb{N}} \frac{1}{2^{z_i}}$ for all $(x, y) \in]0, 1]^2$, where $x = \sum_{i\in\mathbb{N}} \frac{1}{2^{x_i}}$, $y = \sum_{i\in\mathbb{N}} \frac{1}{2^{y_i}}$, and $\{x_i\}_{i\in\mathbb{N}}$, $\{y_i\}_{i\in\mathbb{N}}$, $\{z_i\}_{i\in\mathbb{N}}$ are increasing sequences of natural numbers, and $z_i = x_i + y_i - i$. For the second result it is enough to take an ordinal sum of the Krause t-norm on $[0, a]$ for some $a \in]0, 1[$. Then this ordinal sum is border continuous, non-Archimedean and non-continuous.

In the final part of this chapter we will focus on uninorms that are continuous on $[0, e[^2 \cup]e, 1]^2$. Assume such a uninorm U and focus on T_U (similar observations can be obtained for S_U by duality). We will now show that T_U can be obtained from a continuous t-subnorm ([16]) by redefining its values on the border of the unit square. First, let us recall that a binary operation $M\colon [0,1]^2 \longrightarrow [0,1]$ is a *t-subnorm* if it is commutative, associative, non-decreasing in both variables and $M(x, y) \leq \min(x, y)$ for all $(x, y) \in [0,1]^2$. From each t-subnorm we can obtain a t-norm T given by $T(x, y) = M(x, y)$ for all $(x, y) \in [0, 1[^2$, $T(x, y) = \min(x, y)$ otherwise. This process is called *lifting* of a t-subnorm to a t-norm. Then $T = M$ if and only if M is a t-norm. Vice versa in [15] a border-continuous projection $M_T\colon [0,1]^2 \longrightarrow [0,1]$ of a t-norm was defined by

$$M_T(x, y) = \begin{cases} T(x, y) & \text{if } (x, y) \in [0, 1[^2, \\ T(x^-, y^-) & \text{if } \max(x, y) = 1. \end{cases}$$

The idea of this border-continuous projection was to obtain a reverse process to the lifting of a t-subnorm to a t-norm. However, such a border-continuous projection need not to be monotone.

Example 4 Let $T\colon [0,1]^2 \longrightarrow [0,1]$ be given by

$$T(x, y) = \begin{cases} \min(x, y) & \text{if } \max(x, y) = 1, \\ \frac{2}{3}(x + y) - \frac{5}{6} & \text{if } (x, y) \in [\frac{1}{2}, 1]^2, \max(x, y) < 1, \\ 0 & \text{otherwise.} \end{cases}$$

Then $M_T(\frac{1}{2}, 1) = 0$ and $M_T(\frac{1}{2}, \frac{7}{8}) = \frac{1}{12}$.

Thus the proper definition of a border-continuous projection is rather $M_T(x, y) = T(x, y)$ if $(x, y) \in [0, 1[^2$, $M_T(x, y) = T(x^-, y)$ if $x = 1, y < 1$, $M_T(x, y) = T(x, y^-)$ if $y = 1, x < 1$, and $M_T(x, y) = T(x^-, y^-)$ if $x = y = 1$. In this case a border-continuous projection of a t-norm is commutative, bounded by minimum and non-decreasing in both variables. Note that a border-continuous projection is not border-continuous in the sense that each point from the border of the unit square is a point of continuity of M_T, but in the sense that $M_T(x, 1^-) = M_T(x, 1)$, while the function $M_T(\cdot, 1)$ can be non-continuous. However, if we take the t-norm from Example 4 then M_T is not associative, i.e., not a t-subnorm. Indeed, in this case

$$M_T(x, y) = \begin{cases} \frac{2}{3}(x + y) - \frac{5}{6} & \text{if } (x, y) \in [\frac{1}{2}, 1]^2, \\ 0 & \text{otherwise,} \end{cases}$$

and $M_T(M_T(1, 1), x) = \frac{2 \cdot x}{3} - \frac{1}{2}$ for all $x \geq \frac{3}{4}$, however, $M_T(1, M_T(1, x)) = 0$ for all $x < 1$. Thus we have to characterize all t-norms such that their border-continuous projection is a t-subnorm. This problem was solved in [39]. It is evident that if T is border-continuous then $M_T = T$. Further we have the following result.

Proposition 4 ([39]) *For a t-norm $T\colon [0,1]^2 \longrightarrow [0,1]$ its border-continuous projection $M_T\colon [0,1]^2 \longrightarrow [0,1]$ is a t-subnorm if and only if the following two conditions are satisfied:*

(i) *for all* $x, y \in [0, 1[$ *either* $T(u_0, x) = \lim_{u \longrightarrow 1^-} T(u, x)$ *for some* $u_0 \in [0, 1[$, *or* $T(a, y) = \lim_{v \longrightarrow a^-} T(v, y)$, *where* $a = \lim_{u \longrightarrow 1^-} T(u, x)$,

(ii) *either* $\lim_{u \longrightarrow 1^-} T(u, u) = 1$, *or* $T(u_0, v_0) = \lim_{u \longrightarrow 1^-} T(u, u)$ *for some* $u_0, v_0 \in [0, 1[$, *or for all* $x \in [0, 1[$ *there is* $T(b, x) = \lim_{v \longrightarrow b^-} T(v, x)$, *where* $b = \lim_{u \longrightarrow 1^-} T(u, u)$.

As an easy corollary of the previous result we see that if T is left-continuous on $[0, 1[^2$ then M_T is a t-subnorm: this is for example the case of the Krause t-norm.

Since we focus on t-norms T_U continuous on $[0, 1[^2$ then M_{T_U} is always a continuous t-subnorm and a similar result can be obtained also for S_U. From [30] we know that a continuous t-subnorm is an ordinal sum of continuous Archimedean t-norms and a continuous Archimedean t-subnorm. The set of continuous Archimedean t-subnorms can be divided into three parts: continuous cancellative t-subnorms, continuous nilpotent t-subnorms, continuous t-subnorms with no nilpotent element which are not cancellative. Here continuous nilpotent t-subnorms are such that they posses a nilpotent element $x \in]0, 1]$ with $M(x, x) = 0$. Although the structure of uninorms such that M_{T_U} (M_{S_U}) is not cancellative is quite complicated, in the case that M_{T_U} (M_{S_U}) is a continuous cancellative t-subnorm (t-superconorm) several results similar as in the case of uninorms with strict underlying functions can be shown. Note that a continuous cancellative t-subnorm (t-superconorm) is always Archimedean.

Proposition 5 ([39]) *Let* $U : [0, 1]^2 \longrightarrow [0, 1]$ *be a uninorm with neutral element* $e \in]0, 1[$, *such that* M_{T_U} (M_{C_U}) *is a continuous cancellative t-subnorm (t-superconorm). Then* U *has one of the seven forms from Theorem 5.*

We can use this result to characterize also non-Archimedean uninorms continuous on $[0, e[^2 \cup]e, 1]^2$, such that their underlying t-norm (t-conorm) is an ordinal sum of continuous Archimedean t-norms (t-conorms) and the last (the first) cancellative continuous t-subnorm (t-superconorm).

We will denote by $\mathcal{U}_{lcc}$ be the set of all uninorms continuous on $[0, e[^2 \cup]e, 1]^2$ such that there exists an idempotent point $b_0 \in [0, 1]$, $b_0 < e$, such that U is continuous and cancellative on $]b_0, e[^2$, and by $\mathcal{U}_{rcc}$ be the set of all uninorms continuous on $[0, e[^2 \cup]e, 1]^2$ such that there exists an idempotent point $c_0 \in [0, 1]$, $c_0 > e$, such that U is continuous and cancellative on $]e, c_0[^2$. In [39] it was shown that if $U \in \mathcal{U}_{lcc} \cap \mathcal{U}_{rcc}$ then U is an extended ordinal sum of a uninorm with continuous underlying functions on $[0, b_0[\cup \{U(b_0, c_0)\} \cup]c_0, 1]$ and an Archimedean uninorm cancellative and continuous on $]0, e[^2 \cup]e, 1[^2$ acting on $[b_0, c_0]$.

Further, if $U \in \mathcal{U}_{lcc}$ and $e = \inf\{c \in]e, 1] \mid c \text{ is an idempotent point}\}$ then U is an extended ordinal sum of a uninorm with continuous underlying functions on $[0, b_0] \cup]e, 1]$ and an Archimedean t-norm cancellative and continuous on $]0, 1[^2$ acting on $[b_0, e]$.

Finally, if $U \in \mathcal{U}_{rcc}$ and $e = \sup\{b \in [0, e[\mid b \text{ is an idempotent point}\}$ then U is an extended ordinal sum of a uninorm with continuous underlying functions on $[0, e[\cup [c_0, 1]$ and an Archimedean t-conorm cancellative and continuous on $]0, 1[^2$ acting on $[e, c_0]$.

5 Conclusions and Further Perspectives

We have discussed uninorms with continuous underlying functions, shown several partial results that were achieved, and then demonstrated their representation via the extended ordinal sum construction. We have also characterized the set on which such a uninorm is discontinuous. We have further shown how these results can be used also for uninorms for which only parts of the underlying functions are known.

We have also discussed a generalization of this problem, i.e., the case when a uninorm is continuous on $[0, e[^2 \cup]e, 1]^2$. We have shown the characterization of such uninorms in the case when the underlying functions are related to a continuous cancellative t-subnorm and a continuous cancellative t-superconorm, respectively. An open problem for future research is the full characterization of uninorms continuous on $[0, e[^2 \cup]e, 1]^2$.

We have also shown the characterization of all t-norms such that their border-continuous projection is associative, i.e., a t-subnorm.

As we have seen in Sect. 2, recently a number of papers was focused on uninorms on complete lattices. Therefore future research on uninorms will be related to uninorms on more abstract scales.

Acknowledgments This work was supported by grants VEGA 2/0049/14, APVV-0178-11 and Program Fellowship of SAS.

References

1. Aczél, J.: Lectures on Functional Equations and their Applications. Academic Press, New York (1966)
2. Akella, P.: Structure of n-uninorms. Fuzzy Sets Syst. **158**, 1631–1651 (2007)
3. De Baets, B.: Idempotent uninorms. Eur. J. Oper. Res. **118**, 631–642 (1998)
4. De Baets, B., Fodor, J., Ruiz, D., Torrens, J.: Idempotent uninorms on finite ordinal scales. Int. J. Uncertain. Fuzziness, Knowl.-Based Syst. **107**, 1–14 (2009)
5. Clifford, A.H.: Naturally totally ordered commutative semigroups. Am. J. Math. **76**, 631–646 (1954)
6. Dombi, J.: A general class of fuzzy operators, the De Morgan class of fuzzy operators and fuzziness measures induced by fuzzy operator. Fuzzy Sets Syst. **8**, 149–163 (1982)
7. Drewniak, J., Drygaś, P.: On a class of uninorms. Int. J. Uncertain. Fuzziness, Knowl.-Based Syst. **10**, 5–10 (2002)
8. Drygaś, P.: Discussion of the structure of uninorms. Kybernetika **41**, 213–226 (2005)
9. Drygaś, P.: On monotonic operations which are locally internal on some subset of their domain. Proc. EUSFLAT Conf. **2**, 185–191 (2007)

10. Drygaś, P.: On properties of uninorms with underlying t-norm and t-conorm given as ordinal sums. Fuzzy Sets Syst. **161**, 149–157 (2010)
11. Fodor, J., De Baets, B.: A single-point characterization of representable uninorms. Fuzzy Sets Syst. **202**, 89–99 (2012)
12. Fodor, J., Yager, R.R., Rybalov, A.: Structure of uninorms. Int. J. Uncertain. Fuzziness, Knowl.-Based Syst. **5**, 411–427 (1997)
13. Hájek, P.: Observations on the monoidal t-norm logic. Fuzzy Sets Syst. **132**, 107–112 (2002)
14. Hu, S., Li, Z.: The structure of continuous uninorms. Fuzzy Sets Syst. **124**, 43–52 (2001)
15. Jayaram, B., Baczyński, M., Mesiar, R.: R-implications and the exchange principle: a complete characterization. In: Galichet, S., Montero, J., Mauris, G. (eds.) Proceedings of EUSFLAT-2011 and LFA-2011, pp. 223–229. Aix-les-Bains, France (2011)
16. Jenei, S.: A note on the ordinal sum theorem and its consequence for the construction of triangular norms. Fuzzy Sets Syst. **126**, 199–205 (2002)
17. Kimberling, C.: On a class of associative functions. Publ. Math. Debrecen **20**, 21–39 (1973)
18. Klement, E.P., Mesiar, R., Pap, E.: On the relationship of associative compensatory operators to triangular norms and conorms. Int. J. Uncertainty, Fuzziness, Knowl.-Based Syst. **4**, 129–144 (1996)
19. Klement, E.P., Mesiar, R., Pap, E.: Triangular Norms. Kluwer Academic Publishers, Dordrecht (2000)
20. Krause, G.M.: The Devil's Terraces: a Discontinuous Associative Function, personal communication (2015)
21. Li, G., Liu, H.W., Fodor, J.: Single-point characterization of uninorms with nilpotent underlying t-norm and t-conorm. Int. J. Unc. Fuzz. Knowl. Based Syst. **22**, 591–604 (2014)
22. Li, Y.M., Shi, Z.K.: Remarks on uninorm aggregation operators. Fuzzy Sets Syst. **114**, 377–380 (2000)
23. Liu, H.W.: Semi-uninorms and implications on a complete lattice. Fuzzy Sets Syst. **191**, 72–82 (2012)
24. Martín, J., Mayor, G., Torrens, J.: On locally internal monotonic operations. Fuzzy Sets Syst. **137**, 27–42 (2003)
25. Mas, M., Mayor, G., Torrens, J.: t-operators and uninorms on a finite totally ordered set. Int. J. Intell. Syst. **14**, 909–922 (1999)
26. Mas, M., Monserrat, M., Torrens, J.: On left and right uninorms. Int. J. Uncertainty, Fuzziness Knowl.-Based Syst. **9**(4), 491–507 (2001)
27. Mesiar, R., Navara, M.: Diagonals of continuous triangular norms. Fuzzy Sets Syst. **104**, 35–41 (1999)
28. Mesiar, R.: Triangular norms—an overview. In: Reusch, B., Temme, K.H. (eds.) Computational Intelligence in Theory and Practice, pp. 35–54. Physica-Verlag, Heidelberg (2001)
29. Mesiarová, A.: Wild T-norms. J. Electr. Eng. **12/s**, 36–40 (2000)
30. Mesiarová, A.: Continuous triangular subnorms. Fuzzy Sets Syst. **142**, 75–83 (2004)
31. Mesiarová-Zemánková, A.: Multi-polar t-conorms and uninorms. Inf. Sci. **301**, 227–240 (2015)
32. Mesiarová-Zemánková, A.: Ordinal sum of uninorms and generalized uninorms, Int. J. Approximate Reasoning, under Rev. (2015)
33. Mesiarová-Zemánková, A.: Ordinal sums of representable uninorms, Fuzzy Sets Syst., under Rev. (2015)
34. Mesiarová-Zemánková., A.: Characterization of uninorms with continuous underlying t-norm and t-conorm by their set of discontinuity points. IEEE Trans. Fuzzy Syst., under Rev. (2015)
35. Mesiarová-Zemánková., A.: Characterization of uninorms with continuous underlying t-norm and t-conorm by means of the ordinal sum construction. Int. J. Approximate Reasoning, under Rev. (2015)
36. Mesiarová-Zemánková., A.: T-norms and t-conorms continuous around diagonals. Fuzzy Sets Syst., under Rev. (2015)
37. Mesiarová-Zemánková A.: Continuous completions of triangular norms known on a subregion of the unit interval. Fuzzy Sets Syst., under Rev. (2015)

38. Mesiarová-Zemánková A.: Extremal completions of triangular norms known on a subregion of the unit interval. In: Torra, V., Narukawa, Y. (eds.), Proceedings MDAI 2015 Conference, LNAI 9321, pp. 21–32. Springer (2015)
39. Mesiarová-Zemánková A.: Uninorms continuous on $[0, e[^2 \cup]e, 1]^2$. Inf. Sci., under Rev. (2015)
40. Petrík, M., Mesiar, R.: On the structure of special classes of uninorms. Fuzzy Sets Syst. **240**, 22–38 (2014)
41. Ruiz-Aguilera, D., Torrens, J., De Baets, B., Fodor, J.: Some remarks on the characterization of idempotent uninorms. In: Hüllermeier, E., Kruse, R., Hoffmann, F. (eds.) Computational Intelligence for Knowledge-Based Systems Design, Proceedings of the 13th IPMU 2010 Conference, LNAI 6178, pp. 425–434. Springer, Berlin (2010)
42. Ruiz, D., Torrens, J.: Distributivity and conditional distributivity of a uninorm and a continuous t-conorm. IEEE Trans. Fuzzy Syst. **14**(2), 180–190 (2006)
43. Schweizer, B., Sklar, A.: Probabilistic Metric Spaces. North-Holland, New York (1983)
44. Smutná, D.: Non-Continuous t-norms with continuous diagonal. J. Electr. Eng. **12/s**, 51–53 (2000)
45. Smutná, D.: On a peculiar t-norm. BUSEFAL **75**, 60–67 (1998)
46. Su, Y., Wang, Z., Tang, K.: Left and right semi-uninorms on a complete lattice. Kybernetika **49**(6), 948–961 (2013)
47. Tkadlec, J.: Triangular norms with continuous diagonals. Tatra Mt. Math. Publ. **16**, 187–195 (1999)
48. Wang, Z., Fang, J.X.: Residual operations of left and right uninorms on a complete lattice. Fuzzy Sets Syst. **160**, 22–31 (2009)
49. Yager, R.R., Rybalov, A.: Uninorm aggregation operators. Fuzzy Sets Syst. **80**, 111–120 (1996)
50. Yager, R.R., Rybalov, A.: Bipolar aggregation using the uninorms. Fuzzy Optim. Decis. Making **10**, 59–70 (2011)

The Notions of Overlap and Grouping Functions

Humberto Bustince, Edurne Barrenechea, Miguel Pagola and Javier Fernandez

Abstract In this work, we make a review of the concepts of overlap and grouping functions, mainly from a theoretical point of view. In particular, we summarize some of the most relevant works that have been published in recent years about this topic.

1 Introduction

In many situations it is necessary to assign a given element or object to one out of several available classes. If a clear boundary among the classes does not exist, it may be difficult to carry out such an assignation. Even more, the classes may be fuzzy in nature, and experts may realize that elements are simply in between several classes (see, e.g., [3]). But it may be also the case that even if the boundaries are clear, we do not have enough precision about them [26]. In any of these situations the concept of overlap arises (see [5–7, 14, 15, 32]).

The concept of overlap as a bivariate aggregation operator was first introduced in [9] to measure the degree of overlap of an object in a fuzzy classification system with two classes, with an eye specially kept in image processing applications [22]. However, overlap function have has been succesfully applied to many other situations, when it is necessary to know the degree of overlap of objects in two-class classification systems, as the image segmentation problem described in [22] (in which it is

H. Bustince (✉) · E. Barrenechea · M. Pagola · J. Fernandez
Departamento of Automática y Computación and with the Institute of Smart Cities, Universidad Publica de Navarra, 31006 Navarra, Spain
e-mail: bustince@unavarra.es

E. Barrenechea
e-mail: edurne.barrenechea@unavarra.es

M. Pagola
e-mail: miguel.pagola@unavarra.es

J. Fernandez
e-mail: fcojavier.fernandez@unavarra.es

S. Saminger-Platz and R. Mesiar (eds.), *On Logical, Algebraic, and Probabilistic Aspects of Fuzzy Set Theory*, Studies in Fuzziness and Soft Computing 336,
DOI 10.1007/978-3-319-28808-6_8

necessary to discriminate between object and background) or in the framework of preference relations [10].

Mathematically speaking, overlap functions are just a particular instance of bivariate, continuous aggregation functions [11]; that is, increasing functions which are defined in the unit square and which fulfill appropriate boundary conditions. It is therefore worth to consider the possible relationships between overlap functions and other well-known examples of aggregation functions, as t-norms, copulas, semicopulas or quasi-copulas. Notice that although we, in principle, consider that overlap functions are defined in the unit square, this is not an essential requirement, and other domains may be considered too.

Overlap functions can also be extended to situations where more than two classes are involved, but associativity is not a natural requirement [18]. This is the case, for instance, for classification problems which involve rules of the type:

$$\begin{array}{l}\text{Rule } R_j: \text{ If } x_1 \text{ is } \widetilde{A_{j1}} \text{ and } \dots \text{ and } x_n \text{ is } \widetilde{A_{jn}} \\ \text{then Class} = C_j \text{ with } RW_j,\end{array} \tag{1}$$

then in the inference procedure (see [19–21, 28]) an aggregation

$$A_n(\mu_{\widetilde{A_{j1}}}(x_{p1}), \dots, \mu_{\widetilde{A_{jn}}}(x_{pn}))$$

is commonly used, where $x_p = (x_{p1}, \dots, x_{pn})$ is a new example to be classified. In fact, in this kind of procedures, the product t-norm, is usually considered in order to carry out such aggregation. However, in some of these situations associativity is not a natural assumption. This fact leads to the extension of the concept of overlap function to the n-dimensional setting.

Besides, and closely related to the notion of overlap function, it also appears the concept of a grouping function. In this case, instead of being interested in determining whether an input falls into more than one class, we focus on determining up to what extent it belongs to at least one of the considered classes. The main interest of this notion of grouping function lies in the fact that the class of overlap functions may be obtained by duality with respect to any strong negation from the class of grouping functions, and vice-versa. This implies, as a relevant consequence, that overlap and grouping functions can be used to replace t-norms and t-conorms in those situations where associativity is not a natural requirement. This is the case, for instance and under appropriate circumstances, of preferences structures, see [10].

In this chapter we present an overview of the main concepts and results related to overlap and grouping, following the papers [4, 9, 10, 13, 18, 22, 25]. The structure of the chapter is as follows. We start recalling some preliminary notions and results. In Sect. 3 we introduce the notion of a bivariate overlap functions. In Sect. 4 we discuss additive generators for overlap functions. In Sect. 5 we recall the concept of grouping functions and its relation with overlap functions. Section 6 is devoted to the n-dimensional extension of overlap and grouping functions. We finish with some conclusions and references.

2 Preliminaries

We start recalling some concepts and results which are well-known, in order to fix notations.We start with the concept of automorphism of the unit interval.

Definition 1 An automorphism of the unit interval is any continuous and strictly increasing function $\varphi : [0, 1] \to [0, 1]$ such that $\varphi(0) = 0$ and $\varphi(1) = 1$.

The definition of aggregation functions is a very well-known one. Here we follow the approach and definitions given in [8, 11, 17, 24] (but see also [2, 12]).

Definition 2 An aggregation function of dimension n (n-ary aggregation function) is an increasing mapping $M : [0, 1]^n \to [0, 1]$ such that $M(0, \dots, 0) = 0$ and $M(1, \dots, 1) = 1$.

For the particular case of bivariate aggregation functions we remind the following definitions.

Definition 3 Let M be a bivariate aggregation function.

(i) M is said to be symmetric if $M(x, y) = M(y, x)$ for any $x, y \in [0, 1]$.
(ii) M is said to be associative if $M(M(x, y), z) = M(x, M(y, z))$ for any $x, y, z \in [0, 1]$.

One of the most relevant example of aggregation function is provided by t-norms. These function play a key role, for instance, to model conjunctions in fuzzy logics or intersections in fuzzy set theory.

Definition 4 A triangular norm (t-norm for short) is an associative, symmetric bivariate aggregation function $T : [0, 1]^2 \to [0, 1]$ such that $T(1, x) = x$ for all $x \in [0, 1]$. A strictly increasing (in $]0, 1]^2$) continuous t-norm T is called a strict t-norm.

A particular type of continuous t-norms are Archimedean t-norms (see, e.g., [23]).

Definition 5 A continuous t-norm T is said to be Archimedean if $T(x, x) < x$ for all $x \in]0, 1[$.

It is worth to note that, although this is not the usual definition of Archimedean t-norm that can be found in the literature, both are equivalent when dealing with continuous t-norms [23]. It is also important to recall that every strict t-norm (i.e., any continuous and strictly increasing t-norm) is necessarily Archimedean. In fact, any strict t-norm can be obtained perturbing the product t-norm $T_P(x, y) = xy$ by means of an appropriate automorphism (see [23, 29]).

Theorem 1 *A t-norm T is strict if and only if there exists an automorphism φ of the unit interval such that*

$$T(x, y) = \varphi^{-1}(\varphi(x)\varphi(y)), \qquad x, y \in [0, 1].$$

It is possible to classify t-norms using the notion of ordinal sum, that we recall now.

Definition 6 Suppose that $\{[a_m, b_m]\}$ is a countable family of non-overlapping, closed, non-trivial, proper subintervals of [0, 1], To each $[a_m, b_m]$ in the family associate a t-norm T_m. The ordinal sum of the family $\{([a_m, b_m], T_m)\}$ is the mapping $T : [0, 1]^2 \to [0, 1]$ given by

$$T(x, y) = \begin{cases} a_m + (b_m - a_m)T_m(\frac{x-a_m}{b_m-a_m}, \frac{y-a_m}{b_m-a_m}) \text{ if } (x, y) \in [a_m, b_m]^2 \\ \min(x, y) \text{ otherwise} \end{cases}$$

Each T_m is called a summand.

By means of these ordinal sums, t-norms can be classified as follows [16].

Theorem 2 *Assume that T is a continuous t-norm. Then, one of the following three cases is valid for T:*

1. $T(x, y) = \min(x, y)$;
2. T is Archimedean;
3. there exists a family $\{([a_m, b_m], T_m)\}$ such that T is the ordinal sum of this family and each T_m is a continuous Archimedean t-norm.

A general study on t-norms and their properties may be found, for instance, in [1, 15, 16, 23, 29].

2.1 Copulas, Semicopulas and Quasi-copulas

Definition 7 A mapping $S : [0, 1]^2 \to [0, 1]$ is called a semicopula if it is nondecreasing in each coordinate and 1 is its neutral element, i.e., $S(x, 1) = S(1, x) = x$ for all $x \in [0, 1]$.

Definition 8 A quasi-copula is a semicopula Q which is also a 1-Lipschitz function (with respect to the L^1-norm).

Definition 9 A copula is a semicopula C which is 2-increasing, i.e.,

$$C(x, y) + C(x', y') - C(x', y) - C(x, y') \geq 0 \text{ for all } 0 \leq x \leq x' \leq 1, 0 \leq y \leq y' \leq 1.$$

Observe that, as stated previously, each copula is a quasi-copula. More generally, all copulas, semicopulas and quasi-copulas are aggregation functions.

3 Overlap Functions

In this section, we mainly follows the developments in [9]. As we have already said, overlap functions are a particular instance of bivariate aggregation functions. The formal definition reads as follows.

Definition 10 [9] A mapping $G_O : [0, 1]^2 \to [0, 1]$ is an overlap function if it satisfies the following conditions:
(G_O1). G_O is symmetric.
(G_O2). $G_O(x, y) = 0$ if and only if $xy = 0$.
(G_O3). $G_O(x, y) = 1$ if and only if $xy = 1$.
(G_O4). G_O is non-decreasing.
(G_O5). G_O is continuous.

There are many possible examples of overlap functions. For instance, $G_O(x, y) = \min(x, y)$ or $G_O(x, y) = x^p y^p$ for $p > 0$. Note that, in the latter example the resulting overlap function is not associative unless $p = 1$, whereas, if $p = 1$, we recover the product, which is also a t-norm (as it is also the case of the minimum).

Let's denote by $\mathcal{O}$ the set of all overlap functions. Then:

Theorem 3 *$(\mathcal{O}, \leq_{\mathcal{O}})$ with the ordering $\leq_{\mathcal{O}}$ defined for $G_1, G_2 \in \mathcal{O}$ by*

$$G_1 \leq_{\mathcal{O}} G_2 \text{ if and only if } G_1(x, y) \leq G_2(x, y)$$

for all $x, y \in [0, 1]$, is a lattice.

The lattice $(\mathcal{O}, \leq_{\mathcal{O}})$ is not complete (no top neither bottom elements, for example). On the other hand, it is closed with respect to appropriate aggregation functions. In particular, it holds that:

Theorem 4 *Let $M : [0, 1] \times [0, 1] \to [0, 1]$ be a mapping. For $G_1, G_2 \in \mathcal{O}$, define the mapping $\mathcal{M}(G_1, G_2) : [0, 1] \times [0, 1] \to [0, 1]$ as*

$$\mathcal{M}(G_1, G_2)(x, y) = M(G_1(x, y), G_2(x, y)) \text{ for all } x, y \in [0, 1] .$$

Then, $\mathcal{M}(G_1, G_2) \in \mathcal{O}$ for any $G_1, G_2 \in \mathcal{O}$ if and only if there is a continuous aggregation function $M^* : [0, 1] \times [0, 1] \to [0, 1]$ with no zero divisors and such that also its dual $(M^*)^d$ (that is, the mapping $(M^*)^d(x, y) = 1 - M^*(1 - x, 1 - y)$) has no zero divisors (i.e., if $M^*(x, y) = 1$ then necessarily either $x = 1$ or $y = 1$) so that $M|_E = M^*|_E$, where $E =]0, 1[^2 \cup \{(0, 0), (1, 1)\}$.

An important consequence of the previous result is the convexity of the class of overlap functions.

Corollary 1 *Let $G_1, \ldots, G_m$ be overlap functions and $w_1, \ldots, w_m$ be non negative weights with $\sum w_i = 1$. Then the convex sum $G = \sum w_i G_i$ is also an overlap function.*

Overlap functions can be characterized in terms of rational expressions as follows.

Theorem 5 *The mapping $G_O : [0, 1]^2 \to [0, 1]$ is an overlap function if and only if*

$$G_O(x, y) = \frac{f(x, y)}{f(x, y) + h(x, y)}$$

for some $f, h : [0, 1]^2 \to [0, 1]$ such that

1. *f and h are symmetric;*
2. *f is non decreasing and h is non increasing;*
3. *$f(x, y) = 0$ if and only if $xy = 0$;*
4. *$h(x, y) = 0$ if and only if $xy = 1$;*
5. *f and h are continuous functions.*

Example 1 Take $f(x, y) = \sqrt{xy}$ and $h(x, y) = \max(1 - x, 1 - y)$, then we have that by the construction given in Theorem 5 we get an overlap function

$$G_O(x, y) = \frac{\sqrt{xy}}{\sqrt{xy} + \max(1 - x, 1 - y)} .$$

Corollary 2 *In the setting of Theorem 5, $G_O(x, x) = x$ for some $x \in (0, 1)$ if and only if*

$$f(x, x) = \frac{x}{1 - x} h(x, x) .$$

The rational expressions which are provided in Theorem 5 need not be unique. However, the lack of uniqueness allows us to recover a whole family of overlap functions as follows.

Corollary 3 *Let f and h be two functions in the setting of the previous theorem. Then, for $k_1, k_2 \in]0, \infty[$, the mappings*

$$G_O^{k_1,k_2}(x, y) = \frac{f^{k_1}(x, y)}{f^{k_1}(x, y) + h^{k_2}(x, y)}$$

define a parametric family of overlap functions.

Corollary 4 *In the same setting of Theorem 5, let us assume that G_O can be expressed in two different ways:*

$$G_O(x, y) = \frac{f_1(x, y)}{f_1(x, y) + h_1(x, y)} = \frac{f_2(x, y)}{f_2(x, y) + h_2(x, y)}$$

for any $x, y \in [0, 1]$ *and let* M *be a bivariate continuous aggregation function that is homogeneous of order one. Then, if we define* $f(x, y) = M(f_1(x, y), f_2(x, y))$ *and* $h(x, y) = M(h_1(x, y), h_2(x, y))$ *it also holds that*

$$G_O(x, y) = \frac{f(x, y)}{f(x, y) + h(x, y)} .$$

3.1 Overlap Functions and t-Norms

Overlap functions do not require associativity in their definition. If the associativity property is required, we recover a t-norm.

Theorem 6 *Let* G_O *be and associative overlap function. Then* G_O *is a t-norm.*

Besides, Theorem 2 of classification of t-norms can be extended to cover those t-norms which are also overlap functions as follows.

Theorem 7 *If a t-norm* T *is an overlap function, then* T *belongs to one of the following three types:*

(1) $T = T_M$;
(2) T *is strict;*
(3) T *is the ordinal sum of the family* $\{([a_m, b_m], T_m)\}$, *with all the* T_m *continuous Archimedean and such that if for some* m_0 $a_{m_0} = 0$, *then necessarily* T_{m_0} *is a strict t-norm.*

Example 2 1. In the construction of the following overlap function we use item (3) of Theorem 7 taking as t-norm the product, (which is strict, continuous and Archimedean), for the corresponding interval [0, 0.5].

$$G_O(x, y) = \begin{cases} 2xy \text{ if } (x, y) \in [0, 0.5]^2 \\ \min(x, y) \text{ otherwise} \end{cases}$$

2. In the construction of the following overlap function we take the product and the Lukasiewicz t-norms (see p. 84 of [23]). Nevertheless, in this overlap function we do not consider any interval of the type $[0, b_m]$.

$$G_O(x, y) = \begin{cases} 0.1 + 2.5(x - 0.1)(y - 0.1) \text{ if } (x, y) \in [0.1, 0.5]^2 \\ 0.7 + max(x + y - 1.6, 0) \text{ if } (x, y) \in [0.7, 0.9]^2 \\ \min(x, y) \text{ otherwise} \end{cases}$$

3. The following t-norm satisfies all the properties required to overlap functions, except (G_O2). This is due to the fact that in $[0, 0.25]^2$ we consider the Lukasiewicz t-norm which is continuous and Archimedean but not strict.

$$T(x, y) = \begin{cases} max(x + y - 0.25, 0) \text{ if } (x, y) \in [0, 0.25]^2 \\ \min(x, y) \text{ otherwise} \end{cases}$$

3.2 Overlap Functions and Semicopulas, Quasi-copulas and Copulas

As a first result we have the following.

Proposition 1 *Let S be a symmetric semicopula. Then S is an overlap function if and only if S is continuous and has not zero divisors.*

Corollary 5 *Let Q be a symmetric quasicopula without zero divisors. Then Q is also an overlap function.*

It is also possible to recover copulas from overlap functions.

Theorem 8 *Let G_O be an overlap function being homogeneous of order $k+1$, with $k \in [0, 1]$. Suppose that there exists $e \in [0, 1]$ such that $G_O(x, e) = G_O(e, x) = x$ for all $x \in [0, 1]$, that is, that G_O has a neutral element e. Then G_O is also a copula.*

Notice that the family $(\min(x^k y, xy^k)$ for $k \in [0, 1]$ is the so called Cuadras-Augé family of copulas [27].

4 Additive Generators of Overlap Functions

As we have already discussed, there exists a close relation between overlap functions and other relevant classes of operators such as t-norms. Moreover, in the same way as it is done for the latter, it is possible to study the use additive generators for overlap functions [13].

We start recalling the notion of pseudo-inverse of a given function.

Definition 11 [31] Let $f : [a, b] \to [c, d]$ be an increasing or decreasing function.[1] function $f^{(-1)} : [c, d] \to [a, b]$ defined by

$$f^{(-1)}(y) = \begin{cases} \sup\{x \in [a, b] \mid f(x) < y\} & \text{if } f(a) < f(b), \\ \sup\{x \in [a, b] \mid f(x) > y\} & \text{if } f(a) > f(b), \\ a & \text{if } f(a) = f(b) \end{cases} \quad (2)$$

is called the *pseudo-inverse* of f.

Let's denote the *range* or *image* of a function $f : A \to B$ by $Ran(f)$. Note that, if a function $f : [a, b] \to [c, d]$ is increasing (decreasing) then $f^{(-1)}$ is also increasing (decreasing). If f is strictly increasing (decreasing) then $f^{(-1)}$ is continuous, $f^{(-1)} \circ f = Id_{[a,b]}$ and $f \circ f^{(-1)}(x) = x$ if and only if $x \in Ran(f)$.

Our approach here to additive generators for overlap functions follows is based on the work by Viceník [30, 31]. We start with some auxiliary lemmas.

[1] In this paper, an increasing (decreasing) function does not need to be strictly increasing (decreasing).

Lemma 1 *Let* $\theta : [0, 1] \to [0, \infty]$ *be a decreasing function such that*

1. $\theta(x) + \theta(y) \in Ran(\theta)$, *for* $x, y \in [0, 1]$ *and*
2. *if* $\theta(x) = \theta(0)$ *then* $x = 0$.

Then $\theta(x) + \theta(y) \geq \theta(0)$ *if and only if* $x = 0$ *or* $y = 0$.

Lemma 2 *Consider functions* $\theta : [0, 1] \to [0, \infty]$ *and* $\vartheta : [0, \infty] \to [0, 1]$ *such that, for each* $x_0 \in [0, 1]$, *if it holds that*

$$\vartheta(\theta(x)) = x_0 \text{ if and only if } x = x_0, \tag{3}$$

then $\theta(x) = \theta(x_0)$ *if and only if* $x = x_0$.

Then we arrive at the main result regarding additive generators of overlap functions.

Theorem 9 *Let* $\theta : [0, 1] \to [0, \infty]$ *and* $\vartheta : [0, \infty] \to [0, 1]$ *be continuous and decreasing functions such that*

1. $\theta(x) + \theta(y) \in Ran(\theta)$, *for* $x, y \in [0, 1]$ *;*
2. $\vartheta(\theta(x)) = 0$ *if and only* $x = 0$*;*
3. $\vartheta(\theta(x)) = 1$ *if and only* $x = 1$*;*
4. $\theta(x) + \theta(y) = \theta(1)$ *if and only* $x = 1$ *and* $y = 1$.

Then, the function $O_{\theta,\vartheta} : [0, 1]^2 \to [0, 1]$, *defined by*

$$O_{\theta,\vartheta}(x, y) = \vartheta(\theta(x) + \theta(y)), \tag{4}$$

is an overlap function.

Corollary 6 *Let* $\theta : [0, 1] \to [0, \infty]$ *and* $\vartheta : [0, \infty] \to [0, 1]$ *be continuous and decreasing functions such that*

1. $\theta(x) = \infty$ *if and only if* $x = 0$*;*
2. $\theta(x) = 0$ *if and only if* $x = 1$*;*
3. $\vartheta(x) = 1$ *if and only if* $x = 0$*;*
4. $\vartheta(x) = 0$ *if and only if* $x = \infty$.

Then, the function $O_{\theta,\vartheta} : [0, 1]^2 \to [0, 1]$, *defined by*

$$O_{\theta,\vartheta}(x, y) = \vartheta(\theta(x) + \theta(y)), \tag{5}$$

is an overlap function.

Proposition 2 *Let* $\theta : [0, 1] \to [0, \infty]$ *and* $\vartheta : [0, \infty] \to [0, 1]$ *be continuous and decreasing functions such that*

1. $\vartheta(x) = 1$ *if and only if* $x = 0$;
2. $\vartheta(x) = 0$ *if and only if* $x = \infty$;
3. $0 \in Ran(\theta)$;
4. $O_{\theta,\vartheta}(x, y) = \vartheta(\theta(x) + \theta(y))$ *is an overlap function.*
   Then, the following conditions also hold:
5. $\theta(x) = \infty$ *if and only if* $x = 0$;
6. $\theta(x) = 0$ *if and only if* $x = 1$;

(θ, ϑ) is called an *additive generator pair* of the overlap function $O_{\theta,\vartheta}$, and $O_{\theta,\vartheta}$ is said to be additively generated by the pair (θ, ϑ).

Example 3 Consider the functions $\theta : [0, 1] \to [0, \infty]$ and $\vartheta : [0, \infty] \to [0, 1]$, defined, respectively by:

$$\theta(x) = \begin{cases} -2\ln x & \text{if } x \neq 0 \\ \infty & \text{if } x = 0 \end{cases}$$

and

$$\vartheta(x) = \begin{cases} e^{-x} & \text{if } x \neq \infty \\ 0 & \text{if } x = \infty, \end{cases}$$

which are continuous and decreasing functions, satisfying the conditions 1–4 of Corollary 6. Then, whenever $x \neq 0$ and $y \neq 0$, one has that:

$$O_{\theta,\vartheta}(x, y) = \vartheta(\theta(x) + \theta(y)) = e^{-(-2\ln x - 2\ln y)} = e^{\ln x^2 y^2} = x^2 y^2.$$

Otherwise, if $x = 0$, it holds that

$$O_{\theta,\vartheta}(0, y) = \vartheta(\theta(0) + \theta(y)) = \vartheta(\infty + \theta(y)) = 0,$$

and, similarly, if $y = 0$, then $O_{\theta,\vartheta}(x, 0) = 0$. It follows that

$$O_{\theta,\vartheta}(x, y) = x^2 y^2,$$

and so we recover ae non associative overlap function for which 1 is not a neutral element.

Corollary 7 *Considering the same conditions as in Theorem 9, whenever* $\vartheta = \theta^{(-1)}$ *then* $O_{\theta,\vartheta}$ *is a positive t-norm (i.e., without divisors of zero).*

Theorem 10 *Let* $G_O : [0, 1]^2 \to [0, 1]$ *be an overlap function having 1 as neutral element. Then, if* G_O *is additively generated by a pair* (θ, ϑ)*, with* $\theta : [0, 1] \to [0, \infty]$ *and* $\vartheta : [0, \infty] \to [0, 1]$ *satisfying the conditions of Theorem 9, then* G_O *is associative.*

The following result is straight:

Corollary 8 *Let $G_O : [0,1]^2 \to [0,1]$ be an overlap function additively generated by a pair (θ, ϑ). G_O is a t-norm if and only if 1 is a neutral element of G_O.*

Notice that whenever T is a positive continuous t-norm (that is, an overlap function) that is additively generated by a function $t : [0,1] \to [0,\infty]$, then it is also additively generated by a pair (θ, ϑ) in the sense of Theorem 9, where $\theta = t$ and $\vartheta = t^{(-1)}$, and vice-versa.

A more detailed study on additive generators for overlap functions can be found in [13].

Now let's recall the definition of pseudo-automorphism.

Definition 12 A function $\mathscr{F} : [0,1] \to [0,1]$ is said to be a pseudo-automorphism if the following conditions hold:

(PA1) $\mathscr{F}$ is increasing;
(PA2) $\mathscr{F}$ is continuous;
(PA3) $\mathscr{F}(x) = 1$ if and only if $x = 1$;
(PA3) $\mathscr{F}(x) = 0$ if and only if $x = 0$.

An automorphism $\varphi : [0,1] \to [0,1]$ is a strictly increasing pseudo-automorphism.

Then we have the following important result

Theorem 11 *Let $\varphi : [0,1] \to [0,1]$ be an automorphism and $T : [0,1]^2 \to [0,1]$ be a t-norm. The function $O_{\varphi,T} : [0,1]^2 \to [0,1]$, defined by*

$$O_{\varphi,T}(x, y) = \varphi(T(x, y)), \tag{6}$$

is an overlap function if and only if T is positive and continuous.

$O_{\varphi,T}$ is called the overlap obtained by the distortion of the t-norm T by the automorphism φ, or the overlap function obtained by a (φ, T)-distortion.

However, it is important to remark that not every overlap is obtained as a distortion of a t-norm.

Proposition 3 *There exists an overlap function $G_O : [0,1]^2 \to [0,1]$ which is not a $(\mathscr{F}, T)$-distortion of a positive and continuous t-norm $T : [0,1]^2 \to [0,1]$ for any automorphism $\mathscr{F}$.*

In the same way, it is also true that not every overlap function $G_O : [0,1]^2 \to [0,1]$ can be obtained by means of a $(\mathscr{F}, T)$-distortion, for some pseudo-automorphism $\mathscr{F} : [0,1] \to [0,1]$ and some positive and continuous t-norm $T : [0,1]^2 \to [0,1]$.

5 Overlap Functions and Grouping Functions

The notion of grouping function [10] arises in the applied field when it is necessary to measure up to what extent a given element belongs to at least one of two given classes. The formal definition reads as follows.

Definition 13 A mapping $G_G : [0, 1]^2 \to [0, 1]$ is a grouping function if it satisfies the following conditions:

(G_G1) G_G is symmetric;
(G_G2) $G_G(x, y) = 0$ if and only if $x = y = 0$;
(G_G3) $G_G(x, y) = 1$ if and only if $x = 1$ or $y = 1$;
(G_G4) G_G is non-decreasing;
(G_G5) G_G is continuous.

There exists a close link between overlap and grouping functions. In fact, both concepts are dual to each other.

Definition 14 [22] Let G be an overlap function (resp. a grouping function) and let n_1 and n_2 be two continuous negation operators such that:

1. $n_1(x) = 0$ (resp. $n_2(x) = 0$) if and only if $x = 1$, and
2. $n_1(x) = 1$ (resp. $n_2(x) = 1$) if and only if $x = 0$

Then, the operator
$$\overline{G}^{n_1,n_2}(x, y) = n_1(G(n_2(x), n_2(y)))$$

is called the dual grouping (resp. overlap) of G with respect to n_1 and n_2.

In fact, every grouping function can be obtained as the dual through a strong negation of an overlap function. Besides, note that this duality implies that many of the results we have discussed for overlap functions can be translated straightforwardly for grouping functions just making use of duality. So, in some sense, it seems that overlap and grouping functions are related in the same way as t-norms are t-conorms are. However, the drop of associativity in the definition of both overlap and grouping functions gives raise to dramatic differences from the case of t-norms and t-conorms. In particular, it does not hold that overlap functions are always smaller than or equal to grouping functions.

Example 4 Consider the overlap function given by $G_O(x, y) = \min(\sqrt{x}, \sqrt{y})$ and the grouping function given by $G_G(x, y) = \max(x^2, y^2)$. Then:

$$G_O(0.5, 0.25) = \min(\sqrt{0.5}, \sqrt{0.25}) > \max((0.5)^2, (0.25)^2) = G_G(0.5, 0.25)$$

So in this case $G_O \nleq G_G$.

The problem of when an overlap function is smaller than or equal to a grouping function is very relevant from the point of view of applications and was deeply analyzed in [25], and we now summarize the results from that paper.

The first important concept at this regard is that of f-bounding.

Definition 15 Let f be a mapping from [0, 1] to [0, 1]. An overlap function G_O (resp. a grouping function G_G) is called f-bound if the equality $G_O(x, 1) = f(x)$ (resp. $G_G(x, 0) = f(x)$) holds for all $x \in [0, 1]$

We denote by $\mathscr{O}_f$ ($\mathscr{G}_f$) the class of f-bound overlap functions (grouping functions). Note that the class of overlap functions (resp.grouping functions) can be obtained as the union of these classes $\mathscr{O}_f$ (resp. $\mathscr{G}_f$).

Then the following result is straight.

Proposition 4 *Let G be either an f-bound overlap or an f-bound grouping. Then:*

1. *f is continuous,*
2. *f is increasing,*
3. *$f(x) = 0$ if and only if $x = 0$,*
4. *$f(x) = 1$ if and only if $x = 1$.*

Let Ω denote the set of mappings satisfying the four properties given in Proposition 4. Then

Proposition 5 *Let G_O and G_G be an overlap and a grouping function, respectively, and $f \in \Omega$. Then:*

1. *if $G_O \in \mathscr{O}_f$, then $G_O(x, y) \leq \min\{f(x), f(y)\}$ for all $x, y \in [0, 1]$.*
2. *if $G_G \in \mathscr{G}_f$, then $G_G(x, y) \geq \max\{f(x), f(y)\})$ for all $x, y \in [0, 1]$.*

The relevance of the notion of f-bounding is clear in the following result.

Proposition 6 *Let G_O and G_G be an f_1-overlap function and an f_2-grouping function, respectively, with $f_1, f_2 \in \Omega$. If $f_1 \leq f_2$, then $G_O(x, y) \leq G_G(x, y)$ for all $x, y \in [0, 1]$.*

Corollary 9 *Let G_O be an overlap function and G_G a grouping function. If both, G_O and G_G, are f-bound then $G_O \leq G_G$.*

However, Proposition 6 provides a necessary but not sufficient condition to ensure that a given overlap function is less than or equal to a given grouping function. Nevertheless, the following result holds.

Proposition 7 *Let G_O be an overlap function (resp. let G_G be a grouping function), then for each mapping $f \in \Omega$ there exists $G_G \in \mathscr{G}_f$ (resp. $G_O \in \mathscr{O}_f$) such that $G_O \leq G_G$.*

Besides, the choice of a bounding function imposes some restrictions to the size of the corresponding overlap and grouping functions.

Proposition 8 *Let $f \in \Omega$, then*

1. *the operator $G_O(x, y) = f(\min(x, y))$ is the greatest f-bound overlap and*
2. *the operator $G_G(x, y) = f(\max(x, y))$ is the least f-bound grouping.*

It is relevant to note that there is neither least element in $\mathcal{O}_f$ nor greatest element in $\mathcal{G}_f$.

Proposition 9 *Let $f \in \Omega$, then the infimum of $\mathcal{O}_f$ is:*

$$\bigwedge_{G_O \in \mathcal{O}_f} G_O(x, y) = \begin{cases} f(x) & \text{if } y = 1 \\ f(y) & \text{if } x = 1 \\ 0 & \text{Otherwise} \end{cases}$$

and the supremum of $\mathcal{G}_f$ is:

$$\bigvee_{G_G \in \mathcal{G}_f} G_G(x, y) = \begin{cases} f(x) & \text{if } y = 0 \\ f(y) & \text{if } x = 0 \\ 1 & \text{Otherwise} \end{cases}$$

Corollary 10 *Let $f \in \Omega$. Then the set $\mathcal{O}_f$ (resp. $\mathcal{G}_f$) does not have a least element (resp. a greatest element).*

Another interesting property of $\mathcal{O}_f$ and $\mathcal{G}_f$ is that they are dense as posets.

Proposition 10 *Let G_1 and G_2 be two overlaps in $\mathcal{O}_f$ (resp. two groupings in $\mathcal{G}_f$) such that $G_1 < G_2$. Then there exists $G \in \mathcal{O}_f$ (resp. $G \in \mathcal{G}_f$) such that $G_1 < G < G_2$.*

The following definition establishes a restriction on negations used in the dual construction in order to maintain the f-bound condition.

Let's recall now the notion of f-duality.

Definition 16 Let G be an overlap (resp. grouping), $f \in \Omega$ an automorphism on the unit interval (i.e., a bijective mapping), n a bijective negation[2] and let us consider the negations:

- $n_1(x) = f(n(x))$ and
- $n_2(x) = f^{-1}(n^{-1}(x))$,

Then the operator

$$\overline{G}^{n,f}(x, y) = n_1\big(G(n_2(x), n_2(y))\big)$$

is called the *f-dual grouping* (resp. *f-dual overlap*) of G with respect to n.

[2] This kind of negation is called in the Literature *strict*.

Proposition 11 *Let $f \in \Omega$, $G \in \mathcal{O}_f$ (resp. $G \in \mathcal{G}_f$) be a bijection and n a bijective negation, then the f-dual grouping function (resp. overlap function) of G w.r.t. n is f-bound.*

From Proposition 11, we can ensure that any f-bound overlap functions is less than or equal to any of its f-dual grouping functions.

Corollary 11 *Let $f \in \Omega$, $G_O \in \mathcal{O}_f$ and let $\overline{G_O}^{n,f}$ be an f-dual grouping function of G_O. Then the inequality $G_O \leq \overline{G_O}^{n,f}$ holds.*

And reciprocally:

Corollary 12 *Let $f \in \Omega$, $G_G \in \mathcal{G}_f$ and let $\overline{G_G}^{n,f}$ be an f-dual overlap function of G_G. Then the inequality $\overline{G_G}^{n,f} \leq G_G$ holds.*

In [25] this study is made deeper by considering also the so-called f-diagonal overlap and grouping functions, that is, overlap and grouping functions whose values at the diagonal are given by a fixed function f.

6 n-Dimensional Overlap and Grouping Functions

Since overlap functions are not assumed to be associative, its extensions to the n-dimensional case with $n > 2$ requires of a specific definition. From now on, we follow the developments in [18].

Definition 17 An n-dimensional aggregation function $G_O : [0,1]^n \longrightarrow [0,1]$ is an n-dimensional overlap function if and only if:

1. G_O is symmetric.
2. $G_O(x_1, \ldots, x_n) = 0$ if and only if $\prod_{i=1}^{n} x_i = 0$.
3. $G_O(x_1, \ldots, x_n) = 1$ if and only if $x_i = 1$ for all $i \in \{1, \ldots, n\}$.
4. G_O is increasing.
5. G_O is continuous.

Note that, taking into account this definition, an object c that belongs to three classes C_1, C_2 and C_3 with degrees $x_1 = 1$, $x_2 = 1$ and $x_3 = 0.3$ will not have the maximum degree of overlap since condition (3) of the previous definition is not satisfied. Even more, if the degrees are $x_1 = 1$, $x_2 = 1$ and $x_3 = 0$, from the second condition we will conclude that the n-dimensional degree of overlapping of this object into the classification system given by the classes C_1, C_2 and C_3 will be zero. This is the reason why this first extension of the original idea of overlap proposed in [9] has been called *n-dimensional* overlap. Let us observe that this definition is closely related with the idea of intersection of n classes.

Example 5 It is easy to see that the following aggregation functions are n-dimensional overlap functions:

1. The minimum powered by p. $G_O(x_1, \ldots, x_n) = \min_{1 \le i \le n} \{x_i^p\} = \left[\min_{1 \le i \le n} \{x_i\}\right]^p$ with $p > 0$.
2. The geometric mean. $G_O(x_1, \ldots, x_n) = (\prod_{i=1}^{n} x_i)^{\frac{1}{n}}$.
3. The Einstein product aggregation operator. $EP(x_1, \ldots, x_n) = \dfrac{\prod_{i=1}^{n} x_i}{1 + \prod_{i=1}^{n}(1 - x_i)}$
4. The sinus induced overlap $G_O(x_1, \ldots, x_n) = \sin \frac{\pi}{2}(\prod_{i=1}^{n} x_i)^p$ with $p > 0$.

The characterization results already introduced for bivariate overlap functions may be extended in a straight way for n-dimensional overlap functions. In particular, we remark the following result.

Proposition 12 *Let $A_n : [0, 1]^n \longrightarrow [0, 1]$ be an aggregation function. If A_n is averaging, then A_n is an n-dimensional overlap function if and only if it is symmetric, continuous, has zero as absorbing element and satisfies $A_n(x, 1, \ldots, 1) \neq 1$ for any $x \neq 1$.*

In fact, We have the following theorem.

Theorem 12 *Let $G_1, \ldots, G_m$ be n-dimensional overlap functions and let $M : [0, 1]^m \longrightarrow [0, 1]$ be a continuous and symmetric aggregation function such that if $M(x) = 0$ then $x_i = 0$ for some i and $M(x) = 1$ only if $x_i = 1$ for some i. Then the aggregation function $G : [0, 1]^n \longrightarrow [0, 1]$ defined as $G(x) = M(G_1(x), \ldots, G_m(x))$ is an n-dimensional overlap function.*

Remark 1 Notice that, since any averaging aggregation function M being symmetric and continuous satisfies the conditions of the previous Theorem, it is possible to conclude that any continuous symmetric averaging aggregation of n-dimensional overlap functions is also an n-dimensional overlap function.

As an illustrative case, consider that of OWA operators.

Proposition 13 *Let $W = (w_1, \ldots, w_n) \in [0, 1]^n$ be a weighting vector. The following statements are equivalent:*

1. *The OWA operator defined by the weighting vector W is an n-dimensional overlap function.*
2. $w_n = 1$

Also for the case of Kolmogorov-Nagumo means the following result holds.

Proposition 14 *Let $f : [0, 1] \to [-\infty, 0]$ be a continuous increasing bijection, that is, $f :]0, 1] \to]-\infty, 0]$ is an increasing bijection such that $\lim_{x \to 0} f(x) = -\infty$, so by abuse of notation we define $f(0) = -\infty$. Then the function: $G_O(x_1, \ldots, x_n) = f^{-1}(f(x_1) + \cdots + f(x_n))$ is an n-dimensional overlap function.*

This corresponds to the n-ary extension of a strict t-norm generated by an additive generator f. Note than an analogous extension may be done for grouping functions.

Definition 18 An n-dimensional function

$$G_G : [0, 1]^n \longrightarrow [0, 1]$$

is an n-dimensional grouping function if and only if it satisfies the following conditions:

1. G_G is symmetric.
2. $G_G(x) = 0$ if and only if $x_i = 0$, for all $i = 1, \ldots, n$.
3. $G_G(x) = 1$ if and only if there exist $i \in \{1, \ldots, n\}$ with $x_i = 1$.
4. G_G is non-decreasing.
5. G_G is continuous.

Again, some particular continuous t-conorms (their n-ary forms) and their convex combinations are prototypical examples of n-ary grouping functions.

Example 6 The following aggregation functions are examples of n-dimensional grouping functions:

- The maximum powered by p. $G_G(x_1, \ldots, x_n) = \max\limits_{1 \le i \le n} \{x_i^p\}$ with $p > 0$.
- The Einstein sum aggregation operator. $ES(x_1, \ldots, x_n) = \dfrac{\sum_{i=1}^n x_i}{1 + \prod_{i=1}^n x_i}$

Theorem 13 *Let G_O be an n-dimensional overlap function and let n_1 and n_2 be two continuous negation operators such that:*

1. *$n_1(x) = 0$ (resp. $n_2(x) = 0$) if and only if $x = 1$, and*
2. *$n_1(x) = 1$ (resp. $n_2(x) = 1$) if and only if $x = 0$*

Then the function $G : [0, 1]^n \longrightarrow [0, 1]$ defined as

$$G(x_1, \ldots, x_n) = n_1(G_O(n_2(x_1), \ldots, n_2(x_n))$$

is an n-dimensional grouping aggregation function.

On the other hand, it is possible to build an n-dimensional overlap function from a grouping aggregation function and a negation function using duality, as in the case of bivariate overlap and grouping functions.

Theorem 14 *Let G_G be an n-dimensional grouping function and let n_1 and n_2 be two continuous negation operators such that:*

1. $n_1(x) = 0$ *(resp.* $n_2(x) = 0$*) if and only if* $x = 1$*, and*
2. $n_1(x) = 1$ *(resp.* $n_2(x) = 1$*) if and only if* $x = 0$.

Then the function $G : [0, 1]^n \longrightarrow [0, 1]$ *defined as*

$$G(x_1, \ldots, x_n) = n_1(G_G(n_2(x_1), \ldots, n_2(x_n)))$$

is an n-dimensional overlap aggregation function.

Theorem 15 *Let* $G_1, \ldots G_m$ *be n-dimensional grouping functions and let* $M : [0, 1]^m \longrightarrow [0, 1]$ *be a continuous aggregation function such that* $M(x) = 0$ *if and only if* $x_i = 0$ *for some i and* $M(x) = 1$ *if and only if* $x_i = 1$ *for all i. Then the aggregation function* $G : [0, 1]^n \longrightarrow [0, 1]$ *defined as* $G(x) = M(G_1(x), \ldots, G_m(x))$ *is an n-dimensional grouping function.*

Corollary 13 *Let* $G_1, \ldots, G_m$ *be n-dimensional grouping functions and let* $w_1, \ldots, w_m$ *be nonnegative weights with* $\sum_{i=1}^{m} w_i = 1$*. Then the convex sum* $G(x) = \sum_{i=1}^{m} w_i G_i(x)$ *is also an n-dimensional grouping function.*

A more detailed analysis of n-dimensional overlap and grouping functions can be found in [18].

7 Conclusions

In this chapter we have made a review of the main definitions and results related to overlap and grouping functions. These functions have shown themselves very useful in applications where associativity is not a natural requirement. Although a detailed explanation of such applications would require a too large amount of space, it is worth to mention the use of overlap functions in decision making to define preference structures [10], in classification to get generalizations of fuzzy rule based algorithms which improve the results of those based on the use t-norms [28], or in image processing [22]. In all these cases it is worth to mention that classical algorithms may be improved with the use of these overlap and grouping functions.

Acknowledgments The authors have been supported by project TIN2013-40765-P.

References

1. Alsina, C., Frank, M.J., Schweizer, B.: Associative Functions. Triangular Norms and Copulas. World Scientific, Hackensack (2006)
2. Amo, A., Montero, J., Molina, E.: Representation of consistent recursive rules. Eur. J. Oper. Res. **130**, 29–53 (2001)
3. Amo, A., Montero, J., Biging, G., Cutello, V.: Fuzzy classification systems. Eur. J. Oper. Res. **156**, 459–507 (2004)
4. Bedregal, B., Dimuro, G.P., Bustince, H., Barrenechea, E.: New results on overlap and grouping functions. Inf. Sci. **249**, 148–170 (2013)
5. Bustince, H., Barrenechea, E., Pagola, M., et al.: Weak fuzzy S-subsethood measures. Overlap index. Int. J. Uncertainty Fuzziness Knowl.-Based Syst. **14**(5), 537–560 (2006)
6. Bustince, H., Mohedano, V., Barrenechea, E., et al.: Definition and construction of fuzzy DI-subsethood measures. Inf. Sci. **176**(21), 3190–3231 (2006)
7. Bustince, H., Pagola, M., Barrenechea, E.: Construction of fuzzy indices from fuzzy DI-subsethood measures: application to the global comparison of images. Inf. Sci. **177**(3), 906–929 (2007)
8. Bustince, H., Montero, J., Barrenechea, E., Pagola, M.: Semiautoduality in a restricted family of aggregation operators. Fuzzy Sets Syst. **158**(12), 1360–1377 (2007)
9. Bustince, H., Fernandez, J., Mesiar, R., Montero, J., Orduna, R.: Overlap functions. Nonlin. Anal.-Theory Meth. Appl. **72**, 1488–1499 (2010)
10. Bustince, H., Pagola, M., Mesiar, R., Hullermeier, E., Herrera, F., Montero, J., Orduna, R.: Grouping, overlap, and generalized bientropic functions for fuzzy modeling of pairwise comparisons. IEEE Trans. Fuzzy Syst. **20**, 405–415 (2012)
11. Calvo, T., Kolesárová, A., Komorníkova, M., Mesiar, R.: Aggregation operators: properties, classes and construction methods. In: Aggregation Operators New Trends and Applications. Physica-Verlag, Heidelberg (2002)
12. Cutello, V., Montero, J.: Recursive connective rules. Int. J. Intell. Syst. **14**, 3–20 (1999)
13. Dimuro, G.P., Bedregal, B., Bustince, H., Asiaín, M.J., Mesiar, R.: On additive generators of overlap functions. Fuzzy Sets Syst. (in press) doi:10.1016/j.fss.2015.02.008
14. Dubois, D., Koning, J.L.: Social choice axioms for fuzzy set aggregation. Fuzzy Sets Syst. **58**, 339–342 (1991)
15. Dubois, D., Ostasiewicz, W., Prade, H.: Fuzzy Sets: History and Basic Notions. In: Fundamentals of Fuzzy Sets. Kluwer, Boston (2000)
16. Fodor J., Roubens M., Fuzzy preference modelling and multicriteria decision support. In: Theory and Decision Library, Kluwer Academic Publishers (1994)
17. Gómez, D., Montero, J.: A discussion on aggregation operators. Kybernetika **40**, 107–120 (2004)
18. Gómez, D., Rodríguez, J.T., Montero, J., Bustince, H., Barrenechea, E.: n-dimensional overlap functions. Fuzzy Sets Syst. (in press). doi:10.1016/j.fss.2014.11.023
19. Hosseini, M.S., Eftekhari-Moghadam, A.M.: Fuzzy rule-based reasoning approach for event detection and annotation of broadcast soccer video. Appl. Soft Comput. **13**(2), 846–866 (2013)
20. Ishibuchi, H., Nakashima, T., Nii, M.: Classification and Modeling with Linguistic Information Granules: Advanced Approaches to Linguistic Data Mining. Springer, Berlin (2004)
21. Ishibuchi, H., Yamamoto, T.: Rule weight specification in fuzzy rule-based classification systems. IEEE Trans. Fuzzy Syst. **13**, 428–435 (2005)
22. Jurio, A., Bustince, H., Pagola, M., Pradera, A., Yager, R.: Some properties of overlap and grouping functions and their application to image thresholding. Fuzzy Sets Syst. **229**, 69–90 (2013)
23. Klement, E.P., Mesiar, R., Pap, E.: Triangular Norms, Trends in Logic, Studia Logica Library, vol. 8. Kluwer Academic Publishers, Dordrecht (2000)
24. Klir, G.J., Folger, T.A.: Fuzzy Sets. Uncertainty and Information, Prentice Hall, Englewood Cliffs (1988)

25. Madrid, N., Burusco, A., Bustince, H., Fernandez, J., Perfilieva, I.: Upper bounding overlaps by groupings. Fuzzy Sets Syst. **264**, 76–99 (2015)
26. Montero, J., Gomez, D., Bustince, H.: On the relevance of some families of fuzzy sets. Fuzzy Sets Syst. **158**, 2429–2442 (2007)
27. Nelsen, R.B.: An introduction to Copulas. Lecture Notes in Statistics, vol. 139. Springer, New York (1999)
28. Sanz, J., Galar, M., Jurio, A., Brugos, A., Pagola, M., Bustince, H.: Medical diagnosis of cardiovascular diseases using an interval-valued fuzzy rule-based classification system. Appl. Soft Comput. J. **20**, 103–111 (2014)
29. Schweizer, B., Sklar, A.: Probabilistic Metric Spaces. North-Holland, Amsterdam (1983)
30. Viceník, P.: Additive generators of non-continuous triangular norms, Topological an Algebraic Structures in Fuzzy Sets. Kluwer (2003)
31. Viceník, P.: Additive generators of associative functions. Fuzzy Sets Syst. **153**, 137–160 (2005)
32. Zadeh, L.A.: Fuzzy sets as a basis for a theory of possibility. Fuzzy Sets Syst. **1**, 3–28 (1978)

Asymmetric Copulas and Their Application in Design of Experiments

Fabrizio Durante and Elisa Perrone

Abstract We present an overview on definitions and properties of asymmetric copulas, i.e. copulas whose values are not invariant under any permutation of their arguments. In particular, we review an axiomatic approach in the definition of a measure of asymmetry (non–exchangeability) for copulas, starting with the seminal contributions by Klement and Mesiar [45] and Nelsen [56]. Then we discuss how asymmetric copulas may be useful also in the optimal design of experiments and how they may provide additional insights into these problems.

1 Introduction

The problem of describing relationships among random variables has attracted a lot of attention during the years, especially for inference and prediction. Nowadays, it can benefit of the use of *copulas*, which provide a convenient tool to construct and estimate a multivariate stochastic model. In fact, following a copula approach, we may determine the joint probability law of a random vector $\mathbf{X} = (X_1, \dots, X_d)$ in two steps: first, we fix the marginal behavior of each component X_i, then we describe the relationships among the X_i's by means of a suitable copula, which is a multivariate distribution function whose univariate marginals are uniformly distributed on $[0, 1]$. For more details about copulas, see, for instance, [28, 41, 55, 62].

In order to determine possible models for a random vector $\mathbf{X}$, it is then important to have at disposal a variety of copulas that may cover different situations that arise in practice. To this end, several investigations in the literature considered the problem of determining families of copulas and studying their properties with particular emphasis on their range of association (as measured by Spearman's correlation coef-

F. Durante
Faculty of Economics and Management, Free University of Bozen–Bolzano, Bolzano, Italy
e-mail: fabrizio.durante@unibz.it

E. Perrone (✉)
Institute for Applied Statistics, Johannes Kepler University Linz, Linz, Austria
e-mail: elisa.perrone@jku.at

S. Saminger-Platz and R. Mesiar (eds.), *On Logical, Algebraic, and Probabilistic Aspects of Fuzzy Set Theory*, Studies in Fuzziness and Soft Computing 336,
DOI 10.1007/978-3-319-28808-6_9

ficient and Kendall's tau, for instance), tail dependence and so on. For an overview on different constructions, we refer the reader to [41, 55].

Interestingly, many of these constructions overlap with families of *triangular norms* (see, for instance, [46, 62]). In fact, in dimension $d = 2$, copulas can be seen as binary operations on $[0, 1]$ that are supermodular and have neutral element 1. In particular, when they are associative and commutative, they form an interesting class of triangular norms, known as *Archimedean copulas* (see, for instance, [33, 53]).

Archimedean copulas have become quite popular in applications because of several interesting properties that make them tractable especially for inferential purposes (see, for instance, [35]). However, their main practical limitation is that they are *exchangeable*, i.e. the values assumed by the copula does not change under any permutation of its arguments.

To get more flexibility, a popular approach is to consider (multivariate) models that exploit bivariate Archimedean (or, generally, exchangeable) copulas for determining the marginal or conditional dependence and, hence, combining such pieces into an high–dimension framework. Approaches of such type include vine–pair copula constructions [7], factor copula models [20, 49, 52] or nested Archimedean copulas.

Another possibility is to find methods for providing constructions that, starting with a given (exchangeable) copula C, produce another copula C' (usually, with an additional number of parameters) that need not be exchangeable. See, for instance, Khoudraji's device [31, 43] and related extensions [11, 50, 51].

It should also be stressed that some classes of (semi–)parametric bivariate copulas can explicitly handle the non–exchangeable case. Consider, for instance, extreme–value and Archimax copulas [5, 6, 47], Liouville copulas [54], copulas that are invariant under (univariate) truncation [17, 18], asymmetric semilinear copulas [8] among others.

Methods and tools for coping with non–exchangeable (asymmetric) copulas have become quite popular in the recent years; see, for instance, the excellent overview provided in [36]. In this contribution, we shortly review some of the recent literature on this topic by emphasizing earlier contributions provided in [45, 56], which have suggested possible ways to quantify non–exchangeability in copula models. Then, we consider a novel application of (asymmetric) copulas in optimal designs of experiments and show how their use may provide a convenient tool in this framework too.

2 A Glimpse of Non–exchangeable Copulas

We start by recalling the basic definition of exchangeability in dimension 2.

Definition 1 A (bivariate) copula C is said to be *exchangeable* (or *symmetric*) if $C(u, v) = C(v, u)$ for all $(u, v) \in [0, 1]^2$.

In particular, if (U, V) is a random pair distributed according to an exchangeable copula C, then

$$\mathbb{P}(V \leq v \mid U \leq u) = \mathbb{P}(U \leq v \mid V \leq u).$$

Thus, the conditional distribution of $(V \mid U \leq u)$ is equal to the conditional distribution of $(U \mid V \leq u)$. In particular, it implies that non–exchangeability occurs when there is a causality relationship between U and V.

Below, we illustrate several practical examples where non–exchangeability may appear.

- In the study of financial time series, it is often the case that high losses in one (large) market often imply high losses in other (smaller) markets, while the opposite direction rarely appears. This phenomenon can be interpreted in terms of financial contagion that should be captured with copulas that should describe this asymmetric behavior. See, for instance, [15, 16, 40].
- In reliability theory, the failure of a system can depend (at least) on two characteristics: the usage history and the age. As such, the warranty policy for certain types of products specifies the limits of coverage in terms of both age and usage. However, it is often the case that, in warranty claim data analysis, the dependence between these two features is related to a non–exchangeable copula, as addressed for instance in [68].
- In environmental applications, consider, for instance, a spatial analysis of a river basin where data are collected at different gauge stations. Then, due to the physical and geographic structure of the river networks (for instance, one river can be a tributary of another one), non–exchangeable dependencies may occur at various levels. For more details, see for instance [1, 36].

Remark 1 In fuzzy logic (in a broad sense), instead, the adoption of a symmetric function (e.g., triangular norm) as conjunctive operation is sometimes considered quite strong. Therefore, several investigations have stressed the importance of introducing also asymmetric (i.e. non-commutative) conjunctions [3, 19, 30, 37, 42].

Graphically, exchangeability can be interpreted in the sense that the level sets associated with an exchangeable copula C are symmetric with respect to the main diagonal of the unit square (see Fig. 1). Analogously, the 3D plot of an exchangeable copula (which is a surface) is symmetric with respect to the vertical plane passing through the line $\{x = y\}$.

These graphical interpretations also suggest possible ways to transform an exchangeable copula into another copula that does not share this property. Examples are illustrated below.

Example 1 Let C be an exchangeable copula that is different from the Fréchet-Hoeffding upper bound copula $M_2(u, v) = \min\{u, v\}$. Then C can be modified to a non–exchangeable copula by means of a suitable patchwork construction [12, 13, 26]. Without loss of generality, let R be a rectangle contained in $\{(u, v) \in [0, 1]^2 : u \geq v\}$ such that $V_C(R) > 0$. Then, one can construct another copula C' whose induced

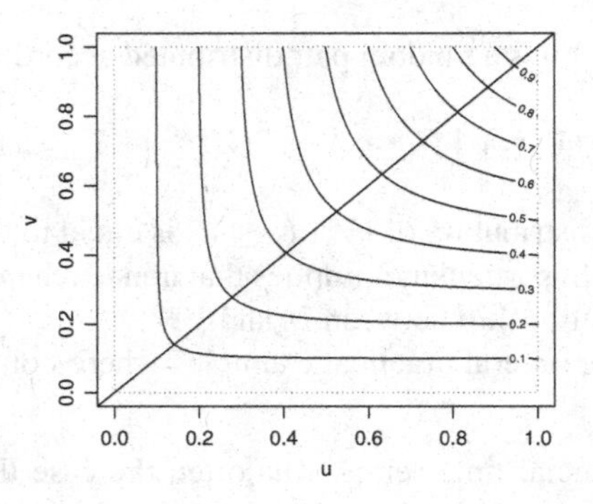

Fig. 1 Level curves of a Clayton copula with parameter $\alpha = 2$

measure $\mu_{C'}$ coincides with μ_C on the Borel sets of $[0, 1]^2 \setminus R$, while the probability mass is spread in a different way on R (see Fig. 2). The copula C' obtained in this way is, in general, non–exchangeable.

For instance, consider the independence copula $\Pi_2(u, v) = uv$ and let $a \in [0, \frac{1}{2}]$. Consider the rectangular patchwork

$$\widetilde{C} = (\langle [0, a] \times [1 - a, 1], C_1 \rangle)^{\Pi},$$

where $C_1 \neq \Pi$. Then, $\widetilde{C}$ is not exchangeable and its expression is given by:

$$\widetilde{C}(u, v) = \begin{cases} a^2 C_1\left(\dfrac{u}{a}, \dfrac{v - \overline{a}}{a}\right) + \overline{a}u, & (u, v) \in [0, a] \times [1 - a, 1], \\ uv, & \text{otherwise}, \end{cases}$$

where $\overline{a} = 1 - a$.

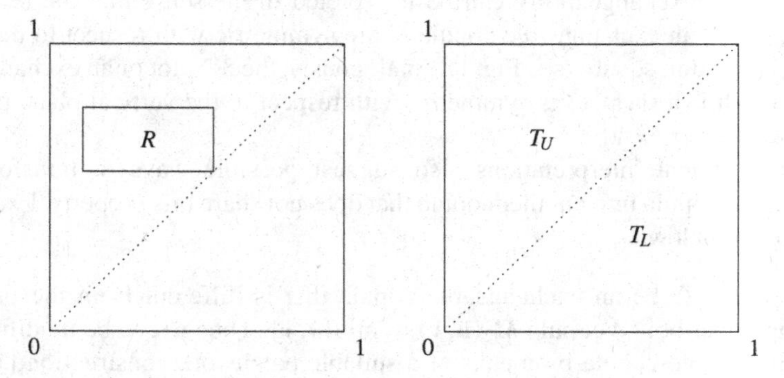

Fig. 2 Graphical illustration of a rectangular (*left*) and diagonal (*right*) patchwork construction

Example 2 Let C be an exchangeable copula, $C \neq M_2$. For every $t \in [0, 1]$, let $\delta_C(t) := C(t, t)$ be the diagonal section of C. In view of the results in [12], there exists another exchangeable copula $C_1 \neq C$ whose diagonal section coincides with C. Moreover, as a consequence of [22, Theorem 1], the following non–exchangeable copula can be given:

$$\widetilde{C}(u, v) = \begin{cases} C(u, v), & (u, v) \in T_L, \\ C_1(u, v), & (u, v) \in T_U, \end{cases}$$

where $T_L = \{(u, v) \in [0, 1]^2 \colon u \geq v\}$, $T_U = [0, 1]^2 \setminus T_L$. This type of construction is known as diagonal patchwork of two copulas (see also [25, 57]).

Remark 2 In previous examples, we start with a copula $C \neq M_2$, However, if $C = M_2$ it can be easily transformed into a non–exchangeable copula by means of a push–forward of its induced measure under a suitable measure–preserving transformation. For more details, see [27, 67].

Example 1 can be also used to check that, for every exchangeable copula $C \neq M_2$, a non–exchangeable copula $\widetilde{C}$ exists that is sufficiently close to C (in the L^∞ norm). Instead, given a non–exchangeable copula, there is no general way to approximate it via an exchangeable one.

At a more abstract level, the sets of exchangeable and non–exchangeable copulas are quite different, as the following general results hold:

- Exchangeable copulas form a closed set in the class $\mathscr{C}$ of all copulas endowed with the L^∞ norm, i.e. a sequence of exchangeable copulas uniformly converges to an exchangeable copula. In particular, exchangeable copulas are not dense in $\mathscr{C}$, while non-exchangeable copulas are so.
- In some sense (namely, in the sense of Baire category), the class of exchangeable copulas is a small set (i.e. a set of first category) in $\mathscr{C}$ endowed with the L^∞ norm, while a typical copula is non-exchangeable (see [14] for more details).

These facts also provide additional motivations to consider in depth non–exchangeable copulas.

3 Measures of Non–exchangeability

As noted in [56], "*in a sense, the relationship between exchangeability and non–exchangeability is analogous to the relationship between independence and dependence*" for random variables. In particular, this analogy could be exploited to construct a measure of non–exchangeability, mimicking the axiomatic approach by Rényi [61]. Following this viewpoint, Klement et al. [21] have introduced a set of axioms for a measure of non–exchangeability for identically distributed and continuous random variables, which turned out to depend only on the associated copula. This definition is reported below.

Definition 2 A function $\mu : \mathscr{C} \to \mathbb{R}_+$ is a *measure of non–exchangeability* for $\mathscr{C}$ if it satisfies the following properties:

- $\mu(C) \leq K$ for some $K \in \mathbb{R}_+$ and for all $C \in \mathscr{C}$;
- $\mu(C) = 0$ if, and only if, C is symmetric;
- $\mu(C) = \mu(C^t)$ for every $C \in \mathscr{C}$, with $C^t(u, v) = C(v, u)$;
- $\mu(C) = \mu(\hat{C})$ for every $C \in \mathscr{C}$, with $\widehat{C}$ survival copula related to C;
- If $C, (C_n)_{n \in \mathbb{N}} \in \mathscr{C}$, $C_n \stackrel{n}{\to} C$ (pointwise) implies $\mu(C_n) \stackrel{n}{\to} \mu(C)$.

The most popular measures of non–exchangeability can be derived from the L_p distance d_p in $\mathscr{C}$ ($p \in [1, +\infty]$), given by

$$d_p(A, B) := \left(\int_0^1 \int_0^1 |A(u, v) - B(u, v)|^p \, du \, dv \right)^{1/p},$$

when p is in $[1, +\infty[$, and, for $p = +\infty$ (see [56]),

$$d_\infty(A, B) := \max_{(u,v) \in [0,1]^2} |A(u, v) - B(u, v)|.$$

In particular, it was proved in [21] the following result.

Theorem 1 *For every $p \in [1, +\infty]$, $\mu_p : \mathscr{C} \to \mathbb{R}_+$ defined by $\mu_p(C) := d_p(C, C^t)$ is a measure of non–exchangeability.*

Moreover, given μ_p defined as above, there exists a constant $K_p \in \mathbb{R}_+$ and (at least) a copula $C_p \in \mathscr{C}$ such that

$$\mu_p(C_p) = K_p \quad \text{and} \quad \forall C \in \mathscr{C} \quad \mu_p(C_p) \geq \mu_p(C).$$

Such C_p is called *maximally non-exchangeable* copula with respect to μ_p. As a consequence, we may always suppose that μ_p takes values on $[0, 1]$.

It was proved in [45, 56] that

$$\forall C \in \mathscr{C} \qquad \max_{(u,v) \in [0,1]^2} |C(u, v) - C(v, u)| \leq \tfrac{1}{3}.$$

Moreover, the copulas that are maximally non–exchangeable with respect to μ_∞ have been derived in [45, 56] and their support is depicted in Fig. 3.

Example 3 The measure of non-exchageability for $\widetilde{C}$ of Example 1 is given by

$$\mu_{+\infty}(\widetilde{C}) = 3a^2 \max_{(x,y) \in [0,1]^2} |C_1(x, y) - xy| \, .$$

Maximum asymmetry for such a $\widetilde{C}$ is, hence, obtained when C_1 belongs to $\{W, M\}$. For such a case, $\mu_{+\infty}(\widetilde{C}) = \frac{3a^2}{4}$.

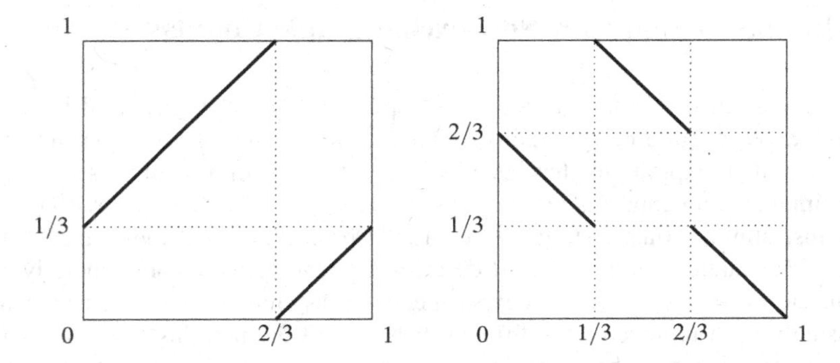

Fig. 3 Support of maximally non-exchangeable copulas with respect to μ_∞

Remark 3 In [4], the authors determined best possible lower and upper copulas C under the constraint that, for a fixed t, $\mu_\infty(C) = t$. Maximally non–exchangeable copulas in d dimensions ($d \geq 2$) have been provided, instead, in [38]. An alternative measure of non–exchangeability for copulas has been proposed in [64].

There is a strong nexus between non-exchangeability measures and dependence measures, as considered also in [56, Sect. 3]. Consider, for instance, the *Schweizer–Wolff* measure of dependence given by

$$\kappa(C) = 4 \max_{(u,v)\in[0,1]^2} |C(u, v) - uv|.$$

For more details, see [63]. It is easily shown that, for every $C \in \mathscr{C}$,

$$\mu_\infty(C) \leq \frac{3}{2}\kappa(C).$$

Moreover, if we consider other (qualitative) dependence notions like positive quadrant dependence, stochastic increasingness, etc., we may notice that:

- positive (resp. negative) dependence plays in favor of exchangeability;
- stronger positive (resp. negative) dependence implies smaller non–exchangeability.

Results of this type can be found, for instance, in [23, 24].

Various possible techniques for measuring asymmetry in copula models have been also employed in developing statistical tests, as done for instance in [34, 36] (see also [48, 60]), or to test special copula-based stochastic processes [2].

4 Optimal Design with Non–exchangeable Copulas

As largely discussed in the introduction, copulas are substantially used in several applied areas. A quite new domain is related to the design of the experiments, an area where their exploitation has been just recently taken into account (see [9, 58]).

Optimal experimental design is a statistical area mainly applied to environmental problems, clinical trials testing and industrial procedures. All these applications relate to the challenging problem of describing phenomena that are generally non-symmetric. However, in design of experiments, the asymmetry of a specific situation has usually been reflected by a different behavior of the margins, only. Moreover, the possible impact of the dependence structure on the design has received little attention.

A first step in including this latter aspect has been made in [58], where a copula–based approach has been introduced. In particular, through several examples it has been shown how taking into account a dependence between the random variables involved in the phenomenon leads to relevant changes on the obtained design. Nevertheless, it has not yet been considered how the impact of the asymmetry of the copula influences the optimal design. In the following, we will investigate this latter aspect by assuming a bivariate stochastic model with identically distributed margins, but (eventually) asymmetric dependence structures. Specifically, the examined asymmetry that will be considered is controlled by adding parameters to an exchangeable copula. Therefore, its impact can be measured by comparing the designs obtained for the symmetric model with the ones obtained for the asymmetric one. Such a study might be of interest because it highlights the robustness of the obtained design against various dependency scenarios.

In the following, we first introduce some basics about experimental design. Then, we illustrate the impact of the asymmetrization on the designs for different values of the model parameters.

4.1 Optimal Experimental Design: Some Basics

From now on, we shall consider a vector $\mathbf{x}^T = (x_1, \dots, x_r) \in \mathscr{X}$ of control variables, where $\mathscr{X} \subset \mathbb{R}^r$ is a compact set. We focus directly on the bivariate case.

The results of the observations and of the expectations in a regression experiment are the vectors:

$$\mathbf{y}(\mathbf{x}) = (y_1(\mathbf{x}), y_2(\mathbf{x})),$$

$$\mathbf{E}[\mathbf{Y}(x)] = \mathbf{E}[(Y_1, Y_2)] = \eta(\mathbf{x}, \beta) = (\eta_1(\mathbf{x}, \beta), \eta_2(\mathbf{x}, \beta)),$$

where $\beta = (\beta_1, \dots, \beta_k)$ is an unknown parameter vector to be estimated and $\eta_i (i = 1, 2)$ are known functions. We denote by $F_{Y_i}(y_i(\mathbf{x}, \beta))$ the margins of each Y_i for $i \in \{1, 2\}$. According to Sklar's theorem [66], we assume that the dependence between

Y_1 and Y_2 is modeled by a copula function C_α, depending on $\alpha = \{\alpha_1, \dots, \alpha_l\}$, an unknown (copula) parameter vector. Hence, the joint model can be expressed in the form

$$C_\alpha(F_{Y_1}(y_1(\mathbf{x}, \beta)), F_{Y_2}(y_2(\mathbf{x}, \beta))).$$

The Fisher Information Matrix for a single observation, i.e., $m(x, \gamma)$, is a $(k+l) \times (k+l)$ matrix whose elements are given by

$$\mathbf{E}\left(-\frac{\partial^2}{\partial\gamma_i\partial\gamma_j}\log\left[\frac{\partial^2}{\partial y_1\partial y_2}C_\alpha(F_{Y_1}(y_1(\mathbf{x}, \beta)), F_{Y_2}(y_2(\mathbf{x}, \beta)))\right]\right) \tag{1}$$

where $\gamma = \{\gamma_1, \dots, \gamma_{k+l}\} = \{\beta_1, \dots, \beta_k, \alpha_1, \dots, \alpha_l\}$.

The aim of design theory is to quantify the amount of information on both sets of parameters α and β, respectively, from the regression experiment embodied in the Fisher Information Matrix.

For N independent observations at $x_1, \dots, x_N$, the corresponding Information matrix is

$$\mathbf{M}(\xi, \gamma) = \sum_{i=1}^{N} w_i m(x_i, \gamma), \ \sum_{i=1}^{N} w_i = 1 \text{ and } \xi = \begin{Bmatrix} x_1 & \dots & x_N \\ w_1 & \dots & w_N \end{Bmatrix}.$$

The approximate design theory is concerned with finding $\xi^*(\gamma)$ such that it maximizes some scalar function $\phi(M(\xi, \gamma))$, i.e., the so-called *design criterion*. Hereinafter, we consider only *D-optimality*, i.e., $\phi(M) = \log\det M$, provided M is nonsingular.

The formulation of a Kiefer-Wolfowitz type equivalence relation (see [44]) is the cornerstone of the theoretical foundation of optimal design. The following theorem of such type is a generalized version of a result formulated in [39] and based on the findings in [65]. A detailed proof of this result can be found in [58].

Theorem 2 *For a local parameter vector* $(\bar{\gamma})$*, the following properties are equivalent:*

- ξ^* *is D-optimal;*
- $\operatorname{tr}[M(\xi^*, \bar{\gamma})^{-1} m(x, \bar{\gamma})] \leq (k+l), \forall x \in \mathscr{X}$*;*
- ξ^* *minimize* $\max_{x \in \mathscr{X}} \operatorname{tr}[M(\xi^*, \bar{\gamma})^{-1} m(x, \bar{\gamma})]$*, over all* $\xi \in \Xi$*, where* Ξ *is the design space.*

Theorem 2 allows one to implement standard design algorithms such as of the Fedorov-Wynn type (see [29, 69]). It also provides simple checks for D-optimality through the maxima of $d(x, \xi^*) = \operatorname{tr}[M(\xi^*, \bar{\gamma})^{-1} m(x, \bar{\gamma})]$, which is usually called *sensitivity function*. The sensitivity function plays an important role in the convex design theory. In fact, its maxima determine the location of the points that are the most informative with respect to the optimality criterion.

The next definition is important for the comparison of two different designs.

Definition 3 Let $(k+l)$ be the number of the model parameters. The ratio

$$D(\xi, \xi') = \left(\frac{|M(\xi, \gamma)|}{|M(\xi', \gamma)|} \right)^{1/(k+l)} \tag{2}$$

is called *D-Efficiency* of the design ξ with respect to the design ξ'.

As it is now evident from previous definitions and results, the optimal designs depend upon the trend model structure and the chosen copula. Furthermore, such designs might also be influenced by the unknown parameter values for γ through the induced nonlinearities (see [59]). Thence, we are resorting to localized designs around the values $\bar{\gamma}$.

4.2 A Model for Binary Outcomes

In this section we present an example with potential applications in clinical trials.

Let us formally introduce the model. We assume a bivariate binary response (Y_{i1}, Y_{i2}), $i = 1, \ldots, n$, with four possible outcomes $\{(0,0), (0,1), (1,0), (1,1)\}$ where 1 usually represents a success and 0 a failure of a given treatment. We denote the joint probabilities of Y_1 and Y_2 by $p_{y_1,y_2} = \mathbb{P}(Y_1 = y_1, Y_2 = y_2)$ where $(y_1, y_2) \in \{0,1\}^2$. In a clinical trial context, Y_1 and Y_2 could represent, for instance, efficacy and toxicity of a tested drug (see [9, 10]).

Now, define

$$\begin{aligned} p_{11} &= C_\alpha(\pi_1, \pi_2), \; p_{10} = \pi_1 - p_{11}, \\ p_{01} &= \pi_2 - p_{11}, \quad p_{00} = 1 - \pi_1 - \pi_2 + p_{11}, \end{aligned} \tag{3}$$

where π_1 and π_2 are the marginal probabilities of success and C_α is a given copula. Following [39], we assume that

$$\log\left(\frac{\pi_i}{1-\pi_i}\right) = \beta_{i1} + \beta_{i2}x, \qquad i = 1, 2 \tag{4}$$

with $x \in [0, 10]$. As shown in [10], in such a model the Fisher information matrix for a single observation can be written as:

$$m(x, \gamma) = \frac{\partial \mathbf{p}}{\partial \gamma}^T \left(P^{-1} + \frac{1}{1 - p_{11} - p_{10} - p_{01}} \mathbf{e}\mathbf{e}^T \right) \frac{\partial \mathbf{p}}{\partial \gamma}, \tag{5}$$

where $\mathbf{p} = (p_{11}, p_{10}, p_{01})$, $P = diag(\mathbf{p})$ and $\mathbf{e} = (1, 1, 1)^T$.

Since our focus is on the effects of asymmetric dependence, here we consider $\pi_1 = \pi_2$, i.e., the same marginal behavior of Y_1 and Y_2. Specifically, we restrict the parameters related to the margins to $\beta_1 = (\beta_{11}, \beta_{12})$ with 'localized' initial values $\bar{\boldsymbol{\beta}}_1 = [-1, 1]$.

We start with an exchangeable model, where the probability of success p_{11} is given by

$$p_{11} = F_{Y_1,Y_2}(\pi_1, \pi_1; \alpha_1) = C(\pi_1, \pi_1; \alpha_1),$$

with C being an exchangeable copula. The idea is to take into account a transformation of the copula C in order to obtain an asymmetric dependence for the random vector (Y_1, Y_2). In practice, this is obtained by bringing two additional parameters into the model that may induce the asymmetry. Then a natural way how to inquire the effect of the applied transformation is simply to compare the designs obtained for C and those obtained for the asymmetric version of C, by calculating the corresponding D-efficiencies.

The copula transformation we take into account consists of modifying a given exchangeable copula $C = C_{\alpha_1}$, with parameter α_1, into the copula $\tilde{C} = \tilde{C}_{\alpha_1,\alpha_2,\alpha_3}$ defined, for every $(u, v) \in [0, 1]^2$, by

$$\tilde{C}(u, v) = u^{\alpha_2} v^{\alpha_3} C_{\alpha_1}(u^{1-\alpha_2}, v^{1-\alpha_3}), \tag{6}$$

where $\alpha_2, \alpha_3 \in [0, 1]$. For $\alpha_2 \neq \alpha_3$, $\tilde{C}$ is non–exchangeable. Transformation (6) is the well-known Khoudraji's asymmetrization, described in [43]. Notice that the dependence in $\tilde{C}$ of (6) is limited since, as shown in [32],

$$\tau(\tilde{C}) \leq \frac{(1-\alpha_2)(1-\alpha_3)}{(1-\alpha_2)+(1-\alpha_3)-(1-\alpha_2)(1-\alpha_3)} =: \tau_{\max}(\tilde{C}).$$

It is worth to stress that, while in the symmetric model the estimated parameter vector is $(\beta_{11}, \beta_{12}, \alpha_1)$, in the asymmetric case we bring the focus on the asymmetrization by considering as vector of the estimated parameters $(\beta_{11}, \beta_{12}, \alpha_1, \alpha_2, \alpha_3)$.

In our example, the symmetric copula C_{α_1} is assumed belonging to the Clayton family and the following scheme is carried on. For some values of Kendall's τ for the copula C_{α_1} of (3), we find the D-Efficiency of the optimal design ξ, which corresponds to the exchangeable Clayton copula with the given τ, with respect to the optimal design $\bar{\xi}^*$, which corresponds to the asymmetric model of Eq. (6) with (suitable) parameters $(\tilde{\alpha_1}, \tilde{\alpha_2}, \tilde{\alpha_3})$ in order to ensure that $\tau(C) = \tau(\tilde{C})$. The losses in D-efficiency (in percentage) are hence used to quantify the impact of the asymmetrization on the design.

In this small numerical illustration we assume that $\tau \in \{0.10, 0.25, 0.50\}$, while $(\tilde{\alpha_2}, \tilde{\alpha_3})$ may vary in $\{0, 0.2, 0.4, 0.6, 0.8\}$ (without loss of generality, we consider $\alpha_2 > \alpha_3$). However, it should be noted that some combinations of $(\tilde{\alpha_2}, \tilde{\alpha_3})$ may not provide the given values of Kendall's measure of association.

Table 1 shows the losses in D-Efficiency for various localized parameter vectors. Analyzing the results, one may notice that the losses are generally quite substantial. Such high losses prove that the symmetric model and the asymmetric one provide reasonably different optimal designs. An evidence of the difference between the two

Table 1 Losses in D-efficiency (in percentage) between Clayton and asymmetric Clayton models

τ	α_1	$(\tilde{\alpha_2}, \tilde{\alpha_3})$	$\tilde{\alpha_1}$	$\tau_{\max}(\tilde{C}_{(\tilde{\alpha_1},\tilde{\alpha_2},\tilde{\alpha_3})})$	Loss in D–efficiency (in %)
0.50	2.00	(0.2, 0.0)	3.5	0.80	43.99
0.25	0.66	(0.2, 0.0)	0.9	0.80	49.49
0.10	0.22	(0.2, 0.0)	0.3	0.80	55.70
0.50	2.00	(0.4, 0.0)	11	0.60	75.70
0.25	0.66	(0.4, 0.0)	1.5	0.60	51.74
0.10	0.22	(0.4, 0.0)	0.4	0.60	56.27
0.25	0.66	(0.6, 0.0)	3.6	0.40	63.80
0.10	0.22	(0.6, 0.0)	0.7	0.40	57.90
0.10	0.22	(0.8, 0.0)	2.2	0.20	67.46
0.25	0.66	(0.4, 0.2)	2.0	0.52	51.98
0.10	0.22	(0.4, 0.2)	0.5	0.52	56.52
0.25	0.66	(0.6, 0.2)	5.3	0.36	64.86
0.10	0.22	(0.6, 0.2)	0.9	0.36	58.07
0.10	0.22	(0.8, 0.2)	3.5	0.19	70.37
0.25	0.66	(0.6, 0.4)	9.5	0.31	65.83
0.10	0.22	(0.6, 0.4)	1.4	0.31	57.42

obtained designs can also be seen by simply comparing Fig. 4a or b with Fig. 5. In fact, the optimal design found for the asymmetric model has three support points, while the one related to the symmetric model just has two support points. Moreover, Table 1 also indicates that the losses depend upon the chosen transformation. As a matter of fact, by looking at the values corresponding to the same fixed τ, a considerable increase of loss relative to stronger asymmetries can be observed. This aspect is also observable from Fig. 4, where the different geometry of two optimal designs corresponding to two distinct asymmetrizations is shown.

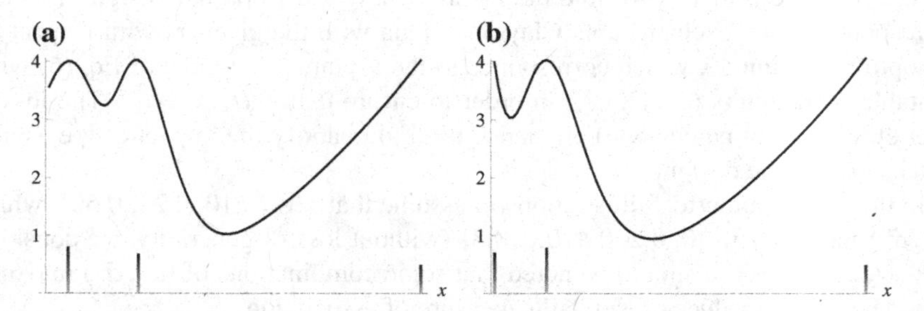

Fig. 4 Sensitivity functions (*continuous lines*) and design weights (*bars*) for asymmetric Clayton with $(\tilde{\alpha_1}, \tilde{\alpha_2}, \tilde{\alpha_3}) = (11, 0.4, 0)$ (Fig. **a**) and $(\tilde{\alpha_1}, \tilde{\alpha_2}, \tilde{\alpha_3}) = (3.5, 0.2, 0)$ (Fig. **b**), respectively

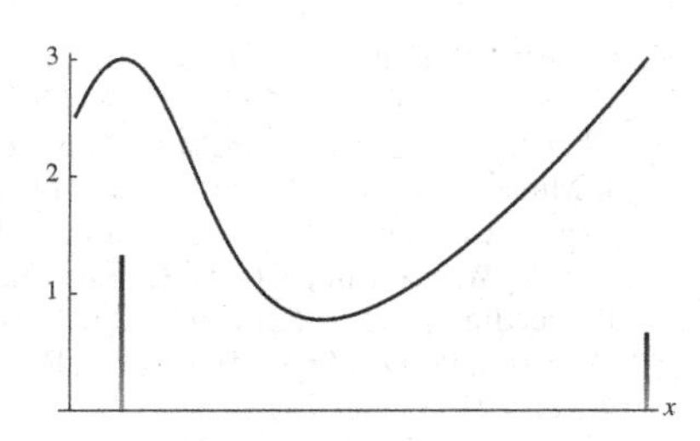

Fig. 5 Sensitivity function (*continuous line*) and design weights (*bars*) for symmetric Clayton with $\tau = 0.5$

5 Conclusions

This paper provides an overview on basics and motivations of asymmetric copulas. The work deals with both theoretical and practical aspects of the usage of such functions. With the aim of highlighting the importance of asymmetric copula models in applications, an example on optimal experimental design is carried on. The study is conducted by comparing designs obtained for an initially symmetric model with optimal designs for asymmetric transformations of the original (symmetric) model. Results for stronger and weaker asymmetrizations are reported. Overall, in the presented illustration, symmetric and asymmetric dependence structures affect the optimal design in a different way. Moreover, the effect of the transformation seems to increase for stronger asymmetrizations. As a conclusion, this work shows that, even by assuming equal marginal behavior, the non-exchangeability of a specific phenomenon may be caught by the dependence structure thanks to the use of flexible copula models.

Acknowledgments This article is devoted to Prof. Erich Peter Klement on the occasion of his retirement.
The first author has been supported by the Faculty of Economics and Management of Free University of Bozen-Bolzano, Italy, via the project "Model Uncertainty and Dependence".
The second author would like to thank Prof. Werner Müller, for his support and useful suggestions. She was supported by the project ANR-2011-IS01-001-01 "DESIRE" and Austrian Science Fund (FWF) I 883-N18.

References

1. Bacigal, T., Jágr, V., Mesiar, R.: Non-exchangeable random variables, Archimax copulas and their fitting to real data. Kybernetika (Prague) **47**(4), 519–531 (2011)
2. Beare, B.K., Seo, J.: Time irreversible copula-based Markov models. Econom. Theory **30**(5), 923–960 (2014)
3. Běhounek, L., Bodenhofer, U., Cintula, P., Saminger-Platz, S., Sarkoci, P.: Graded dominance and related graded properties of fuzzy connectives. Fuzzy Sets Syst. **262**, 78–101 (2015)
4. Beliakov, G., De Baets, B., De Meyer, H., Nelsen, R.B., Úbeda-Flores, M.: Best-possible bounds on the set of copulas with given degree of non-exchangeability. J. Math. Anal. Appl. **417**(1), 451–468 (2014)

5. Capéraà, P., Fougères, A.L., Genest, C.: Bivariate distributions with given extreme value attractor. J. Multivar. Anal. **72**(1), 30–49 (2000)
6. Charpentier, A., Fougères, A.L., Genest, C., Nešlehová, J.G.: Multivariate Archimax copulas. J. Multivar. Anal. **126**, 118–136 (2014)
7. Czado, C.: Pair-copula constructions of multivariate copulas. In: Jaworski, P., Durante, F., Härdle, W., Rychlik, T. (eds.) Copula Theory and Its Applications. Lecture Notes in Statistics—Proceedings, vol. 198, pp. 93–109. Springer, Berlin (2010)
8. De Baets, B., De Meyer, H., Mesiar, R.: Asymmetric semilinear copulas. Kybernetika (Prague) **43**(2), 221–233 (2007)
9. Denman, N.G., McGree, J.M., Eccleston, J.A., Duffull, S.B.: Design of experiments for bivariate binary responses modelled by Copula functions. Comput. Stat. Data Anal. **55**(4), 1509–1520 (2011)
10. Dragalin, V., Fedorov, V.: Adaptive designs for dose-finding based on efficacytoxicity response. J. Stat. Plan. Infer. **136**(6), 1800–1823 (2006)
11. Durante, F.: Construction of non-exchangeable bivariate distribution functions. Stat. Pap. **50**(2), 383–391 (2009)
12. Durante, F., Fernández-Sánchez, J.: On the classes of copulas and quasi-copulas with a given diagonal section. Int. J. Uncertain. Fuzziness Knowl.-Based Syst. **19**(1), 1–10 (2011)
13. Durante, F., Fernández-Sánchez, J., Sempi, C.: Multivariate patchwork copulas: a unified approach with applications to partial comonotonicity. Insur. Math. Econ. **53**, 897–905 (2013)
14. Durante, F., Fernández-Sánchez, J., Trutschnig, W.: Baire category results for exchangeable copulas. Fuzzy Sets Syst. **284**, 146–151 (2016)
15. Durante, F., Foscolo, E., Jaworski, P., Wang, H.: A spatial contagion measure for financial time series. Expert Syst. Appl. **41**(8), 4023–4034 (2014)
16. Durante, F., Jaworski, P.: Spatial contagion between financial markets: a copula-based approach. Appl. Stoch. Models Bus. Ind. **26**(5), 551–564 (2010)
17. Durante, F., Jaworski, P.: Invariant dependence structure under univariate truncation. Statistics **46**(2), 263–277 (2012)
18. Durante, F., Jaworski, P., Mesiar, R.: Invariant dependence structures and Archimedean copulas. Stat. Probab. Lett. **81**(12), 1995–2003 (2011)
19. Durante, F., Klement, E., Mesiar, R., Sempi, C.: Conjunctors and their residual implicators: characterizations and construction methods. Mediterr. J. Math. **4**(3), 343–356 (2007)
20. Durante, F., Klement, E., Quesada-Molina, J., Sarkoci, P.: Remarks on two product-like constructions for copulas. Kybernetika (Prague) **43**(2), 235–244 (2007)
21. Durante, F., Klement, E., Sempi, C., Úbeda-Flores, M.: Measures of non-exchangeability for bivariate random vectors. Stat. Pap. **51**(3), 687–699 (2010)
22. Durante, F., Kolesárová, A., Mesiar, R., Sempi, C.: Copulas with given diagonal sections: novel constructions and applications. Int. J. Uncertain. Fuzziness Knowl.-Based Syst. **15**(4), 397–410 (2007)
23. Durante, F., Papini, P.L.: Componentwise concave copulas and their asymmetry. Kybernetika (Prague) **45**(6), 1003–1011 (2009)
24. Durante, F., Papini, P.L.: Non-exchangeability of negatively dependent random variables. Metrika **71**(2), 139–149 (2010)
25. Durante, F., Rodríguez-Lallena, J.A., Úbeda-Flores, M.: New constructions of diagonal patchwork copulas. Inf. Sci. **179**(19), 3383–3391 (2009)
26. Durante, F., Saminger-Platz, S., Sarkoci, P.: Rectangular patchwork for bivariate copulas and tail dependence. Comm. Stat. Theory Methods **38**(15), 2515–2527 (2009)
27. Durante, F., Sarkoci, P., Sempi, C.: Shuffles of copulas. J. Math. Anal. Appl. **352**(2), 914–921 (2009)
28. Durante, F., Sempi, C.: Principles of Copula Theory. CRC/Chapman and Hall, Boca Raton (2015)
29. Fedorov, V.V.: The design of experiments in the multiresponse case. Theory Probab. Appl. **16**(2), 323–332 (1971)

30. Fodor, J.C., Keresztfalvi, T.: Nonstandard conjunctions and implications in fuzzy logic. Int. J. Approx. Reason. **12**(2), 69–84 (1995)
31. Genest, C., Ghoudi, K., Rivest, L.P.: Understanding relationships using copulas, by E. Frees and E. Valdez, January 1998. N. Am. Actuar. J. **2**(3), 143–149 (1998)
32. Genest, C., Kojadinovic, I., Nešlehová, J., Yan, J.: A goodness-of-fit test for bivariate extreme-value copulas. Bernoulli **17**(1), 253–275 (2011)
33. Genest, C., MacKay, R.J.: Copules archimédiennes et familles de lois bidimensionnelles dont les marges sont données. Can. J. Stat. **14**(2), 145–159 (1986)
34. Genest, C., Nešlehová, J., Quessy, J.F.: Tests of symmetry for bivariate copulas. Ann. Inst. Stat. Math. **64**(4), 811–834 (2012)
35. Genest, C., Nešlehová, J., Ziegel, J.: Inference in multivariate Archimedean copula models. TEST **20**(2), 223–256 (2011)
36. Genest, C., Nešlehová, J.: Assessing and modeling asymmetry in bivariate continuous data. In: Jaworski, P., Durante, F., Härdle, W. (eds.) Copulae in Mathematical and Quantitative Finance. Lecture Notes in Statistics, pp. 91–114. Springer, Berlin (2013)
37. Hájek, P., Mesiar, R.: On copulas, quasicopulas and fuzzy logic. Soft Comput. **12**(12), 123–1243 (2008)
38. Harder, M., Stadtmüller, U.: Maximal non-exchangeability in dimension d. J. Multivar. Anal. **124**, 31–41 (2014)
39. Heise, M.A., Myers, R.H.: Optimal designs for bivariate logistic regression. Biometrics **52**(2), 613–624 (1996)
40. Jaworski, P., Pitera, M.: On spatial contagion and multivariate GARCH models. Appl. Stoch. Models Bus. Ind. **30**, 303–327 (2014)
41. Joe, H.: Dependence Modeling with Copulas. Chapman and Hall/CRC, London (2014)
42. Kawaguchi, M.F., Miyakoshi, M.: Composite fuzzy relational equations with noncommutative conjunctions. Inf. Sci. **110**(1–2), 113–125 (1998)
43. Khoudraji, A.: Contributions à l'étude des copules et à la modélisation des valeurs extrêmes bivariées. Ph.D. thesis, Université de Laval, Québec (Canada) (1995)
44. Kiefer, J., Wolfowitz, J.: The equivalence of two extremum problems. Can. J. Math. **12**, 363–366 (1960)
45. Klement, E.P., Mesiar, R.: How non-symmetric can a copula be? Comment. Math. Univ. Carolin. **47**(1), 141–148 (2006)
46. Klement, E.P., Mesiar, R., Pap, E.: Triangular norms. Trends in Logic-Studia Logica Library, vol. 8. Kluwer Academic Publishers, Dordrecht (2000)
47. Klement, E.P., Mesiar, R., Pap, E.: Archimax copulas and invariance under transformations. C. R. Math. Acad. Sci. Paris **340**(10), 755–758 (2005)
48. Kojadinovic, I., Yan, J.: A non-parametric test of exchangeability for extreme-value and left-tail decreasing bivariate copulas. Scand. J. Stat. **39**(3), 480–496 (2012)
49. Krupskii, P., Joe, H.: Factor copula models for multivariate data. J. Multivar. Anal. **120**, 85–101 (2013)
50. Liebscher, E.: Construction of asymmetric multivariate copulas. J. Multivar. Anal. **99**(10), 2234–2250 (2008)
51. Liebscher, E.: Erratum to Construction of asymmetric multivariate copulas. J. Multivar. Anal. **102**(4), 869–870 (2011)
52. Mazo, G., Girard, S., Forbes, F.: A flexible and tractable class of one-factor copulas. Stat. Comput., in press (2015)
53. McNeil, A.J., Nešlehová, J.: Multivariate Archimedean copulas, d-monotone functions and ℓ_1-norm symmetric distributions. Ann. Stat. **37**(5B), 3059–3097 (2009)
54. McNeil, A.J., Nešlehová, J.: From Archimedean to Liouville copulas. J. Multivar. Anal. **101**(8), 1772–1790 (2010)
55. Nelsen, R.B.: An Introduction to Copulas. Springer Series in Statistics, 2nd edn. Springer, New York (2006)
56. Nelsen, R.B.: Extremes of nonexchangeability. Stat. Pap. **48**(2), 329–336 (2007)

57. Nelsen, R.B., Quesada-Molina, J.J., Rodríguez-Lallena, J.A., Úbeda-Flores, M.: On the construction of copulas and quasi-copulas with given diagonal sections. Insur. Math. Econom. **42**(2), 473–483 (2008)
58. Perrone, E., Müller, W.G.: Optimal design for copula models. Statistics, in press (2015). doi:10.1080/02331888.2015.1111892
59. Pronzato, L., Pázman, A.: Design of Eperiments in Nonlinear Models. Springer Lecture Notes in Statistics, vol. 212 (2013)
60. Quessy, J.F., Bahraoui, T.: Graphical and formal statistical tools for assessing the symmetry of bivariate copulas. Can. J. Stat. **41**(4), 637–656 (2013)
61. Rényi, A.: On measures of dependence. Acta Math. Acad. Sci. Hungar. **10**, 441–451 (1959)
62. Schweizer, B., Sklar, A.: North-Holland Series in Probability and Applied Mathematics. Probabilistic metric spaces. North-Holland Publishing Co., New York (1983)
63. Schweizer, B., Wolff, E.F.: On nonparametric measures of dependence for random variables. Ann. Stat. **9**(4), 879–885 (1981)
64. Siburg, K.F., Stoimenov, P.A.: Symmetry of functions and exchangeability of random variables. Stat. Pap. **52**(1), 1–15 (2011)
65. Silvey, S.D.: Optimal Design (Science Paperbacks). Chapman and Hall (1980)
66. Sklar, A.: Fonctions de répartition à n dimensions et leurs marges. Publications de l'Institut de Statistique de Paris **8**, 229–231 (1959)
67. Trutschnig, W., Fernández Sánchez, J.: Some results on shuffles of two-dimensional copulas. J. Stat. Plan. Infer. **143**(2), 251–260 (2013)
68. Wu, S.: Construction of asymmetric copulas and its application in two-dimensional reliability modelling. Eur. J. Oper. Res. **238**, 476–485 (2014)
69. Wynn, H.P.: The sequential generation of D-optimum experimental designs. Ann. Math. Stat. **41**(5), 1655–1664 (1970)

Copulæ of Processes Related to the Brownian Motion: A Brief Survey

Carlo Sempi

Abstract The copulas of a few stochastic processes related to the Brownian motion are derived; specifically, if (X_t) is one such process, the copula of the pair (X_s, X_t) is determined for $s < t$.

1 Introduction

The study of the copulas of stochastic processes is rapidly becoming important. However, not all aspects of stochastic processes can be dealt with through copulas. In general, one may safely say that, given a stochastic process (X_t) and n times $t_1 < t_2 < \ldots < t_n$, the distribution function(=d.f.) of the random variables X_{t_1}, ..., X_{t_n} can in principle be amenable to calculation via copulas. Here we survey known results on the deduction of copulas $C_{s,t}$ of pairs (X_s, X_t) with $s < t$ where $(X_t)_{t \geq 0}$ is a stochastic process related to the Brownian motion, in a sense to be made precise in the following. Also, following [6], we establish the copula of a Brownian motion (B_t) and its supremum S_t. Although the present paper is a survey a few new results are presented, notably in Sects. 3 and 4. Of the growing literature on these aspects only the items that are strictly relevant for the present chapter will be quoted.

We briefly recall the definition of a (bivariate) copula and the main properties that will be needed.

Definition 1 A *copula* is a function $C : \mathbb{I}^2 \to \mathbb{I}$, where $\mathbb{I} = [0, 1]$ such that:

(a) for every $u \in \mathbb{I}$, $C(u, 0) = C(0, u) = 0$ and $C(u, 1) = C(1, u) = u$;
(b) C is 2–increasing: for all u, u', v and v' in $\mathbb{I}$ with $u \leq u'$ and $v \leq v'$

$$C(u', v') - C(u, v') - C(u', v) + C(u, v) \geq 0 .$$

C. Sempi (✉)
Dipartimento di Matematica e Fisica "Ennio De Giorgi",
Università del Salento, Lecce 73100, Italy
e-mail: carlo.sempi@unisalento.it

S. Saminger-Platz and R. Mesiar (eds.), *On Logical, Algebraic, and Probabilistic Aspects of Fuzzy Set Theory*, Studies in Fuzziness and Soft Computing 336,
DOI 10.1007/978-3-319-28808-6_10

In other words, a copula C is (the restriction to the unit square $\mathbb{I}^2$ of) a d.f. that concentrates all the probability mass on $\mathbb{I}^2$ and that has uniform margins. The main result is provided by Sklar's theorem [7].

Theorem 1 *Let a random vector* $\mathbf{X} = (X, Y)$ *be given on a probability space* $(\Omega, \mathscr{F}, \mathbb{P})$, *let* $H(x, y) := \mathbb{P}(X \leq x, Y \leq y)$ *be the joint d.f. of* $\mathbf{X}$, *and let* $F(x) = \mathbb{P}(X \leq x)$ *and* $G(y) = \mathbb{P}(Y \leq y)$ *be its marginals. Then there exists a copula* $C = C_{\mathbf{X}}$ *such that, for every point* $(x, y) \in \mathbb{R}^2$,

$$H(x, y) = C\left(F(x), G(y)\right) . \tag{1}$$

If the marginals F and G are continuous, then the copula C is uniquely defined.

Under the assumptions of Theorem 1, if F and G are continuous, then there exists a unique copula C associated with $\mathbf{X}$ that is determined, for all $(u, v) \in [0,1[^2$, via the formula

$$C(u, v) = H\left(F^{(-1)}(u), G^{(-1)}(v)\right) , \tag{2}$$

where $F^{(-1)}(t) := \inf\{x \in \mathbb{R} : F(x) \geq t\}$ is the right–continuous quasi–inverse of F; similarly for $G^{(-1)}$. When both F and G are continuous one may speak of *the* copula of the random vector $\mathbf{X}$ and denote it by $C_{\mathbf{X}}$.

Theorem 2 ([1]) *In the probability space* $(\Omega, \mathscr{F}, \mathbb{P})$ *let the continuous random variables X and Y have d.f.'s F and G and copula C. Then, one has a.e.*

$$\mathbb{P}(X \leq x \mid Y)(\omega) = \mathbb{E}\left(\mathbf{1}_{\{X \leq x\}} \mid Y\right)(\omega) = \partial_2 C\left(F_X(x), F_Y(Y(\omega))\right) , \tag{3}$$

$$\mathbb{P}(Y \leq y \mid X)(\omega) = \mathbb{E}\left(\mathbf{1}_{\{Y \leq y\}} \mid X\right)(\omega) = \partial_1 C(F_X(X(\omega)), F_Y(y)) . \tag{4}$$

Here we have set

$$\partial_1 f(s, t) := \frac{\partial f(s, t)}{\partial s} , \quad \text{and} \quad \partial_2 f(s, t) := \frac{\partial f(s, t)}{\partial t} .$$

Theorem 3 *Let X and Y be continuous random variables defined on the probability space* $(\Omega, \mathscr{F}, \mathbb{P})$ *and consider the continuous mappings*

$$f : \operatorname{Ran} X \to \mathbb{R} \quad \textit{and} \quad g : \operatorname{Ran} Y \to \mathbb{R} .$$

(a) *If both f and g are strictly increasing, then, for all* $(u, v) \in \mathbb{I}^2$,

$$C_{f(X),g(Y)}(u, v) = C_{XY}(u, v) ;$$

(b) *if both f and g are strictly decreasing, then, for all* $(u, v) \in \mathbb{I}^2$,

$$C_{f(X),g(Y)}(u, v) = C^{\sigma_1\sigma_2}(u, v) = u + v - 1 + C_{XY}(1 - u, 1 - v)\,.$$

We refer to [4] and to the forthcoming monograph [2] for the properties of copulas and to [3, 5, 8] for those of stochastic processes.

2 The Copula of a Brownian Motion

With the exception of the last remark, the results of this section were established in [1].

Given a standard Brownian motion (=BM, for short) $(B_t)_{t\geq 0}$ on the probabilitiy space $(\Omega, \mathscr{F}, \mathbb{P})$, we wish to find the copula $C^B_{s,t}$ of the pair (B_s, B_t) with $s < t$.

The transition probabilities for the BM (B_t) are given, for $s \in]0, t[$ and x and y in $\mathbb{R}$, by

$$\mathbb{P}(x, s; y, t) := \mathbb{P}(B_t \leq y \mid B_s = x) = \Phi\left(\frac{y - x}{\sqrt{t - s}}\right),$$

where Φ denotes the distribution function of the standard normal law $N(0, 1)$. By Theorem 2 one has

$$\mathbb{P}(x, s; y, t) = \partial_1 C^B_{s,t}\,(F_s(x), F_t(y))\,,$$

where F_t is the d.f. of B_t. Therefore

$$\begin{aligned} C^B_{s,t}\,(F_s(x), F_t(y)) &= \int_0^{F_s(x)} \partial_1 C^B_{s,t}\,(w, F_t(y))\, dw \\ &= \int_0^{F_s(x)} \Phi\left(\frac{y - x}{\sqrt{t - s}}\right) dw\,. \end{aligned} \tag{5}$$

As is well known (see, e.g., [3] or [5]) the d.f. F_t of B_t is given, for $t > 0$, by

$$F_t(x) = \Phi\left(\frac{x}{\sqrt{t}}\right).$$

This is both continuous and strictly increasing. As a consequence the copula of (B_s, B_t) is uniquely determined; moreover, F_t has an inverse, which will be needed shortly and which is easily calculated

$$F_t^{-1}(u) = \sqrt{t}\,\Phi^{-1}(u) \qquad (u \in]0, 1[)\,.$$

Replacing this result into (5) yields the copula $C^B_{s,t}$

$$C^B_{s,t}(u,v) = \int_0^u \Phi\left(\frac{\sqrt{t}\,\Phi^{-1}(v) - \sqrt{s}\,\Phi^{-1}(w)}{\sqrt{t-s}}\right) dw\,. \tag{6}$$

This copula has partial derivatives given, for $(u,v) \in\,]0,1[^2$ by

$$\partial_1 C^B_{s,t}(u,v) = \Phi\left(\frac{\sqrt{t}\,\Phi^{-1}(v) - \sqrt{s}\,\Phi^{-1}(u)}{\sqrt{t-s}}\right), \tag{7}$$

$$\partial_2 C^B_{s,t}(u,v) = \frac{1}{\varphi\left(\Phi^{-1}(v)\right)}\sqrt{\frac{t}{t-s}}\int_0^u \varphi\left(\frac{\sqrt{t}\,\Phi^{-1}(v) - \sqrt{s}\,\Phi^{-1}(w)}{\sqrt{t-s}}\right) dw\,,$$

where φ is the density of the standard normal law $N(0,1)$. A further derivation of (7) provides the density of $C^B_{s,t}$

$$c^B_{s,t}(u,v) = \sqrt{\frac{t}{t-s}}\,\frac{1}{\varphi\left(\Phi^{-1}(v)\right)}\,\varphi\left(\frac{\sqrt{t}\,\Phi^{-1}(v) - \sqrt{s}\,\Phi^{-1}(u)}{\sqrt{t-s}}\right). \tag{8}$$

Notice that, because of Theorem 3(a), the copula of the Brownian bridge $B^*_t = B_t - t\,B_1$ coincides with the copula (6) of the Brownian motion (B_t).

3 The Copula of the Geometric Brownian Motion

A geometric Brownian motion (X_t) satisfies the stochastic integral Eq. (see, e.g., [3])

$$X_t = X_0 + \mu \int_0^t X_s\, ds + \nu \int_0^t X_s\, dB_s\,, \tag{9}$$

where $\mu \in \mathbb{R}$, $\nu > 0$, $X_0 \neq 0$. The solution of (9) is given by

$$X_t = X_0 \exp\left(\left(\mu - \frac{1}{2}\nu^2\right) t + \nu\, B_t\right).$$

Thus if $X_0 > 0$, then X_t is a strictly increasing function of B_t, $X_t = f_t(B_t)$, where

$$f_t(x) := X_0 \exp\left(\left(\mu - \frac{1}{2}\nu^2\right) t + \nu\, x\right).$$

As a consequence of Theorem 3 (a) the copula $C^{GB+}_{s,t}$ of the random vector (X_s, X_t) with $s \in\,]0,t[$ coincides with the copula $C^B_{s,t}$ of the BM given by (6) and has the same density (8).

If, on the other hand, $X_0 < 0$, then X_t is a strictly decreasing function of B_t, $X_t = f_t(B_t)$, with the same function f_t as above, which is now strictly decreasing. Therefore, by Theorem 3(b) the copula of (X_s, X_t) is given by

$$\begin{aligned} C_{s,t}^{GB-}(u,v) &= u + v - 1 + C_{s,t}^{B}(1-u, 1-v) \\ &= u + v - 1 + \int_0^{1-u} \Phi\left(\frac{\sqrt{t}\,\Phi^{-1}(1-v) - \sqrt{s}\,\Phi^{-1}(w)}{\sqrt{t-s}}\right) dw\,, \end{aligned} \tag{10}$$

which differs from the copula of Eq. (6). When $X_0 < 0$, the density of $C_{s,t}^{GB-}(u,v)$ is given by

$$c_{s,t}^{GB-}(u,v) = \sqrt{\frac{t}{t-s}}\,\frac{1}{\varphi\left(\Phi^{-1}(1-v)\right)}\,\varphi\left(\frac{\sqrt{t}\,\Phi^{-1}(1-v) - \sqrt{s}\,\Phi^{-1}(1-u)}{\sqrt{t-s}}\right).$$

4 The Copula of the Ornstein–Uhlenbeck Process

The Ornstein–Uhlenbeck process U_t solves the stochastic differential equation

$$dU_t = -U_t\,dt + \sqrt{2}\,dB_t\,, \tag{11}$$

where (B_t) is a Brownian motion. The connexion between the Ornstein–Uhlenbeck process (U_t) and the Brownian motion (B_t) is given by

$$U_t = 2^{-t}\,B_{e^{2t}}\,, \tag{12}$$

so that the distribution function $F_t^{\mathbf{OU}}$ of U_t is

$$\begin{aligned} F_t^{\mathbf{OU}}(x) &= \mathbb{P}\left(U_t \le x\right) = \mathbb{P}\left(e^{-t}\,B_{e^{2t}} \le x\right) = \mathbb{P}\left(B_{e^{2t}} \le e^t\,x\right) \\ &= \Phi\left(\frac{e^t\,x}{\sqrt{e^{2t}}}\right) = \Phi(x)\,. \end{aligned}$$

Thus

$$\left(F_t^{\mathbf{OU}}\right)^{-1}(u) = \Phi^{-1}(u)\,.$$

The joint distribution function of (U_s, U_t) is given, for $s \in\,]0, t[$ and for x and y in $\mathbb{R}$ by

$$\begin{aligned} F_{s,t}^{\mathbf{OU}}(x,y) &= \mathbb{P}(U_s \le x, U_t \le y) = C_{s,t}^{\mathbf{OU}}\left(F_s^{\mathbf{OU}}(x), F_t^{\mathbf{OU}}(y)\right) \\ &= C_{s,t}^{\mathbf{OU}}\left(\Phi(x), \Phi(y)\right)\,, \end{aligned} \tag{13}$$

so that

$$C^{\mathbf{OU}}_{s,t}(u, v) = F^{\mathbf{OU}}_{s,t}\left(\Phi^{-1}(u), \Phi^{-1}(v)\right) .$$

On the other hand, one has, in view of Eq. (5),

$$\begin{aligned}F^{\mathbf{OU}}_{s,t}(x, y) &= \mathbb{P}(U_s \le x, U_t \le y) = \mathbb{P}\left(e^{-s} B_{e^{2s}} \le x, e^{-t} B_{e^{2t}} \le y\right) \\ &= \mathbb{P}\left(B_{e^{2s}} \le e^s x, B_{e^{2t}} \le e^t y\right) = C^B_{e^{2s},e^{2t}}\left(F^B_{e^{2s}}(e^s x), F^B_{e^{2t}}(e^t y)\right) \\ &= C^B_{e^{2s},e^{2t}}\left(\Phi\left(\frac{e^s x}{\sqrt{e^{2s}}}\right), \Phi\left(\frac{e^t x}{\sqrt{e^{2t}}}\right)\right) = C^B_{e^{2s},e^{2t}}\left(\Phi(x), \Phi(y)\right) . \qquad (14)\end{aligned}$$

By comparing Eqs. (13) and (14), one obtains

$$C^{\mathbf{OU}}_{s,t}(u, v) = \int_0^u \Phi\left(\frac{e^t\,\Phi^{-1}(v) - e^s\,\Phi^{-1}(w)}{\sqrt{e^{2t} - e^{2s}}}\right) dw , \qquad (15)$$

which is the expression of the copula of the Ornstein–Uhlenbeck process.

Its density is given by

$$c^{\mathbf{OU}}_{s,t}(u, v) = \frac{e^t}{\sqrt{e^{2t} - e^{2s}}}\,\varphi\left(\frac{e^t\,\Phi^{-1}(v) - e^s\,\Phi^{-1}(u)}{\sqrt{e^{2t} - e^{2s}}}\right) .$$

5 The Supremum of a Browian Motion

The supremum (S_t) of a Brownian motion (B_t) is defined by

$$S_t := \sup\{B_s : s \le t\} .$$

By recourse to the reflection principle, see, e.g., [5, Sect. 1.3], one has

$$F_{S_t}(y) = 2\,\Phi\left(\frac{y}{\sqrt{t}}\right) - 1 ;$$

since Φ is strictly increasing, F_{S_t} has an inverse given by

$$F^{-1}_{S_t}(v) = \sqrt{t}\,\Phi^{-1}\left(\frac{v+1}{2}\right) . \qquad (16)$$

Let H_t denote the joint distribution function of (B_t, S_t); for $x \le y$, one has, because of the continuity of both B_t and S_t and as a consequence of Eq. (13.4) in [5],

$$
\begin{aligned}
H_t(x,y) &= \mathbb{P}(B_t \le x, S_t \le y) = \mathbb{P}(B_t \le x) - \mathbb{P}(B_t \le x, S_t \ge y) \\
&= \Phi\left(\frac{x}{\sqrt{t}}\right) - \mathbb{P}(B_t \le y - (y-x), S_t \ge y) \\
&= \Phi\left(\frac{x}{\sqrt{t}}\right) - \mathbb{P}(B_t \ge 2y - x) \\
&= \Phi\left(\frac{x}{\sqrt{t}}\right) - \left(1 - \Phi\left(\frac{2y-x}{\sqrt{t}}\right)\right) \\
&= \Phi\left(\frac{x}{\sqrt{t}}\right) - \Phi\left(\frac{x-2y}{\sqrt{t}}\right), \qquad (17)
\end{aligned}
$$

while, if $x > y$, then

$$
H_t(x,y) = \mathbb{P}(S_t \le y) = 2\,\Phi\left(\frac{y}{\sqrt{t}}\right) - 1. \qquad (18)
$$

Then the unique copula C_t^{BS} of the random vector (B_t, S_t) can be obtained by

$$
C_t^{BS}(u,v) = H_t\left(F_{B_t}^{-1}(u), F_{S_t}^{-1}(v)\right).
$$

From this, one has, if $u \le (v+1)/2$, or, equivalently, if $F_{B_t}^{-1}(u) \le F_{S_t}^{-1}(v)$,

$$
\begin{aligned}
C_t^{BS}(u,v) &= \Phi\left(\sqrt{t}\,\frac{1}{\sqrt{t}}\,\Phi^{-1}(u)\right) \\
&\quad - \Phi\left(\frac{1}{\sqrt{t}}\left[\sqrt{t}\,\Phi^{-1}(u) - 2\sqrt{t}\,\Phi^{-1}\left(\frac{v+1}{2}\right)\right]\right) \\
&= u - \Phi\left(\Phi^{-1}(u) - 2\,\Phi^{-1}\left(\frac{v+1}{2}\right)\right). \qquad (19)
\end{aligned}
$$

If, on the other hand, $u > (v+1)/2$, then

$$
C_t^{BS}(u,v) = 2\,\Phi\left(\sqrt{t}\,\frac{1}{\sqrt{t}}\,\Phi^{-1}\left(\frac{v+1}{2}\right)\right) - 1 = v.
$$

Notice that in either case the result does not depend on time: the copula of B_t and S_t is independent of t and is given by

$$
C^{BS}(u,v) = \begin{cases} u - \Phi\left(\Phi^{-1}(u) - 2\,\Phi^{-1}\left(\frac{v+1}{2}\right)\right), & u \le (v+1)/2, \\ v, & u \ge (v+1)/2. \end{cases}
$$

References

1. Darsow, W.F., Nguyen, B.E., Olsen, T.: Copulas and Markov processes. Illinois J. Math. **36**, 600–642 (1992)
2. Durante, F., Sempi, C.: Principles of Copula Theory. Chapman & Hall/CRC, Boca Raton (2015)
3. Karatzas, I., Shreve, S.E.: Brownian Motion and Stochastic Calculus, 2nd edn. Springer, Berlin (1991)
4. Nelsen, R.B.: An Introduction to Copulas, 2nd edn. Springer, New York (2006)
5. Rogers, L.C.G., Williams, D.: Diffusions, Markov processes and martingales. In: Foundations, 2nd Edn. Cambridge University Press (2000)
6. Schmitz, V.: Copulas and stochastic processes. Ph.D. Dissertation, Rheinische-Westfälische Technische Hochschule, Aachen, Germany (2003)
7. Sklar, A.: Fonctions de répartition à n dimensions et leurs marges. Publ. Inst. Statist. Univ. Paris **8**, 229–231 (1959)
8. Stirzaker, D.: Stochastic processes and models. Oxford University Press (2005)

Extensions of Capacities

Anna Kolesárová and Andrea Stupňanová

Abstract We study extensions of capacities on $N = \{1, \ldots, n\}$ to n-ary aggregation functions acting on $[0, 1]^n$. Besides recalling the universal integral based approaches following the ideas of Klement et al. and generalizations of the Lovász and Owen extensions, we also present some new approaches to extending such capacities which are based on a generalization of the formulas for the discrete Choquet and Sugeno integrals.

1 Introduction

In decision-making processes crisp alternatives are characterized by $\{0, 1\}$-valued score vectors. Given a set $N = \{1, \ldots, n\}$ of criteria, crisp alternatives are described by score vectors $\mathbf{x} = (x_1, \ldots, x_n) \in \{0, 1\}^n$ and evaluated by Boolean utility functions. A Boolean (normed) utility function is a pseudo-Boolean nondecreasing function $u\colon \{0, 1\}^n \to [0, 1]$ satisfying the properties $u(\mathbf{0}) = u(0, \ldots, 0) = 0$ and $u(\mathbf{1}) = u(1, \ldots, 1) = 1$. When criteria are evaluated in the graded scale $[0, 1]$, there is a need to extend Boolean utility functions to utility functions acting on score vectors $\mathbf{x} \in [0, 1]^n$. Utility functions $U\colon [0, 1]^n \to [0, 1]$ should also be nondecreasing to satisfy the Pareto principle. Boolean utility functions $u\colon \{0, 1\}^n \to [0, 1]$ can be identified with capacities on the set N and normed utility functions $U\colon [0, 1]^n \to [0, 1]$ are in a one-to-one correspondence with n-ary aggregation functions on the interval $[0, 1]$. Recall that a capacity m on the set N, [9, 30], is a nondecreasing set function $m\colon 2^N \to [0, 1]$ with the properties $m(\emptyset) = 0$ and $m(N) = 1$. The corresponding m and u are linked by the relation $m(K) = u(\mathbf{1}_K)$, $K \subseteq N$, where $\mathbf{1}_K$ is

---

A. Kolesárová (✉)
Faculty of Chemical and Food Technology, Slovak University of Technology in Bratislava, Radlinského 9, 812 37 Bratislava 1, Slovakia
e-mail: anna.kolesarova@stuba.sk

A. Stupňanová
Faculty of Civil Engineering, Slovak University of Technology in Bratislava, Radlinského 11, 810 05 Bratislava 1, Slovakia
e-mail: andrea.stupnanova@stuba.sk

S. Saminger-Platz and R. Mesiar (eds.), *On Logical, Algebraic, and Probabilistic Aspects of Fuzzy Set Theory*, Studies in Fuzziness and Soft Computing 336,
DOI 10.1007/978-3-319-28808-6_11

the indicator of the set K. An n-ary aggregation function ($n \in \mathbb{N}, n \geq 2$) on the interval $[0, 1]$ is a function $A: [0, 1]^n \to [0, 1]$ which is nondecreasing in each variable and satisfies the boundary conditions $A(\mathbf{0}) = 0$ and $A(\mathbf{1}) = 1$, see [1, 3, 8]. Instead of extensions of Boolean utility functions to utility functions one can investigate extensions of capacities to aggregation functions, i.e., for a given capacity m to look for aggregation functions A satisfying the property $A(\mathbf{1}_K) = m(K)$ for all $K \subseteq N$. Throughout this paper, the set of all capacities on N will be denoted by $\mathscr{M}_n$ and the set of all n-ary aggregation functions on $[0, 1]$ by $\mathscr{A}_n$. The aim of this contribution is to discuss extensions of capacities to n-ary functions acting on $[0, 1]^n$. Note that, in general, these extensions need not be monotone and their range need not be included in $[0, 1]$. In such cases, we will look for the constraints ensuring that for any capacity m on N the corresponding extension is an aggregation function, i.e., it can be seen as a normed utility function satisfying the Pareto principle.

There are several kinds of integrals on N which are based on capacities. Though these integrals are monotone, they are not always extensions of the corresponding capacities. For example, the concave integral introduced by Lehrer [16] is an extension of a capacity m only if m is totally balanced. In particular, if m is supermodular then the concave integral coincides with the Choquet integral [5, 7]. Similarly, in general, the convex integral [22] does not extend the corresponding capacity. It coincides with the Choquet integral if and only if m is a submodular capacity. The Pan-integral of Yang and Klir [31], see also [30], which is based on a semiring $(\mathbb{R}_+, \oplus, \odot)$, is an extension of the considered capacity m only in some particular cases, namely if m is $\oplus$-superadditive, i.e., if $m(K \cup L) \geq m(K) \oplus m(L)$ whenever $K, L \subset N$, $K \cap L = \emptyset$. On the other hand, universal integrals on $[0, 1]$, introduced by Klement et al. in [12], can serve as a positive example of a method how to extend a capacity $m: 2^N \to \{0, 1\}$ into an aggregation function $A: [0, 1]^n \to [0, 1]$ so that $m(K) = A(\mathbf{1}_K)$ for any $K \subseteq N$.

The contribution is organized as follows. In the next section, we recall and exemplify universal integrals and bring several non-classical integrals. In particular, we recall extremal semicopula-based integrals and hierarchical classes of copula-based integrals. We also recall a related extension method proposed by Klement et al. in [11]. In Sect. 3, we discuss the Möbius transform-based extension method generalizing the classical Lovász and Owen extensions [17, 26] and an extension method based on the possibilistic Möbius transform. In Sect. 4, we show new types of extension methods based on a generalization of formulas for the discrete Choquet and Sugeno integrals.

2 Universal Integrals on [0, 1]

A function $\otimes: [0, 1]^2 \to [0, 1]$ is called a semicopula [6] if it is monotone and $e = 1$ is its neutral element, i.e., $x \otimes 1 = 1 \otimes x = x$ for each $x \in [0, 1]$. Observe that semicopulas are not required to be associative or symmetric.

In what follows, we recall the concept of a universal integral on [0, 1], which was introduced by Klement et al. in [12], as a common framework covering the Choquet, Shilkret and Sugeno integrals.

Definition 1 A mapping $I\colon \bigcup_{n\in\mathbb{N}} (\mathcal{M}_n \times [0,1]^n) \to [0,1]$ is called a universal integral on [0, 1] whenever it satisfies the following axioms:

(UI1) For each fixed $n \in \mathbb{N}$, $I|\mathcal{M}_n \times [0,1]^n$ is nondecreasing in both components.
(UI2) There is a semicopula $\otimes\colon [0,1]^2 \to [0,1]$ such that for any $n \in \mathbb{N}, m \in \mathcal{M}_n$, $c \in [0,1]$ and $K \subseteq N$,

$$I(m, c\cdot \mathbf{1}_K) = c \otimes m(K).$$

(UI3) For any $(m_1, \mathbf{x}_1), (m_2, \mathbf{x}_2) \in \bigcup_{n\in\mathbb{N}} (\mathcal{M}_n \times [0,1]^n)$ such that $m_1(\{i \in N_1 \mid x_{1,i} \geq t\}) = m_2(\{j \in N_2 \mid x_{2,j} \geq t\})$ it holds that

$$I(m_1, \mathbf{x}_1) = I(m_2, \mathbf{x}_2).$$

By the axiom (UI2), it holds that

$$I(m, \mathbf{1}_K) = 1 \otimes m(K) = m(K),$$

i.e., universal integrals always extend the considered capacities. The monotonicity of $I(m,\cdot)$ is guaranteed due to (UI1), i.e., $I(m,\cdot)$ is an aggregation function.

Now, we recall three basic universal integrals.

- The *Choquet integral* was introduced in 1953 in [5] as follows

$$Ch_m(\mathbf{x}) = \int_0^1 m(\{i \in N \mid x_i \geq t\})\mathrm{d}t. \tag{1}$$

There are three other equivalent formulas for the discrete Choquet integral. The first one is based on the Möbius transform $M_m : 2^N \to \mathbb{R}$ of a capacity m, given by

$$M_m(K) = \sum_{L\subseteq K} (-1)^{\mathrm{card}(K\setminus L)} m(L), \tag{2}$$

and was introduced in [4]:

$$Ch_m(\mathbf{x}) = \sum_{K\subseteq N} M_m(K) \min\{x_i \mid i \in K\}. \tag{3}$$

Note that the right-hand side of this formula is also known as the Lovász extension of m [17].

Taking into account a geometrical meaning of (1), we can write the following equivalent formulas for the discrete Choquet integral:

$$Ch_m(\mathbf{x}) = \sum_{i=1}^{n} \left(x_{(i)} - x_{(i-1)}\right) m\left(K_{(i)}\right) \tag{4}$$

and

$$Ch_m(\mathbf{x}) = \sum_{i=1}^{n} x_{(i)} \left(m\left(K_{(i)}\right) - m\left(K_{(i+1)}\right)\right), \tag{5}$$

where $(\cdot): N \to N$ is a permutation such that $x_{(1)} \leq \cdots \leq x_{(n)}$ and $K_{(i)} = \{(i), \ldots, (n)\}, i = 1, \ldots, n$. By convention, in (4), $x_{(0)} = 0$ and in (5), $K_{(n+1)} = \emptyset$. Note that if there are some ties between the input values, there exist more permutations satisfying the above constraints, but both formulas (4) and (5) always give the same value of $Ch_m(\mathbf{x})$.

- The *Shilkret integral* was introduced in 1971 in [27], originally for maxitive measures only, as follows

$$Sh_m(\mathbf{x}) = \sup\left\{t \cdot m(\{i \in N \mid x_i \geq t\}) \mid t \in [0, 1]\right\}, \tag{6}$$

  and, equivalently, it can be written as

$$Sh_m(\mathbf{x}) = \sup\left\{x_{(i)} \cdot m\left(K_{(i)}\right) \mid i \in N\right\}. \tag{7}$$

- The *Sugeno integral* was introduced in 1974 in [29] (in Japanese published in 1972) as follows

$$Su_m(\mathbf{x}) = \sup\{\min\{t, m(\{i \in N \mid x_i \geq t\})\} \mid t \in [0, 1]\}. \tag{8}$$

  Equivalently, it holds that

$$Su_m(\mathbf{x}) = \sup\left\{\min\left\{x_{(i)}, m\left(K_{(i)}\right)\right\} \mid i \in N\right\}. \tag{9}$$

Formally, both the Shilkret and Sugeno integrals can be seen as particular examples of the smallest universal integral on [0, 1] related to a semicopula $\otimes$, which was introduced in [12] by the formula

$$I_{\otimes}(m, \mathbf{x}) = \sup\left\{t \otimes m(\{i \in N \mid x_i \geq t\}) \mid t \in [0, 1]\right\}, \tag{10}$$

which can also be written as

$$I_{\otimes}(m, \mathbf{x}) = \sup\left\{x_{(i)} \otimes m\left(K_{(i)}\right) \mid i \in N\right\}. \tag{11}$$

Namely, the Shilkret integral can be obtained for the semicopula $\otimes = \Pi$, $\Pi(x, y) = xy$, and the Sugeno integral for $\otimes = Min$, $Min(x, y) = \min\{x, y\}$.

Now, let us consider copulas [25] as special types of semicopulas. Recall that $C\colon [0, 1]^2 \to [0, 1]$ is a copula if it is a supermodular semicopula, i.e., if for all $\mathbf{x}, \mathbf{y} \in [0, 1]^2$ it holds that

$$C(\mathbf{x} \vee \mathbf{y}) + C(\mathbf{x} \wedge \mathbf{y}) \geq C(\mathbf{x}) + C(\mathbf{y}).$$

Clearly, for $\otimes = C$, the formula (10) gives

$$I_C(m, \mathbf{x}) = \sup\{C(t, m(\{i \in N \mid x_i \geq t\}) \mid t \in [0, 1]\}, \tag{12}$$

which can also be written as

$$I_C(m, \mathbf{x}) = \sup\left\{C\left(x_{(i)}, m\left(K_{(i)}\right)\right) \mid i \in N\right\}.$$

A special class of universal integrals on [0, 1] based on copulas was introduced in [12], see also [11], as follows:

$$I_{[C]}(m, \mathbf{x}) = P_C\left(\left\{(x, y) \in [0, 1]^2 \mid y < m(\{i \in N \mid x_i \geq x\}\right)\right). \tag{13}$$

Here P_C is a probability measure on Borel subsets of the unite square $[0, 1]^2$ generated by the probabilities of the rectangles $[0, x] \times [0, y]$, $P_C([0, x] \times [0, y]) = C(x, y)$. Consequently, using the notation as in (4) and (5), we can write the formula (13) in the form

$$I_{[C]}(m, \mathbf{x}) = \sum_{i=1}^{n} \left(C\left(x_{(i)}, m(K_{(i)})\right) - C\left(x_{(i-1)}, m(K_{(i)})\right)\right) \tag{14}$$

and also as

$$I_{[C]}(m, \mathbf{x}) = \sum_{i=1}^{n} \left(C\left(x_{(i)}, m(K_{(i)})\right) - C\left(x_{(i)}, m(K_{(i+1)})\right)\right). \tag{15}$$

Clearly, if C is the standard product copula Π, which models the independence of random variables, then (14) turns into (4) and (15) turns into (5). Hence, $I_{[\Pi]}$ is the Choquet integral, i.e., $I_{[\Pi]}(m, \mathbf{x}) = Ch_m(\mathbf{x})$ for any $m \in \mathcal{M}_n$ and $\mathbf{x} \in [0, 1]^n$, $n \in \mathbb{N}$.

Similarly, one can show that the comonotone dependence copula Min generates the Sugeno integral, i.e., for any $m \in \mathcal{M}_n$ and $\mathbf{x} \in [0, 1]^n$, $n \in \mathbb{N}$, $I_{[Min]}(m, \mathbf{x}) = Su_m(\mathbf{x})$.

In [13], a family of general copula-based integrals was proposed, compare also [23]. We adopt this proposal to the framework of discrete universal integrals as follows:

Definition 2 Let $n \in \mathbb{N}$ and let $C\colon [0,1]^2 \to [0,1]$ be a fixed copula. The (n, C)-universal integral on $[0,1]$,

$$I_C^{(n)} : \bigcup_{k\in\mathbb{N}} (\mathscr{M}_k \times [0,1]^k) \to [0,1],$$

is given by

$$I_C^{(n)}(m, \mathbf{x}) = \sup \left\{ \sum_{i=1}^{n} \left(C\left(\sum_{j=0}^{i} a_j, m\left(\left\{ p \in \{1,\ldots,k\} \mid x_p \geq \sum_{j=0}^{i} a_j \right\} \right) \right) \right. \right. \\ \left. \left. - C\left(\sum_{j=0}^{i-1} a_j, m\left(\left\{ p \in \{1,\ldots,k\} \mid x_p \geq \sum_{j=0}^{i} a_j \right\} \right) \right) \right) \right\}, \tag{16}$$

where $a_0 = 0, a_1, \ldots, a_n \geq 0$ and $\sum_{j=1}^{n} a_j \leq 1$.

Remark 1 Note that if $n = 1$, for an arbitrary semicopula $S\colon [0,1]^2 \to [0,1]$, the functional $I_S^{(1)} = I_S$ given by (16) is the (weakest) universal integral linked to S, compare (10) and (11). However, $I_S^{(2)}$ does not satisfy the axiom of the monotonicity of universal integrals, in general. To ensure this, S should be supermodular, i.e., a copula.

Example 1 Consider $n = 3$ and the uniform capacity $m \in \mathscr{M}_3$, $m(E) = \frac{\operatorname{card}(E)}{3}$. Let $S\colon [0,1]^2 \to [0,1]$ be a semicopula such that

$$S\left(\frac{1}{3}, \frac{2}{3}\right) = S\left(\frac{2}{3}, \frac{1}{3}\right) = S\left(\frac{2}{3}, \frac{2}{3}\right) = \frac{1}{3}, \text{ and } S\left(\frac{1}{3}, \frac{1}{3}\right) = 0.$$

For example, the function

$$S(x, y) = \operatorname{med}\left(Min(x, y), W(x, y), \max\left\{ x - \frac{1}{3}, y - \frac{1}{3} \right\} \right),$$

where $W\colon [0,1]^2 \to [0,1]$, $W(x, y) = \max\{0, x + y - 1\}$, is the Fréchet-Hoeffding lower bound (the weakest copula), satisfies the given conditions. Then

$$I_S^{(2)}\left(m, \left(\frac{2}{3}, \frac{2}{3}, \frac{2}{3}\right)\right) = \frac{2}{3},$$

$$I_S^{(2)}\left(m, \left(\frac{1}{3}, \frac{2}{3}, \frac{2}{3}\right)\right) = S\left(\frac{1}{3}, 1\right) + S\left(\frac{2}{3}, \frac{2}{3}\right) - S\left(\frac{1}{3}, \frac{2}{3}\right) = \frac{1}{3} + \frac{1}{3} - \frac{1}{3} = \frac{1}{3},$$

$$I_S^{(2)}\left(m, \left(\frac{1}{3}, \frac{1}{3}, \frac{2}{3}\right)\right) = S\left(\frac{1}{3}, 1\right) + S\left(\frac{2}{3}, \frac{1}{3}\right) - S\left(\frac{1}{3}, \frac{1}{3}\right) = \frac{1}{3} + \frac{1}{3} - 0 = \frac{2}{3},$$

which shows that the functional $I_S^{(2)}(m, \cdot)$ is not monotone.

Proposition 1 *Let* $C: [0,1]^2 \to [0,1]$ *be a copula. Then*

(i) $I_C = I_C^{(1)} \leq I_C^{(2)} \leq \cdots \leq I_C^{(n)} \leq \cdots \leq I_{[C]}$.
(ii) *For a fixed n, for any* $m \in \mathcal{M}_n$ *and any copula C, the integrals* $I_C^{(n)}$ *and* $I_{[C]}$ *coincide.*

The integrals $I_C^{(n)}$, $n \in \mathbb{N}$, can be seen as lower approximations of the integral $I_{[C]}$. In a similar way, we can introduce upper approximations of $I_{[C]}$.

Definition 3 Let $n \in \mathbb{N}$ and let $C: [0,1]^2 \to [0,1]$ be a fixed copula. The (C, n)-universal integral on $[0, 1]$,

$$I_{(n)}^C \colon \bigcup_{k \in \mathbb{N}} \left(\mathcal{M}_k \times [0,1]^k\right) \to [0,1],$$

is given by

$$I_{(n)}^C(m, \mathbf{x}) = \inf \left\{ \sum_{i=1}^{n} \left(C\left(\sum_{j=0}^{i-1} a_j, m\left(\left\{ p \in \{1, \ldots, k\} \mid x_p > \sum_{j=0}^{i-1} a_j \right\} \right) \right) \right. \right.$$
$$\left. \left. - C\left(\sum_{j=0}^{i-1} a_j, m\left(\left\{ p \in \{1, \ldots, k\} \mid x_p > \sum_{j=0}^{i-1} a_j \right\} \right) \right) \right) \right\}, \tag{17}$$

where $a_0 = 0$, $a_1, \ldots, a_n \geq 0$ and $\max\{x_1, \ldots, x_n\} \leq \sum_{j=1}^{n} a_j \leq 1$.

Observe that for any copula C it holds that $I_{(1)}^C(m, \mathbf{x}) = (\max x_i, m(\{i \in N \mid x_i > 0\}))$.

Proposition 2 *Let* $C: [0,1]^2 \to [0,1]$ *be a copula. Then*

(i) $I_{(1)}^C \geq I_{(2)}^C \geq \cdots \geq I_{(n)}^C \geq \cdots \geq I_{[C]}$.
(ii) *For a fixed n, for any* $m \in \mathcal{M}_n$ *and any copula C, the integrals* $I_{(n)}^C$ *and* $I_{[C]}$ *coincide.*

Summarizing Propositions 1 and 2, we get the following conclusion.

Corollary 1 *For a fixed n, for any* $m \in \mathcal{M}_n$*, copula C and* $\mathbf{x} \in [0,1]^n$*, it holds that*

$$\begin{aligned} I_C(m, \mathbf{x}) &= I_C^{(1)}(m, \mathbf{x}) \leq I_C^{(2)}(m, \mathbf{x}) \leq \cdots \leq I_C^{(n)}(m, \mathbf{x}) = I_{[C]}(m, \mathbf{x}) \\ &= I_{(n)}^C(m, \mathbf{x}) \leq \cdots \leq I_{(2)}^C(m, \mathbf{x}) \leq I_{(1)}^C(m, \mathbf{x}). \end{aligned}$$

Example 2 Let $n = 3$ and let $m \in \mathcal{M}_3$ be a uniform capacity, $m(E) = \frac{\text{card}(E)}{3}$ and $\mathbf{x} = \left(\frac{1}{3}, \frac{2}{3}, 1\right)$. Then for the product copula Π we have

$$I_\Pi(m, \mathbf{x}) = Sh_m(\mathbf{x}) = I_\Pi^{(1)}(m, \mathbf{x}) = \frac{4}{9},$$
$$I_\Pi^{(2)}(m, \mathbf{x}) = \frac{5}{9},$$
$$I_\Pi^{(3)}(m, \mathbf{x}) = I_{(3)}^\Pi(m, \mathbf{x}) = Ch_m(\mathbf{x}) = \frac{6}{9},$$
$$I_{(2)}^\Pi(m, \mathbf{x}) = \frac{7}{9},$$
$$I_{(1)}^\Pi(m, \mathbf{x}) = \frac{9}{9} = 1.$$

For the Fréchet-Hoeffding lower bound W we get:

$$I_W(m, \mathbf{x}) = I_W^{(1)}(m, \mathbf{x}) = \frac{1}{3},$$
$$I_W^{(2)}(m, \mathbf{x}) = \frac{2}{3},$$
$$I_W^{(3)}(m, \mathbf{x}) = I_{[W]}(m, \mathbf{x}) = I_{(3)}^W(m, \mathbf{x}) = I_{(2)}^W(m, \mathbf{x}) = I_{(1)}^W(m, \mathbf{x}) = \frac{3}{3} = 1,$$

and for the Fréchet-Hoeffding upper bound Min:

$$I_{Min}(m, \mathbf{x}) = Su_m(\mathbf{x}) = I_{Min}^{(1)}(m, \mathbf{x}) = \cdots = I_{(2)}^{Min}(m, \mathbf{x}) = \frac{2}{3},$$
$$I_{(1)}^{Min}(m, \mathbf{x}) = 1.$$

3 Möbius Transform-Based Extensions of Capacities

The formula (3) can be seen as an extension of the capacity m constructed by means of the Möbius transform of m and the aggregation function Min, which inspired us to propose a generalization of this approach. We briefly recall the main idea and the results from [14].

Let $\mathbf{x} = (x_1, \ldots, x_n) \in [0, 1]^n$ be any input n-tuple. To each subset $K \subseteq N$ we assign an n-tuple $\mathbf{x}_K = (u_1, \ldots, u_n)$, where

$$u_i = \begin{cases} x_i & \text{if } i \in K, \\ 1 & \text{otherwise.} \end{cases}$$

Clearly, $\mathbf{x}_\emptyset = (1, \dots, 1) = \mathbf{1}$ and $\mathbf{x}_N = \mathbf{x}$. Let A be an n-ary aggregation function, m a capacity on N and M_m its Möbius transform. Let us define the function $F_{m,A}$: $[0, 1]^n \to \mathbb{R}$ by

$$F_{m,A}(x_1, \dots, x_n) = \sum_{K \subseteq N} M_m(K)\, A(\mathbf{x}_K). \tag{18}$$

Note that, in general, there is a difference between formulas (3) and (18). While in (3) the aggregation function Min, $Min(x_1, \dots, x_k) = \min\{x_1, \dots, x_k\}$, is considered as a k-ary aggregation function for $k = 1, \dots, n$, in (18) only a fixed n-ary aggregation function A is applied. The k-tuples corresponding to the index sets $K \subseteq N$ with $|K| = k$ are always completed into n-tuples by setting to 1 the components with the indices in $N \setminus K$. Moreover, formula (3) defines an extended aggregation function [8], i.e., it can be applied for any arity n. However, for the n-ary aggregation function $A = Min$ for which $e = 1$ is a neutral element, both formulas coincide, i.e., $F_{m,Min}$ given by (18) is the Lovász extension of m [8, 17, 20]. Similarly, for the product aggregation function $A = \Pi$, $\Pi(x_1, \dots, x_n) = x_1 \cdot x_2 \cdot \dots \cdot x_n$, (18) gives the same values as $\sum_{K \subseteq N} M_m(K) \prod_{i \in K} x_i$, i.e., $F_{m,\Pi}$ is the so-called Owen extension of m [26].

In general, the function $F_{m,A}$ defined by (18) is neither an extension of m nor an aggregation function.

Let us first characterize all aggregation functions $A \in \mathscr{A}_n$ such that for each $m \in \mathscr{M}_n$, $F_{m,A}$ is an extension of m.

Theorem 1 *Let $A \in \mathscr{A}_n$. For each $m \in \mathscr{M}_n$, the function $F_{m,A}$ defined by (18) is an extension of m if and only if A is an aggregation function with zero annihilator.*

As mentioned above, our aim is to characterize all aggregation functions $A \in \mathscr{A}_n$ with the property that, for all $m \in \mathscr{M}_n$, the function $F_{m,A}$ is an aggregation function extending m. Let us first focus on the binary case:

Let $A \in \mathscr{A}_2$ be a binary aggregation function with zero annihilator and $m \in \mathscr{M}_2$. If m is determined by the values $a, b \in [0, 1]$, $m(\{1\}) = a$, $m(\{2\}) = b$, then

$$F_{m,A}(x, y) = a\, A(x, 1) + b\, A(1, y) + (1 - a - b) A(x, y). \tag{19}$$

For characterizing the functions $A \in \mathscr{A}_2$ satisfying our requirement, the notion of 2-dimensional quasi-copula (quasi-copula for short) is needed. Recall that a function $Q\colon [0, 1]^2 \to [0, 1]$ is called a quasi-copula if it is a 1-Lipschitz semicopula, i.e., a semicopula satisfying for all $x, x', y, y' \in [0, 1]$ the property

$$|Q(x', y') - Q(x, y)| \leq |x' - x| + |y' - y|.$$

Note that each copula C is also a quasi-copula, but in general, the opposite claim is not true.

Theorem 2 *Let $A \in \mathscr{A}_2$. For each $m \in \mathscr{M}_2$, the function $F_{m,A}$ given by (18) is an aggregation function extending m if and only if for each $(x, y) \in [0, 1]^2$ it holds that*

$$A(x, y) = Q(f(x), g(y)), \tag{20}$$

where Q is a quasi-copula and f, g are nondecreasing $[0, 1] \to [0, 1]$ functions with the properties $f(0) = g(0) = 0$, $f(1) = g(1) = 1$.

Corollary 2 *For aggregation functions $A \in \mathscr{A}_2$ given by $A(x, y) = Q(f(x), g(y))$ with Q, f and g satisfying Theorem 2, it holds that*

$$F_{m,A}(x, y) = F_{m,Q}(f(x), g(y)).$$

Example 3 Consider the Hamacher product (with parameter 0) $C^{H_0}: [0, 1]^2 \to [0, 1]$ given by $C^{H_0}(x, y) = \frac{xy}{x+y-xy}$ whenever $(x, y) \neq (0, 0)$ (and $C^{H_0}(0, 0) = 0$). Note that C^{H_0} is a copula, i.e., also a quasi-copula. Define the functions $f, g: [0, 1] \to [0, 1]$ by $f(x) = g(x) = \frac{2x}{1+x}$. Then, by Theorem 2, we obtain the aggregation function A,

$$A(x, y) = C^{H_0}(f(x), g(y)) = \frac{2xy}{x + y},$$

i.e., the standard harmonic mean.

Let m be any capacity in $\mathscr{M}_2$, with $m(\{1\}) = a$, $m(\{2\}) = b$, a, $b \in [0, 1]$. Then, using the formula (19), see also Corollary 2, we get the aggregation function $F_{m,A}: [0, 1]^2 \to [0, 1]$,

$$F_{m,A}(x, y) = \frac{2ax}{x + 1} + \frac{2by}{y + 1} + (1 - a - b)\frac{2xy}{x + y},$$

which extends the capacity m.

Before characterizing all n-ary aggregation functions $A \in \mathscr{A}_n$ satisfying our requirement, let us recall that for an n-ary aggregation function A the A-volume of an n-box $[\mathbf{a}, \mathbf{b}]$ in $[0, 1]^n$, $[\mathbf{a}, \mathbf{b}] = [a_1, b_1] \times \cdots \times [a_n, b_n]$, is defined by

$$V_A([\mathbf{a}, \mathbf{b}]) = \sum (-1)^{\alpha(\mathbf{c})} A(\mathbf{c}),$$

where the sum is taken over all vertices $\mathbf{c} = (c_1, \ldots, c_n)$ of the n-box $[\mathbf{a}, \mathbf{b}]$ (i.e., each c_k is equal to either a_k or b_k), and $\alpha(\mathbf{c})$ is the number of indices k's such that $c_k = a_k$.

For $n \geq 2$, we have the following general result.

Theorem 3 *For a given aggregation function $A \in \mathscr{A}_n$, $n \geq 2$, the following claims are equivalent.*

(i) *For each fuzzy measure $m \in \mathscr{M}_n$, the function $F_{m,A}$ is an aggregation function extending m.*
(ii) *A is an aggregation function with zero annihilator and for each $[\mathbf{a}, \mathbf{b}] \subseteq [0, 1]^n$ such that $\{0, 1\} \cap \{a_1, \ldots, a_n, b_1, \ldots, b_n\} \neq \emptyset$, the A-volume $V_A([\mathbf{a}, \mathbf{b}])$ is non-negative.*

For example, all n-copulas [25, 28] are suitable aggregation functions for our construction. Recall that *n-copulas* are defined as functions $C\colon [0, 1]^n \to [0, 1]$ satisfying

(C1) the boundary conditions:
if $0 \in \{x_1, \ldots, x_n\}$ then $C(x_1, \ldots, x_n) = 0$,
$C(1, \ldots, 1, x_j, 1, \ldots, 1) = x_j$ for each $j = 1, \ldots, n$ and each $x_j \in [0, 1]$,
(C2) the n-increasing property:
$V_C([\mathbf{a}, \mathbf{b}]) \geq 0$ for each n-box $[\mathbf{a}, \mathbf{b}]$ in $[0, 1]^n$.

It is easy to see that the aggregation functions described in the following proposition also have zero annihilator and that the volumes of all n-boxes in $[0, 1]^n$ are non-negative.

Proposition 3 *Let C be an n-copula, $f_i\colon [0, 1] \to [0, 1]$, $i = 1, \ldots, n$, nondecreasing functions such that for each i, $f_i(0) = 0$, $f_i(1) = 1$. Then the function $A\colon [0, 1]^n \to [0, 1]$ defined by*

$$A(x_1, \ldots, x_n) = C\left(f_1(x_1), \ldots, f_n(x_n)\right),$$

is an n-ary aggregation function such that, for all $m \in \mathscr{M}_n$, the function $F_{m,A}$ is an aggregation function extending m.

However, there are aggregation functions A which are neither copulas nor obtained by a distortion of some copula as in Proposition 3, and in spite of that, for all $m \in \mathscr{M}_n$, $F_{m,A}$ is an aggregation function extending m, see the following example.

Example 4 Consider the function $A\colon [0, 1]^3 \to [0, 1]$, given by

$$A(x, y, z) = xyz \min(1, x + y + z).$$

The function A is a ternary aggregation function with zero annihilator. After a quite tedious computations one obtains that, for each $m \in \mathscr{M}_3$, $F_{m,A}$ is an aggregation function extending m. However, A is not a copula, because, e.g., for $\mathbf{a} = (0.3, 0.3, 0.3)$ and $\mathbf{b} = (0.35, 0.35, 0.35)$ the A-volume of the corresponding 3-box is $V_A([\mathbf{a}, \mathbf{b}]) = -0.0019 < 0$, i.e., the 3-increasing property of A fails. This example can be generalized to any $n > 3$.

Finally, let us mention that in [15] we also studied extension methods based on the possibilistic Möbius transform $M_m^\vee$ of a capacity m, where $M_m^\vee : 2^N \to [0,1]$ is given by

$$M_m^\vee(K) = \begin{cases} 0 & \text{if } m(K) = m(L) \\ & \text{for some } L \subsetneq K, \\ m(K) & \text{otherwise.} \end{cases} \tag{21}$$

More details on the possibilistic Möbius transform can be found, e.g., in [19, 21].

For a semicopula $\otimes : [0,1]^2 \to [0,1]$, a capacity $m \in \mathcal{M}_n$ and an aggregation function $A \in \mathcal{A}_n$, let $F_{m,A}^{\vee,\otimes} : [0,1]^n \to [0,1]$ be a function given by

$$F_{m,A}^{\vee,\otimes}(\mathbf{x}) = \bigvee_{K \subseteq N} \left(M_m^\vee(K) \otimes A(\mathbf{x}_K) \right), \tag{22}$$

where the meaning of $\mathbf{x}_K$ is the same as in (18).

Note that the function $F_{m,A}^{\vee,\otimes}$ can also be written as

$$F_{m,A}^{\vee,\otimes}(\mathbf{x}) = \bigvee_{K \subseteq N} (m(K) \otimes A(\mathbf{x}_K)). \tag{23}$$

Theorem 4 *Let $A \in \mathcal{A}_n$, $n \geq 2$. Then the function $F_{m,A}^{\vee,\otimes}$ is for each $m \in \mathcal{M}_n$ and each semicopula $\otimes$ an aggregation function extending m if and only if A is an aggregation function with zero annihilator.*

Let us note that

- Taking $\otimes = \wedge$, where $\wedge(x, y) = \min\{x, y\}$, and $A = Min$, we obtain
$$F_{m,Min}^{\vee,\wedge}(\mathbf{x}) = Su_m(\mathbf{x}),$$
i.e., $F_{m,Min}^{\vee,\wedge}$ is the Sugeno integral with respect to the capacity m (compare with the relation of $F_{m,Min}$ and the Choquet integral Ch_m).
- In general, for any $n \geq 2$, $F_{m,Min}^{\vee,\Pi}$ leads to the Shilkret integral Sh_m.
- Similarly, for any semicopula $\otimes$, the function $F_{m,Min}^{\vee,\otimes}$ coincides with the weakest universal integral $I_\otimes$ based on $\otimes$ with respect to the capacity m, see (10).

4 Modifications of the Choquet and Sugeno Integrals

4.1 Fusion Function-Based Discrete Choquet-Like Integral

Inspired by the formula (4) for the discrete Choquet integral, we proposed its fusion function-based modification [24], substituting the product in (4) by a binary fusion function. Note that a binary fusion function is any function $F\colon [0,1]^2 \to [0,1]$, see [2].

Definition 4 ([24]) Let $F\colon [0,1]^2 \to [0,1]$ be a fusion function satisfying for each $y \in [0,1]$ the property $F(0,y) = 0$. Then the functional $C_m^F\colon [0,1]^n \to [0,n]$ given by

$$C_m^F(\mathbf{x}) = \sum_{i=1}^{n} F\left(x_{(i)} - x_{(i-1)}, m(K_{(i)})\right). \tag{24}$$

is called a fusion function-based discrete Choquet-like integral (F-based discrete Choquet-like integral for short).

The meaning of symbols in (24) is the same as in (4). Note that due to the property $F(0,y) = 0$ valid for each $y \in [0,1]$, the functional C_m^F is defined correctly, see [24]. Therefore, we will always work with $F \in \mathcal{F}_0$, where

$$\mathcal{F}_0 = \{F\colon [0,1]^2 \to [0,1] \mid \forall y \in [0,1],\ F(0,y) = 0\}.$$

It is clear that for $F = \Pi$, for any capacity $m \in \mathcal{M}_n$, C_m^Π is equal to the Choquet integral, $C_m^\Pi = Ch_m$.

However, in general, functions C_m^F need not be monotone. Moreover, their range need not be contained in $[0,1]$, see the following example.

Example 5 Consider the function

$$F(x,y) = \begin{cases} 0 & \text{if } x = 0, \\ y & \text{if } x = 1, \\ 1 & \text{otherwise.} \end{cases}$$

Then, for a uniform capacity $m \in \mathcal{M}_n, m(K) = \frac{\text{card}(K)}{n}$, and any $\mathbf{x} \in]0,1]^n$, such that $\text{card}(\{x_1, \ldots, x_n\}) = n$, we have $C_m^F(\mathbf{x}) = n$ and $C_m^F(\mathbf{1}_K) = m(K)$ for each $K \subseteq N$. Considering, for example, $n = 3$ and the points $(0,0,0)$, $(0.2, 0.3, 0.7)$ and $(1,1,1)$, we get $C_m^F(0,0,0) = 0$, $C_m^F(0.2, 0.3, 0.7) = 3$ and $C_m^F(1,1,1) = 1$, which shows that the monotonicity of C_m^F is violated and $Ran(C_m^F) \nsubseteq [0,1]$. Note that C_m^F is an extension of m for any $m \in \mathcal{M}_n$.

A deep study of properties of functionals C_m^F can be found in [24]. In this paper we only focus on the properties of fusion functions F which are necessary and sufficient for C_m^F to be an aggregation function and an extension of m for any $m \in \mathcal{M}_n$.

Proposition 4 *Let $F\colon [0,1]^2 \to [0,1]$ be a fusion function. Then, for any capacity $m \in \mathcal{M}_n$, C_m^F extends m, i.e., $C_m^F(\mathbf{1}_K) = m(K)$ for each $K \subseteq N$, if and only if $F(1,y) = y$ for each $y \in [0,1]$.*

Now, we fix $n = 2$. Then any capacity $m \in \mathcal{M}_2$ is completely determined by values $a, b \in [0,1]$, where $a = m(\{1\})$ and $b = m(\{2\})$. For any $F \in \mathcal{F}_0$ and $m \in \mathcal{M}_2$ it holds that

$$C_m^F(x, y) = \begin{cases} F(x, 1) + F(y - x, b) & \text{if } x \leq y, \\ F(y, 1) + F(x - y, a) & \text{otherwise.} \end{cases} \quad (25)$$

Note that always $C_m^F(0, 0) = 0$ and $C_m^F(1, 1) = 1$ whenever $F(1, 1) = 1$.

Let us denote by $\mathscr{F}_{(2)}$ the class of all binary fusion functions $F \in \mathscr{F}_0$ such that for each $m \in \mathscr{M}_2$, C_m^F is a binary aggregation function. The following theorem provides a characterization of all $F \in \mathscr{F}_{(2)}$.

Theorem 5 *Let $F \in \mathscr{F}_0$. Then $F \in \mathscr{F}_{(2)}$ if and only if $F(x, 1) = x$ for all $x \in [0, 1]$ and the function $F(\cdot, y) \colon [0, 1] \to [0, 1]$ is increasing and 1-Lipschitz for each $y \in [0, 1]$, i.e.,*

$$0 \leq F(x_2, y) - F(x_1, y) \leq x_2 - x_1$$

whenever $0 \leq x_1 \leq x_2 \leq 1$.

Note that if C_m^F is an aggregation function for any $m \in \mathscr{M}_2$, then C_m^F is idempotent and translation invariant [24].

Before giving an example, we recall that each quasi-copula Q satisfies the constraints of Theorem 5, hence $Q \in \mathscr{F}_{(2)}$.

Example 6 (i) Consider the greatest quasi-copula Min and a capacity $m \in \mathscr{M}_2$, with $m(\{1\}) = a$, $m(\{2\}) = b$. Then

$$\begin{aligned} C_m^{Min}(x, y) &= \begin{cases} \min\{y, x + b\} & \text{if } x \leq y, \\ \min\{x, y + a\} & \text{otherwise} \end{cases} \\ &= \max\{\min\{y, x + b\}, \min\{x, y + a\}\}, \end{aligned}$$

see Fig. 1.

(ii) Consider the smallest quasi-copula W and $m \in \mathscr{M}_2$, determined by $a, b \in [0, 1]$ as in (i). Then

$$\begin{aligned} C_m^{W}(x, y) &= \begin{cases} \max\{x, y + b - 1\} & \text{if } x \leq y, \\ \max\{y, x + a - 1\} & \text{otherwise} \end{cases} \\ &= \min\{\max\{x, y + b - 1\}, \max\{y, x + a - 1\}\}, \end{aligned}$$

see Fig. 2.

In Theorem 5, we have characterized the class $\mathscr{F}_{(2)}$ of all binary fusion functions $F \in \mathscr{F}_0$ such that the F-based discrete Choquet-like integral C_m^F defined on $[0, 1]^2$ is a binary aggregation for any $m \in \mathscr{M}_2$. Now, we will characterize the class $\mathscr{F}_{(n)}$, $n > 2$, of all binary fusion functions $F \in \mathscr{F}_0$ for which C_m^F defined on $[0, 1]^n$ is an n-ary aggregation function for any $m \in \mathscr{M}_n$.

Theorem 6 *Let $F \in \mathscr{F}_0$. For any $n > 2$, $F \in \mathscr{F}_{(n)}$ if and only if $F \in \mathscr{G}$, where $\mathscr{G}$ is the set of all $F \in \mathscr{F}_0$ which are given by $F(x, y) = xf(y)$, for some increasing function $f \colon [0, 1] \to [0, 1]$ satisfying $f(1) = 1$.*

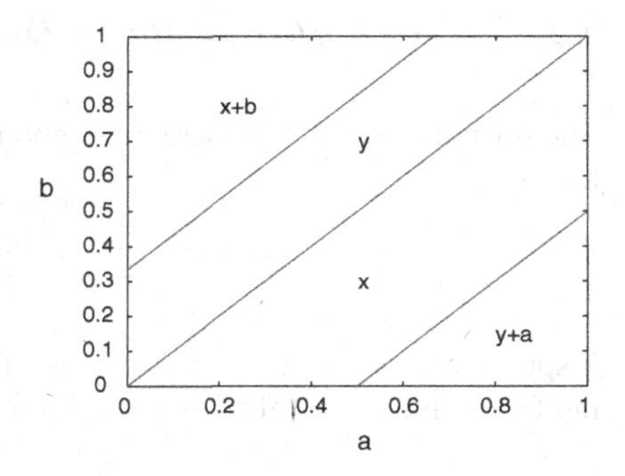

Fig. 1 2D illustration of C_m^{Min} from Example 6(i) for $a = 1/2, b = 1/3$

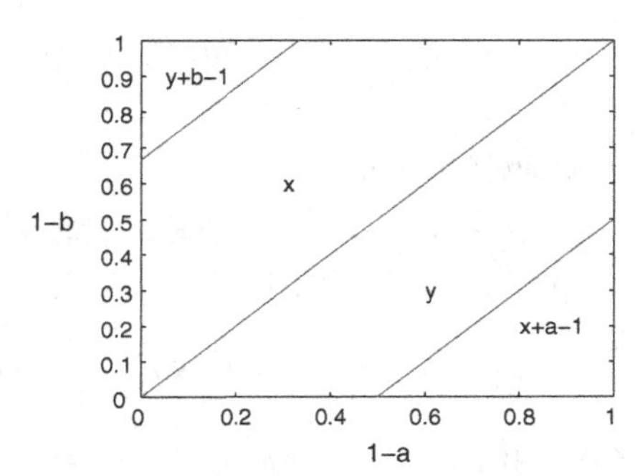

Fig. 2 2D illustration of C_m^W from Example 6(ii) for $a = 1/2, b = 1/3$

Remark 2 In the case of Theorem 6, if $F \in \mathscr{F}_{(n)}$, then $C_m^F = Ch_{f(m)}$, where $Ch_{f(m)}$ is the Choquet integral with respect to the capacity $f(m) \in \mathscr{M}_n$ given by

$$f(m)(K) = \begin{cases} 0 & \text{if } K = \emptyset, \\ f(m(K)) & \text{otherwise.} \end{cases}$$

Indeed, if $F \in \mathscr{F}_{(n)} = \mathscr{G}$, then $F(x, y) = xf(y)$, and

$$C_m^F(\mathbf{x}) = \sum_{i=1}^{n} F\left(x_{(i)} - x_{(i-1)}, m\left(K_{(i)}\right)\right) = \sum_{i=1}^{n} \left(x_{(i)} - x_{(i-1)}\right) f\left(m\left(K_{(i)}\right)\right)$$
$$= Ch_{f(m)}(\mathbf{x}).$$

Summarizing, we have proved that for any $n > 2$, an F-based discrete Choquet-like integral C_m^F is an n-ary aggregation function for any $m \in \mathscr{M}_n$ if and only if it is the Choquet integral with respect to the capacity m distorted by a function f generating F, i.e., $C_m^F = Ch_{f(m)}$.

In our approach, we have generalized the formula (4) of the discrete Choquet integral. Note that the formula (5) can be generalized in a similar way. For more details we refer to [10].

4.2 Fusion Function-Based Discrete Sugeno-Like Integral

The formula (9) for the discrete Sugeno integral $Su_m \colon [0,1]^n \to [0,1]$ can be written as

$$Su_m(\mathbf{x}) = \bigvee_{i=1}^{n} \min\left\{x_{(i)}, m\left(K_{(i)}\right)\right\}.$$

Inspired by this formula, for any fusion function $F \colon [0,1]^2 \to [0,1]$, we can define the function $Su_m^F \colon [0,1]^n \to [0,1]$ by the formula

$$Su_m^F(\mathbf{x}) = \bigvee_{i=1}^{n} F\left(x_{(i)}, m\left(K_{(i)}\right)\right), \tag{26}$$

compare with [18].

Example 7 (i) For fusion functions F given by $F(x,y) = x^\lambda y$, with $\lambda \in]0, \infty[$, the formula (26) gives

$$Su_m^F(\mathbf{x}) = Sh_m\left(x_1^\lambda, \ldots, x_n^\lambda\right).$$

(ii) If $F(x,y) = \min\left\{x^\lambda, y\right\}$, $\lambda \in]0, \infty[$, then

$$Su_m^F(\mathbf{x}) = Su_m\left(x_1^\lambda, \ldots, x_n^\lambda\right).$$

Clearly, for $\lambda = 1$, in (i) we obtain $F = \Pi$ and $Su_m^\Pi(\mathbf{x}) = Sh_m(\mathbf{x})$ for each $\mathbf{x} \in [0,1]^n$ (compare with (7)), i.e., our approach also covers the Shilkret integral.

We first show a sufficient condition for F ensuring that Su_m^F is a pre-aggregation function for any capacity m. A function $A \colon [0,1]^n \to [0,1]$ is said to be a pre-aggregation function [18] if it satisfies the boundary conditions of aggregation functions $A(\mathbf{0}) = 0$ and $A(\mathbf{1}) = 1$, and if there is at least one direction $\mathbf{r} = (r_1, \ldots, r_n) \in [0,1]^n$, $\mathbf{r} \neq \mathbf{0}$, in which A is nondecreasing, i.e.,

$$A(x_1 + cr_1, \ldots, x_n + cr_n) \geq A(x_1, \ldots, x_n)$$

for each $(x_1, \ldots, x_n) \in [0,1]^n$ and $c > 0$ such that $(x_1 + cr_1, \ldots, x_n + cr_n) \in [0,1]^n$. Note that aggregation functions are pre-aggregation functions which are nondecreasing in the direction of any vector $\mathbf{r} \in [0,1]^n \setminus \{\mathbf{0}\}$.

Proposition 5 *Let $F \colon [0,1]^2 \to [0,1]$ be a function increasing in the first variable and let for each $y \in [0,1]$, $F(0,y) = 0$ and $F(1,1) = 1$. Then S_m^F defined in (26) is a pre-aggregation function for any capacity $m \in \mathcal{M}_n$.*

Example 8 Consider the function $F \colon [0,1]^2 \to [0,1]$, $F(x,y) = x|2y - 1|$. Note that F is a proper pre-aggregation function (not an aggregation function) which satisfies the constraints of Proposition 5, and thus, for any $m \in \mathcal{M}_n$, the function $S_m^F \colon [0, 1]^n \to [0, 1]$, $S_m^F(\mathbf{x}) = \bigvee_{i=1}^{n} F\left(x_{(i)}, m\left(A_{(i)}\right)\right)$ is a pre-aggregation function (even an aggregation function). For example, for $n = 2$, $m(\{1\}) = 1/3$, $m(\{2\}) = 3/4$, we get

$$S_m^F(x, y) = \begin{cases} x \vee \frac{y}{2} & if\ x \le y, \\ y \vee \frac{x}{3} & if\ x > y. \end{cases}$$

Note that any function F satisfying the constraints of Proposition 5 is, in fact, a binary $(1, 0)$-nondecreasing pre-aggregation function which satisfies $F(0, y) = 0$ for each $y \in [0, 1]$.

It is not difficult to check that Su_m^F, given by (26), extends an arbitrary capacity m whenever $F(1, y) = y$ for each $y \in [0, 1]$. Moreover, the monotonicity of F ensures the monotonicity of Su_m^F, independently of m. If F is an aggregation function, so is C_m^F. Summarizing these facts, we obtain the following result.

Theorem 7 *Let $F \colon [0, 1]^2 \to [0, 1]$ be an aggregation function such that $F(0, y) = 0$ and $F(1, y) = y$ for each $y \in [0, 1]$. Then, for any $n \ge 2$ and $m \in \mathcal{M}_n$, the function $Su_m^F \colon [0, 1]^n \to [0, 1]$ defined by (26), is an aggregation function extending the capacity m.*

Acknowledgments This work was supported by the project APVV–14–0013.

References

1. Beliakov, G., Pradera, A., Calvo, T.: Aggregation Functions: A Guide for Practitioners. Springer, Heidelberg (2007)
2. Bustince, H., Fernandez, J., Kolesárová, A., Mesiar, R.: Directional monotonicity of fusion functions. Eur. J. Oper. Res. **244**, 300–308 (2015)
3. Calvo, T., Kolesárová, A., Komorníková, M., Mesiar, R.: Aggregation operators: properties, classes and construction methods. In: Calvo, T., Mayor, G., Mesiar, R. (eds.) Aggregation Operators. New Trends and Applications, pp. 3–107. Physica-Verlag, Heidelberg (2002)
4. Chateauneuf, A., Jaffray, J.Y.: Some characterizations of lower probabilities and other monotone capacities through the use of Möbius inversion. Math. Soc. Sci. **17**, 263–283 (1989)
5. Choquet, G.: Theory of capacities. Ann. Inst. Fourier **5**, 131–295 (1953)
6. Durante, F., Sempi, C.: Semicopulae. Kybernetika **41**, 315–328 (2005)
7. Grabisch, M.: Fuzzy integral in multicriteria decision making. Fuzzy Sets Syst. **69**, 279–298 (1995)
8. Grabisch, M., Marichal, J.-L., Mesiar, R., Pap, E.: Aggregation Functions. Cambridge University Press, Cambridge (2009)
9. Grabisch, M., Murofushi, T., Sugeno, M.: Fuzzy measures and integrals. Theory and Applications. Physica Verlag, Heidelberg (2000)
10. Horanská, Ľ., Šipošová, A.: A note on a generalization of the Choquet integral. In: Proceedings of Uncertainty Modelling 2015, STU, Bratislava (2015)
11. Klement, E.P., Mesiar, R., Pap, E.: Measure-based aggregation operators. Fuzzy Sets Syst. **142**(1), 3–14 (2004)

12. Klement, E.P., Mesiar, R., Pap, E.: A universal integral as common frame for Choquet and Sugeno integral. IEEE Trans. Fuzzy Syst. **18**, 178–187 (2010)
13. Klement, E.P., Mesiar, R., Spizzichino, F., Stupňanová, A.: Universal integrals based on copulas. Fuzzy Optim. Decis. Making **13**, 273–289 (2014)
14. Kolesárová, A., Stupňanová, A., Beganová, J.: Aggregation-based extensions of fuzzy measures. Fuzzy Sets Syst. **194**, 1–14 (2012)
15. Kolesárová, A., Stupňanová, A.: On some extensions methods for normed utility functions. In: Proceedings AGOP'2011, pp. 169–174. Benevento, Italy (2011)
16. Lehrer, E.: A new integral for capacities. Econ. Theor. **39**, 157–176 (2009)
17. Lovász, L.: Submodular function and convexity. In: Mathematical Programming: The State of the Art, pp. 235–257. Springer, Berlin (1983)
18. Lucca, G., Sanz, J.A., Pereira Dimuro, G., Bedregal, B., Mesiar, R., Kolesárová, A., Bustince, H.: Pre-aggregation functions: construction and an application. IEEE Trans. Fuzzy Syst. (2015), accepted for publication
19. Marichal, J.-L., Mathonet, P., Tousset, E.: Mesures floues définies sur une échelle ordinale. Working paper (1996)
20. Marichal, J.-L.: Aggregation of interacting criteria by means of the discrete Choquet integral. In: Calvo, T., Mayor, G., Mesiar, R. (eds.) Aggregation Operators. New Trends and Applications, pp. 224–244. Physica-Verlag, Heidelberg (2002)
21. Mesiar, R.: k-Order pan-additive discrete fuzzy measures. In: 7th IFSA World Congress, pp. 488–490. Prague (1997)
22. Mesiar, R., Li, J., Pap, E.: Superdecomposition integral. Fuzzy Sets Syst. **259**, 3–11 (2015)
23. Mesiar, R., Stupňanová, A.: Decomposition integrals. Int. J. Approximate Reasoning **54**, 1252–1259 (2013)
24. Mesiar, R., Kolesárová, A., Bustince, H., Dimuro, G.P., Bedregal, B.: Fusion functions based Choquet-like integrals. Submitted (2015)
25. Nelsen, R.B.: An Introduction to Copulas, 2nd edn. Springer, New York (2006)
26. Owen, G.: Multilinear extensions of games. In: The Shapley value. In: Roth, A.E. (ed.) Essays in Honour of Lloyd S. Shapley, pp. 139–151. Cambridge University Press (1988)
27. Shilkret, N.: Maxitive measure and integration. Indag. Math. **33**, 109–116 (1971)
28. Sklar, A.: Fonctions de répartition à n dimensions et leurs marges. Publ. Inst. Stat. Univ. Paris **8**, 229–231 (1959)
29. Sugeno, M.: Theory of fuzzy integrals and its applications. PhD thesis, Tokyo Institute of Technology (1974)
30. Wang, Z., Klir, G.J.: Fuzzy Measure Theory. Plenum Press, New York (1992)
31. Yang, Q.: The pan integral on the fuzzy measure space. Fuzzy Math. **3**, 107–114 (1985). (in Chinese)

Multi-source Information Fusion Using Measure Representations

Ronald R. Yager

Abstract We first look at the issue of representing information about an uncertain variable using a measure. We focus on some notable measures that can be used. We discuss the role of aggregation functions in the task of combining measures to form new measures. We look at this in the framework of multi-source information fusion. We focus on the fusion of probabilistic and possibilistic information and discuss its role in hard-soft information fusion. We look at some characterizing features associated with measures used to represent uncertain values of variables. We discuss the concepts of assurance and opportunity that play a role in the process of answering questions using information obtained from a measure.

1 Introduction

Intelligent decision-making must take advantage of various types of available information, much of which has some degree of uncertainty. Some of the most important sources of information are statistical data, physical sensors and human observation. Here we are interested in the problem of multi-source uncertain information fusion [10]. We note that statistical data as well physical sensor provided information generally have a probabilistic type of uncertainty whereas the linguistic information provided by humans typically introduces a possibilistic type of uncertainty [19]. We note that many different structures have been suggested for the representation of uncertain information [8, 9, 13]. In order to provide a unified framework for the representation of different types of uncertain information we use a set measure approach for the representation of uncertain information. We discuss a set measure representation of uncertain information. Here we shall we focus on measures over finite universes. In the multi-source fusion problem, we have a collection of pieces of information that must be fused based on expert provided instructions on how to fuse these pieces of information. Generally these instructions can involve a combination of

R. Yager (✉)
Machine Intelligence Institute, Iona College, New Rochelle, NY 10801, USA
e-mail: yager@panix.com

S. Saminger-Platz and R. Mesiar (eds.), *On Logical, Algebraic, and Probabilistic Aspects of Fuzzy Set Theory*, Studies in Fuzziness and Soft Computing 336,
DOI 10.1007/978-3-319-28808-6_12

linguistically and mathematically expressed directions. We began to look at aggregation functions [1, 5] as a way for fusing this information. Some of the most important sources of information are statistical data, physical sensors and human observation. We note that statistical data as well physical sensor provided information generally have a probabilistic type of uncertainty whereas the linguistic information provided by humans typically introduces a possibilistic type of uncertainty [19]. The combination of these two types of information is referred to as hard-soft fusion [10] and generally involves the fusion of probabilistic and possibilistic measures. Two characterizing features of measures that are used to represent uncertain information are the entropy and attitudinal character of a measure, we briefly investigate these concepts. We discuss the ideas of assurance and opportunity [20] that play a fundamental role in the task of answering questions using information represented with a measure.

2 Representing Uncertain Information Using Set Measures

A mapping $\mu\colon 2^X \to [0, 1]$ is called a set function. We now introduce a special type of set function called a fuzzy measure, it is also referred to as a monotonic set measure [6, 12, 14].

Definition 1 A fuzzy measure on X is a mapping $\mu\colon 2^X \to [0, 1]$ such that

1. $\mu(\emptyset) = 0$,
2. $\mu(X) = 1$,
3. $\mu(A) \geq \mu(B)$ if $B \subseteq A$.

We can easily see that for any subsets A and B we have

$$\mu(A \cap B) \leq \min(\mu(A), \mu(B)),$$
$$\mu(A \cup B) \geq \max(\mu(A), \mu(B)).$$

A set function we can associate with any measure is its dual. We define the dual of μ as the set function $\hat{\mu}\colon X \to [0, 1]$ defined such that $\hat{\mu}(A) = 1 - \mu(\overline{A})$ where $\overline{A}$ is the complement of A. We can easily show that if μ is a measure on X then $\hat{\mu}$ is also a measure on X. We note that the dual of the dual is the original measure, $\hat{\hat{\mu}}(A) = 1 - \hat{\mu}(A) = 1 - (1 - \mu(A)) = \mu(A)$. Thus a measure and its dual are unique pairs.

We say an element $x \in X$ is irrelevant with respect to μ if for all A we have $\mu(A \cup \{x\}) = \mu(A)$. We observe that if x is irrelevant then $\mu(\{x\}) = 0$. If we define $E_{-x} = X \setminus \{x\}$ and if x is irrelevant then $\mu(E_{-x}) = 1$. We see this follows since $X = E_{-x} \cup \{x\}$ and hence $\mu(X) = \mu(E_{-x} \cup \{x\}) = \mu(E_{-x})$. We must emphasize that $\mu(\{x\}) = 0$ does not necessarily make x irrelevant. We note that if x is irrelevant to μ it is also irrelevant to its dual $\hat{\mu}$.

A fuzzy measure provides a very general structure for the representation of our knowledge about an uncertain variable. Let V be a variable taking its value in the space X. In using μ to express our knowledge about V we provide the following interpretation. For any subset B of X we have that $\mu(B)$ indicates our **anticipation** that the value of V lies in B. We see that $\mu(\emptyset) = 0$ reflects the fact that the value of V will not be in the null set. The property $\mu(X) = 1$, called the normality condition, indicates the fact that we are completely confident that the value of V lies in X. Finally the monotonicity of μ reflects the fact that you cannot be more confident of finding the value V in the set B than in a set that contains B. Here we say that the set or event A happens if the value of V lies in A, thus $\mu(A)$ is seen as our anticipation that the event A happens.

We shall use the expression V *is* μ to denote the situation where knowledge about V is carried by the set measure μ. In the following we shall restrict ourselves to the case where X is finite. Let us look at the representation of some types of knowledge using this representation. Consider first the case of certainty where we know that the value V is exactly x_1. We can express this using a fuzzy measure μ such that $\mu(B) = 1$ if $x_1 \in B$ and $\mu(B) = 0$ if $x_1 \notin B$. We refer to this as a Dirac measure focused at x_1. Consider now the case of probabilistic uncertainty where we have $Prob(x_i) = p_i$. In this case we define μ such that $\mu(B) = \sum_{x_i \in B} p_i$. Here we see that $\mu(\{x_i\}) = p_i$. Thus we see that the probability measure has the property of additivity. Thus the probability measure is a special fuzzy measure. In this case we refer to $\mu(B)$ by the more specific name of probability of B, $\mu(B) = Prob(B)$. Since $\mu(X) = Prob(X) = 1$ here we require $\sum_{i=1}^{n} p_i = 1$. For this measure $\mu(A \cup B) = \mu(A) + \mu(B) - \mu(A \cap B)$. We see here if $A \cap B = \emptyset$ then $\mu(A \cup B) = \mu(A) + \mu(B)$. We note that this measure requires only knowledge of the p_i to completely determine $\mu(A)$ for all A. A special case of this is where all p_i are the same, here $p_i = 1/n$ where n is the cardinality of X. Another important type of uncertainty representable in this framework is possibilistic uncertainty [4, 21]. We recall that usually this type of uncertainty is obtained from a linguistic description of our knowledge. Here we associate with each x_i a value $\pi_i \in [0, 1]$ called the possibility of x_i. In this case the associated measure μ is called a possibility measure and is defined as $\mu(A) = \max_{x_i \in A} \pi_i$. The condition that $\mu(X) = 1$ requires that at least one $\pi_i = 1$. We see for this measure that $\mu(A \cup B) = \max(\mu(A), \mu(B))$ for any A and B. We note that the Dirac measure is the only measure that is both a probability and possibility measure. Another class of measures are decomposable measures. These generalize the possibility measure. Here if S is any t-conorm [7] then if $\mu(\{x_i\}) = a_i$ we have $\mu(A) = S_{x_i \in A}(a_i)$. A requirement here is that $\mu(X) = S_{x_i \in X}(a_i) = 1$. We note that for these measure if $A \cap B = \emptyset$, then $\mu(A \cup B) = S(\mu(A), \mu(B))$. Since max is a t-conorm we see that the possibility measure is an example of these decomposable measures. Another notable type of measure is one in which $\mu(A \cap B) = \min(\mu(A), \mu(B))$ these are referred to as certainty or necessity measures [2]. In this case we refer to $\mu(A)$ as the necessity of A and use the notation $Nec(A)$ instead of $\mu(A)$. We note that if we let $F_i = X \setminus \{x_i\}$ then we can completely define this measure using $\mu(F_i)$. Here for any A, $\mu(A) = \min_{x_i \in A}(\mu(F_i))$. We shall refer to $\mu(F_i) = \beta_i$. Since $\mu(\emptyset) = \min_{x_j \in X}(\beta_j)$ then we see that fact that $\mu(\emptyset) = 0$ requires at least one $\beta_j = 0$. The case where all β_j are equal implies all $\beta_j = 0$. We note that a necessity measure is the dual of a possibility measure. We also note that these necessity measures are notable in that while for all measures $\mu(A \cap B) \leq \min(\mu(A), \mu(B))$, the necessity measure attains this minimum, $\mu(A \cap B) = \min(\mu(A), \mu(B))$. A generalization of the necessity measure can be obtained by defining $\mu(A \cap B) = T(\mu(A), \mu(B))$ where T is a t-norm [7]. Here again all we need are n values for $\mu(F_i) = \beta_i$. Another important example of fuzzy measures are the cardinality based uncertainty measures. Here our anticipation that the value of the variable lies in a particular subset just depends on the number of elements in the subset. No distinction is made between the individual outcomes elements. Let a_i be a collection of parameters such that $0 = a_0 \leq a_1 \leq \cdots \leq a_n = 1$. Let us denote $|B|$ as the cardinality of the set B, the number of elements in B. We now define $\mu(B) = a_{|B|}$. Here then for all subsets with the same number of elements we have the same of anticipation of finding the value in it. We can show for this measure that no elements are impossible. A useful parameter associated with these measures is $w_i = a_i - a_{i-1}$ which indicates the benefit of adding the ith element to a set. Two important examples of cardinality based uncertainty measures are the optimistic measure μ^* and the pessimistic measure μ_*. For μ^* we have $\mu^*(A) = 1$ for $A \neq \emptyset$. and $\mu^*(\emptyset) = 0$ and for $\mu_*(A) = 0$ for $A \neq X$ and $\mu_*(X) = 1$. For μ^* we have $a_j = 1$ for all $j \geq 1$ and $a_0 = 0$ and for μ_* we have $a_j = 0$ for all $j < n$ and $a_n = 1$. Another example of cardinality-based measure is one where $a_i = i/n$. We can use the w_i to provide a measure of optimism associated with a cardinality based measure μ on X [15]. If $|X| = n$ then

$$Op(\mu) = \sum_{j=1}^{n} \frac{(n-j)w_j}{n-1}.$$

We see when μ is μ^* that $Op(\mu^*) = 1$ and when $\mu = \mu_*$ that $Op(\mu) = 0$. If μ is such that $a_i = i/n$ for all i, then $Op(\mu) = 0.5$.

3 Measures Derived from Other Measures

Assume μ_1 is some fuzzy measure on X and $F\colon\ [0, 1] \to [0, 1]$ is such that $F(0) = 0$, $F(1) = 1$, and $F(a) \geq F(b)$ if $a > b$ then it can be shown that $\mu(A) = F(\mu_1(A))$ is a fuzzy measure [19].

Another class of derived measures are those composed from other measures. Let μ_1 and μ_2 be two fuzzy measures on X. We can show that the set function μ defined such that $\mu(A) = \mu_1(A)\mu_2(A)$ for all $A \subseteq X$ is also a measure. We see that $\mu(\emptyset) = \mu_1(\emptyset)\mu_2(\emptyset)$, $\mu(X) = \mu_1(X)\mu_2(X)$ and if $A \supseteq B$ then $\mu_1(A) \geq \mu_1(B)$ and $\mu_2(A) \geq \mu_2(B)$ and hence $\mu(A) = \mu_1(A)\mu_2(A) \geq \mu_1(B)\mu_2(B) = \mu(B)$. Thus μ as defined above is a fuzzy measure. This result can easily be extended to the fusion of q fuzzy measure $\mu(A) = \prod_{i=1}^{q} \mu_i(A)$.

In the following we shall provide a generalization of this result, which shall form the basis of our approach to the fusion of multi-source information.

Definition 2 An aggregation function G is a function of the integer $q > 1$ arguments $G\colon [0, 1]^q \rightarrow [0, 1]$ having the properties $G(0, 0, \ldots, 0) = 0$, $G(1, 1, \ldots, 1) = 1$ and $G(a_1, \ldots a_q) \geq G(b_1, \ldots, b_q)$ if all $a_j \geq b_j$.

Theorem 1 *Assume for $j = 1$ to q that μ_j are a collection of fuzzy measures on X. Then μ defined such that for all $A \subseteq X$*

$$\mu(A) = G(\mu_1(A), \ldots, \mu_q(A))$$

is a fuzzy measure.

Proof 1. $\mu(\emptyset) = G(\mu_1(\emptyset), \ldots, \mu_q(\emptyset)) = G(0, 0, \ldots, 0) = 0$,
2. $\mu(X) = G(\mu_1(X), \ldots, \mu_q(X)) = G(1, \ldots, 1) = 1$,
3. We have $\mu(A) = G(\mu_1(A), \ldots, \mu_q(A))$ and $\mu(B) = G(\mu_1(B), \ldots, \mu_q(B))$. With $B \subseteq A$, since $\mu_j(A) \geq \mu_j(B)$ for all j, then $\mu(A) \geq \mu(B)$.

In situations where we have a finite space and the μ_j are simple measures the calculation of the aggregate measure is relatively easily. We shall refer to these measures as *quasi-simple*. Let $X = \{x_1, \ldots, x_q\}$. Assume we have q simple measures, μ_j for $j = 1$ to q. For $j = 1$ to q and $i = l$ to n let α_{ij} be the parameters associated with these measures. Thus here $\mu_k(A)$ just depends on α_{ik} for $x_i \in A$. Let G be some aggregation function then with $\mu(A) = G(\mu_1(A), \ldots, \mu_q(A))$ the calculation of $\mu(A)$ just depends on the values of $\mu_j(A)$ which in turn just depends on the parameter α_{ij}. The key point here is that for a measure defined using the aggregation of simple measures, the calculation of the measure of a set is not complex.

4 On the Fusion of Information from Multiple Sources

In the problem of fusing information from multiple sources we have a collection of n sources each of which is providing information about the variable V. Here we shall assume each of these pieces of information can be expressed in terms of a fuzzy measure. Thus our information is a collection V *is* μ_j for $j = 1$ to q where each μ_j is a fuzzy measure defined on the domain X of V.

In addition we must have some expert provided instructions on how to fuse these pieces of information so as to obtain a unified view of the value of V. The basis of this expert provided knowledge can be very diverse. It can be based on a human expert's practical experience in processing multiple-sourced information. It can be based on some formal data mining technology. Most generally these instructions can involve a combination of linguistically and mathematical expressed directions. A fundamental task in multi-source information fusion (MSIF) is the translation of these instructions into formal operations that can be applied. The task of operationalizing these expert provided instructions is generally a very complex one, it often involves a tradeoff between precisely following the instructions and functionality, translating the instruction into implementable operations. Here the capacity of Zadeh's paradigm of computing with words [22, 23] can become very useful for translating these instructions into formal operations.

The type of aggregation operators previously introduced provides a very useful tool for implementing a wide body of expert provided instructions for fusing multiple pieces of information. One of our interests here is to look at the use of these aggregation operators for the fusion of information expressed via fuzzy measures. We shall be particularly concerned with probability and possibility type information as they represent two very important classes of provided information. We note that possibilistic information often arises from a linguistic description of the value of some variable. An example of this is information such as the house is *close*. Here *close* is a linguistic term that can be expressed using fuzzy sets which in turn induces a possibility distribution on the variable "the distance of the house to the river." Probabilistic information often appears because it provides an effective model to represent the accuracy of physical sensing devices. It should be emphasized that the use of probability is not a reflection of randomness in the variable of interest but a reflection of the lack of accuracy of the measuring device.

5 Conjuncting Possibility and Probability Measures: Hard-Soft Fusion

In the following we shall consider the conjunction "anding" fusion of probabilistic and possibilistic information. The t-norm aggregation operator provides an aggregation implementation of an anding or conjunction of multiple pieces of information [1]. This use of "and" usually reflects an instruction to find a fused value that simultaneously satisfies all the multiple pieces of information. Assume we have two pieces of information, $V\ is\ \mu_1$ and $V\ is\ \mu_2$. Consider their fusion $V\ is\ \mu$ where $V\ is\ \mu = V is$ μ_1 **and** V is μ_2. Here μ is a fuzzy measure defined such that for any subset A of the domain X of V we have

$$\mu(A) = T(\mu_1(A), \mu_2(A))$$

where T is a t-norm.

Let us look at this for some notable cases of μ_1 and μ_2. Consider first the case where $V\ is\ \mu_1$, indicates the value of V is exactly x_1 and $V\ is\ \mu_2$, indicates the value of V is exactly x_2. Here μ_1 has $\mu_1(A) = 1$ for all A s.t. $x_1 \in A$ and $\mu_1(A) = 0$ for all A s.t. $x_1 \notin A$. μ_2 has similar information, $\mu_2(A) = 1$ if $x_2 \in A$ and $\mu_2(A) = 0$ if $x_2 \notin A$. In this case we get as our fused value $\mu(A) = 1$ if $\{x_1, x_2\} \subseteq A$ and $\mu(A) = 0$ otherwise. Here we have an anticipation of one of finding the value of V in any set containing both x_1 and x_2. It is interesting to note that in the case where we have two sources giving very different values for the variable we provide a fused solution that tried to unify these two conflicting values rather then reporting a conflict. This is useful, because for some decisions the action taken when V is either x_1 or x_2 is the same so a solution need not differentiate. In a more general situation, if we have

q sources each saying that V has the value x_j then our fused function μ would be such that $\mu(A) = 1$ if $\{x_1, \ldots, x_j\} \subseteq A$ and $\mu(A) = 0$ otherwise. Again here we see a kind of unification of the different pieces of information.

We can now consider the case where μ_1 is as above but where μ_2 is a possibility distribution with $\mu_2(\{x_j\}) = \pi_j$ and $\mu_2(A) = \max_{x_j \in A} \pi_j$. In this case we have for any subset A with $x_1 \in A$ that $\mu(A) = \max_{x_j \in A} \pi_j$ and $\mu(A) = 0$ if $x_1 \notin A$. In particular we note that $\mu(\{x_1\}) = \pi_1$ and $\mu(\{x_j\}) = 0$ for $j \neq 1$. We see that μ is not a pure possibility distribution. We note that all the preceding results hold for any t-norm since at least one of the arguments was always with 1 or 0.

We again consider the case of two pieces of information, V *is* μ_1 and V *is* μ_2 where μ_1 indicates that V is exactly x_1 while μ_2 is any arbitrary measure. Here we get for the fusion of the conjunction of these that V *is* μ where μ is such that $\mu(A) = 0$ if $x_1 \notin A$ and $\mu(A) = \mu_2(A)$ if $x_1 \in A$.

An interesting special case of this occurs when μ_2 is a probability distribution with p_i the probability of x_i. In this case we see that while $\mu(\{x_1\}) = p_1$ we have $\mu(\{x_j\}) = 0$ for $j \neq 1$. Furthermore for any A such that $x_1 \notin A$ we get $\mu(A) = 0$ while for any A such that $x_1 \in A$ we have $\mu(A) = \sum_{x_j \in A} p_j$. We see μ is not a probability measure if $p_1 < 1$. As indicated earlier $\mu(A)$ is our anticipation of finding a value V in the subset A. We observe that $\mu(A) = 0$ or $\mu(A) \geq p_1$.

We now consider the conjunction in the case where μ_1 is a probability distribution and μ_2 is a possibility distribution. In this case $\mu(A) = T(\mu_1(A), \mu_2(A))$ where $\mu_1(A) = \sum_{x_j \in A} p_j$ and $\mu_2(A) = \max_{x_i \in A} \pi_i$. We observe here that for any t-norm T we get $A = \{x_j\}$ that $\mu(\{x_j\}) = T(p_j, \pi_j)$.

Let us look at this fusion of probabilistic and possibilistic information for some notable examples of t-norms. We first consider the product t-norm. In this case $\mu(\{x_j\}) = p_j \cdot \pi_j$ and more generally

$$\mu(A) = \mu_1(A) \cdot \mu_2(A) = \left(\sum_{x_j \in A} p_j \right) \cdot \max_{x_j \in A} \pi_j = \text{Prob}(A) \cdot \text{Poss}(A).$$

We further observe that the product t-norm leads to a nice formulation where

$$\mu(A) = \left(\sum_{x_j \in A} p_j \right) \cdot \max_{x_j \in A} \pi_j = \sum_{x_j \in A} p_j \cdot \left(\max_{x_j \in A} \pi_j \right).$$

Thus we see that each of the p_j is multiplied by the maximum possibility in A. Furthermore we can express this as

$$\mu(A) = \sum_{x_j \in A} p_j(\pi_j + \Delta_j) = \sum_{x_j \in A} (p_j \pi_j + p_j \Delta_j)$$

where $\Delta_j = \pi_A^* - \pi_j$ with $\pi_A^* = \max_{x_k \in A} \pi_k$. With $p_j \pi_j = \mu(\{x_j\})$ we see that we can express

$$\mu(A) = \sum_{x_j \in A} (\mu(\{x_j\}) + p_j \Delta_j).$$

We see also if $\tilde{A} = A \cup \{x_{j+1}\}$ then

$$\mu(\tilde{A}) = \mu(A) + \mu(\{x_{j+1}\}) + p_{j+1}\Delta_j, \qquad \text{if } \pi_{j+1} \leq \pi_A^*$$
$$\mu(\tilde{A}) = \mu(A) + \mu(\{x_{j+1}\}) + \pi_{j+1} - \pi_A^*, \quad \text{if } \pi_{j+1} > \pi_A^*.$$

We further observe that if A has one x_j such that $\pi_j = 1$ then $\mu(\tilde{A}) = \mu(A) + p_{j+1}$.

Example 1 Assume $X = \{x_1, x_2, x_3\}$ and let μ_1 be a probability distribution such that $p_1 = 0.7$, $p_2 = 0.2$ and $p_3 = 0.1$. Let μ_2 be a possibility distribution such that $\pi_1 = 0.3$, $\pi_2 = 1$ and $\pi_3 = 0.8$. In this case we have for any subset A of X that $\mu(A) = \mu_1(A) \cdot \mu_2(A) = \text{Prob}(A) \cdot \text{Poss}(A)$:

$$\mu(\{x_1\}) = 0.7 \cdot 0.3 = 0.21 \qquad \mu(\{x_2\}) = 0.2 \cdot 1 = 0.2$$
$$\mu(\{x_3\}) = 0.1 \cdot 0.8 = 0.8 \qquad \mu(\{x_1, x_2\}) = 0.9 \cdot 1 = 0.9$$
$$\mu(\{x_1, x_3\}) = 0.8 \cdot 0.8 = 0.64 \qquad \mu(\{x_2, x_3\}) = 0.3 \cdot 1 = 0.3$$
$$\mu(\{x_1, x_2, x_3\}) = 1 \cdot 1 = 1$$

We see here that we have introduced a new class of measures, which we shall refer to as P2 measures. Let us understand this measure. This measure is characterized by two sets of parameters $P = [p_1, \ldots, p_n]$ and $\Pi = [\pi_1, \ldots, \pi_n]$ where all p_i and $\pi_i \in [0, 1]$ and $\sum_{i=1}^n p_i = 1$ and $\max_j \pi_j = 1$. Here $\mu(A) = \sum_{x_j \in A} p_i \cdot \max_{x_j \in A} \pi_j$ and in particular $\mu(\{x_j\}) = \pi_j p_j$.

We note that this approach can be easily generalized to the case where we have more than two sources of information some of which are possibilistic and some probabilistic. Thus here if V *is* μ_i are a collection of q pieces of information about V then their fusion under the preceding "anding" imperative is $\mu(A) = \prod_{i=1}^q \mu_i(A)$.

6 Characterizing Features: Entropy and Attitude

Here we now describe some tools useful in the characterization of fuzzy measures that are used to model uncertain variables. These are particularly useful in comparing fuzzy measures with respect to properties such as their overall uncertainty. We refer the reader to [16–18] more details on these. In probability theory an important tool is the Shannon entropy. With this tool we are able to indicate the overall uncertainty of a probability distribution. Assume we have a probability distribution on the space $X = \{x_1, \ldots, x_n\}$ where p_j is the probability of x_j. Here p_j is capturing the distinction in our belief about each of the elements, p_j is the support for x_j as the outcome. We recall the Shannon entropy associated with this distribution is $H(P) = -\sum_j p_j \ln(p_j)$. The Shannon entropy quantifies the overall uncertainty associated with the probability distribution. In [16] we extended this idea to fuzzy measures. In order to extend this to a general measure μ we first introduce the idea of the Shapley index [11]. For any $x_j \in X$ we define its Shapley index S_j as

$$S_j = \sum_{k=0}^{n-1} (r_k \sum_{\substack{K \subset F_j \\ |K|=k}} (\mu(K \cup \{x_j\}) - \mu(K))).$$

In the above K is a subset of cardinality $|K|$, $F_j = X \setminus \{x_j\}$ and $r_k = \frac{(n-k-1)!k!}{n!}$. It can be shown that for any fuzzy measure μ it is always the case that $S_j \in [0, 1]$ and $\sum_{j=1}^{n} S_j = 1$.

In some sense the S_j are reflecting the information about the distinction between the anticipation of each of x_j. As a matter of fact it can be that if μ is a probability measure then $S_j = p_j$. Thus the Shapley index is capturing information about the distinction in the elements.

Using these Shapley values in [16] we extended the idea of entropy to a measure μ by defining the entropy of μ as $H(\mu) = -\sum_j S_j \ln(S_j)$. Since $S_i = p_i$ for a probability measure this definition is compatible with the classic entropy. We note that if μ is a possibility measure and the x_i are indexed such that $\mu(\{x_j\}) \geq \mu(\{x_i\})$ for $j > i$ then $S_j = \sum_{i=1}^{j} \frac{\mu(\{x_i\}) - \mu(\{x_{i-1}\})}{n+1-i}$.

A special interesting case is that of the cardinality based measure. For this measure it can be shown that $S_j = 1/n$ for all x_j. Here then there is no distinction between the x_j and the entropy is the largest, $H(\mu) = \ln(n)$.

If μ is a measure with Shapley indices S_j and Shapley entropy $H(\mu) = -\sum_j S_j \ln(S_j)$ then if $\hat{\mu}$ is the dual of μ we now show that its Shapley indices $\hat{S}_j = S_j$ and hence $H(\hat{\mu}) = H(\mu)$, duals have the same entropy. For the dual we have

$$\hat{S}_j = \sum_{k=0}^{n-1} (r_k \sum_{\substack{k \in F_j, \\ |K|=k}} (\hat{\mu}(K \cup \{x_j\}) - \hat{\mu}(K)))$$

where $F_j = X \setminus \{x_j\}$. Since $\hat{\mu}(A) = 1 - \mu(A)$ then

$$\hat{\mu}(K \cup \{x_j\}) - \hat{\mu}(K) = (\mu(\overline{K}) - \mu(\overline{K \cup \{x_j\}})).$$

We observe that $\overline{K \cup \{x_j\}} = \overline{K} \setminus \{x_j\}$ and hence we get

$$\hat{\mu}(K \cup \{x_j\}) - \hat{\mu}(K) = \mu(\overline{K}) - \mu(\overline{K} \setminus \{x_j\})$$

and therefore $\hat{S}_j = S_j$.

Now we consider another characterizing feature of a fuzzy measure used to convey information about an uncertain variable which is called the attitudinal character of the fuzzy measure [18]. Consider the two cardinality based measures μ^* and μ_*. We recall $\mu^*(A) = 1$ for all $A \neq \emptyset$ and $\mu^*(\emptyset) = 0$ while $\mu_*(A) = 0$ for all $A \neq X$ and $\mu(X) = 1$. As these are both cardinality-based measures they both have no information distinguishing the elements in X regarding their being the value of V. We recall that $\mu(A)$ is the anticipation of finding the value of V in A. We see that these two measures are dealing with our lack of information in very different ways, μ^* deals with the complete lack of knowledge in a very optimistic way, it anticipates finding the value of V in any non-null subset. On the other hand μ_* deals with the complete lack of knowledge in a very pessimistic way it doesnt anticipate finding the value of V in any subset except X. We see that these two measures display polar attitudes about the anticipation of finding V when faced with lack of information. Another cardinality based measure, $\mu(A) = \frac{\text{card}(A)}{n}$, falls between these two extremes and is more neutral.

The preceding fuzzy measures have illustrated different attitudes about the nature of uncertainty. In the following we introduce a characterization of a fuzzy measure that allows us to quantify these differing attitudes. We begin by introducing the cardinality index of a fuzzy measure [18].

Definition 3 Let μ be a measure on X where the cardinality of X, $|X| = n$. For $k = 0$ to $n - 1$ we define C_k as

$$C_k = \lambda_k \sum_{\substack{\text{all } K \\ |K|=k}} \left(\sum_{x \notin K} (\mu(K \cup \{x\}) - \mu(K)) \right)$$

where $\lambda_k = \frac{(n-k-1)!k!}{n!}$. We call C_k the kth cardinality index.

We see that C_k measures the average gain in anticipation we get in going from a subset of cardinality k to one of cardinality $k + 1$. In [18] it was shown that $C_k \in [0, 1]$ and $\sum_{k=0}^{n-1} C_k = 1$.

Yager [18] used the cardinality index to define a characteristic of a fuzzy measure called its attitudinal character. The attitudinal character of a measure m is defined as

$$AC(\mu) = \frac{1}{n-1} \sum_{k=0}^{n-1} (n + k + 1) C_k.$$

We note that since $\sum_{k=0}^{n-1} C_k = 1$ then we can express the attitudinal character as

$$AC(\mu) = 1 - \frac{\sum_{k=0}^{n-1} k C_k}{n-1}.$$

In [18] it was shown that $AC(\mu) \in [0, 1]$. Furthermore the large values of AC indicate a more optimistic type of measure while the small values indicate a more pessimistic type of measure, $AC(\mu) = 1$ being the most optimistic and $AC(\mu) = 0$ being the most pessimistic. Thus the attitudinal character is providing a scale characterizing a measures degree of optimism/pessimism about its anticipation of finding the variables value in a given subset.

In [18] we obtained the cardinality index for cardinality-based measures and probabilistic type measures. We recall for a cardinality based measure μ we have a set $0 \leq a_0 \leq a_i \leq a_n = 1$ such that $\mu(E) = a_{|E|}$, where $|E|$ is the cardinality of E. We showed that for this type of measure the cardinality index is

$$C_k = a_{k+1} - a_k \qquad \text{for } k = 0 \text{ to } n-1.$$

A few notable examples of this are worth pointing out here. In the case of μ^* where $\mu^*(E) = 1$ for all $E \neq \emptyset$ and $\mu^*(\emptyset) = 0$ it was shown that $C_0 = 1$ and $C_k = 0$ for all $k \neq 0$. In the case μ_* where $\mu_*(E) = 0$ for $E \neq X$ and $\mu(X) = 1$ then $C_{n-1} = 1$ and $C_k = 0$ for other k. In the case where $a_j = \frac{j}{n}$ then $C_k = \frac{1}{n}$ for all $k = 0$ to $n-1$.

Another situation studied in [18] was the additive or probabilistic uncertainty. Here we associate with each x_i a value $p_i \in [0, 1]$ so that they sum to one. In [18] it was shown that independent of the value of the p_i the cardinality index for all k is always $C_k = \frac{1}{n}$. Thus the cardinality index doesn't distinguish between probability distributions, all probability distributions have the same cardinality index.

Let us now consider the attitudinal character, $AC(\mu) = 1 - \frac{\sum_{k=0}^{n-1} kC_k}{n-1}$, of these measures. First we consider μ^*, here $C_0 = 1$ and all other $C_k = 0$ and hence $AC(\mu^*) = 1$. This is as expected since μ^* is the most optimistic measure. On the other hand for μ_* since $C_{n-1} = 1$ and all other $C_k = 0$ we have $AC(\mu_*) = 0$. Again this is as expected since μ_* is very pessimistic. Consider now the probabilistic case where $C_k = \frac{1}{n}$ for all k. Here $AC(\mu) = 0.5$, this has a neutral value of 0.5. This is also the case of the cardinality based measure with $a_j = \frac{j}{n}$.

We shall investigate the relationship between the cardinality index of a measure μ and the cardinality index of its dual $\hat{\mu}$. We recall for μ,

$$C_k = \lambda_k \sum_{\substack{\text{all } K \\ |K|=k}} \Big(\sum_{x \notin K} (\mu(K \cup \{x\}) - \mu(K))\Big),$$

it is the average increase in anticipation in going for sets of cardinality k to $k+1$. If $\hat{\mu}$ is the dual then

$$\hat{C}_k = \lambda_k \sum_{\substack{\text{all } K \\ |K|=k}} \Big(\sum_{x \notin K} (\hat{\mu}(K \cup \{x\}) - \hat{\mu}(K))\Big).$$

Since $\hat{\mu}$ is the dual of μ then $\hat{\mu}(E) = 1 - \mu(\overline{E})$. In this case we see that $\hat{\mu}(K) = 1 - \mu(\overline{K})$ and $\hat{\mu}(K \cup \{x\}) = 1 - \mu(\overline{K \cup \{x\}})$ and there then

$$\sum_{x \notin K} (\hat{\mu}(K \cup \{x\}) - \hat{\mu}(K)) = \sum_{x \notin K} (\mu(\overline{K}) - \mu(\overline{K \cup \{x\}})).$$

We note that $\overline{K} = X \setminus K$ and $\overline{K \cup \{x\}} = X \setminus (K \cup \{x\}) = \overline{K} \setminus \{x\}$. Thus we get that

$$\hat{\mu}(K \cup \{x\}) - \hat{\mu}(K) = \mu(\overline{K}) - \mu(\overline{K} \setminus \{x\}).$$

If $|K| = k$ then $|\overline{K}| = n - k$ and $|\overline{K} \setminus \{x\}| = n - k - 1$. Here then we see that $\hat{C}_k$ is the average increase in anticipation of μ in going from sets of cardinality $n - k - 1$ to $n - k$ which is what we denoted as C_{n-k-1}. Thus we see that $\hat{C}_k = C_{k-n-1}$. Using this we can be relate the attitudinal characters of μ and $\hat{\mu}$.

The following important theorem relates the attitudinal characters of dual measures.

Theorem 2 $AC(\hat{\mu}) = 1 - AC(\mu)$.

7 Assurance and Opportunity of Outcomes

Here we shall discuss the concepts of assurance and opportunity which play an important role in the process of answering questions using information obtained from the fusion of measures. Consider now we have a variable V with domain X. Assume our knowledge about the value of V is expressed using a measure μ on X. Thus for any subset A of X the value $\mu(A)$ indicates our anticipation of A occurring, that is of finding the value of V in the set A. Does a value $\mu(A) = 1$ assure us that A will occur? Let us look at this in more detail. Consider the measure $\mu^*(A) = 1$ for all $A \neq \emptyset$. In the case we have $\mu^*(A) = 1$. However in this case we also have $\mu^*(\overline{A}) = 1$ so here we have just as strong an anticipation that A will not occur. In order for us to be assured that A will occur we have to anticipate A will occur and also anticipate that $\overline{A}$ will not occur. The degree to which our anticipation that $\overline{A}$ **will not occur** can be measured by $1 - \mu(\overline{A})$. However we note that this is the dual of μ, $\hat{\mu}(A)$. Using this we introduce a set function called the *assurance* of A which we define as $\lambda(A) = \mu(A) \wedge \hat{\mu}(A)$ [20]. We easily see that λ is a measure. One thing we note about this measure of assurance is that $\lambda(A) \leq \mu(A)$. Thus the measure of assurance of a set is never larger than its measure of anticipation. We also observe that $\lambda(A) \leq \hat{\mu}(A)$.

Consider the two measures μ and $\hat{\mu}$ its dual. We have $\lambda_\mu(A) = \mu(A) \wedge \hat{\mu}(A)$ and $\lambda_{\hat{\mu}}(A) = \hat{\mu}(A) \wedge \hat{\hat{\mu}}(A)$. However since $\hat{\hat{\mu}}(A) = \mu(A)$ we have for measures that are duals we have $\lambda_\mu(A) = \lambda_{\hat{\mu}}(A)$. However it is not the case that $\hat{\mu}(A) = \mu(A)$.

Closely related to the concept of assurance is the concept of *opportunity*. We define this as the set function Ψ where $\Psi(A) = \mu(A) \vee \hat{\mu}(A)$ and we refer to $\Psi(A)$ as the opportunity associated with A. It essentially measures the opportunity of V lying in A. It is clear this $\Psi(A) \geq \lambda(A)$. We can easily show that Ψ is itself a measure. We note that $\Psi(A) \geq \mu(A)$ and $\Psi(A) \geq \hat{\mu}(A)$. As is the case for the measure of assurance we see that if μ and $\hat{\mu}$ are duals then $\Psi_\mu(A) = \Psi_{\hat{\mu}}(A)$. We indicate that $\Psi(A)$ is seen as the opportunity that the value of V will lie in A. We note that $\Psi(A)$ can be seen as an optimistic view of the occurrence of A and $\lambda(A)$ as a pessimistic view.

Actually we see a duality relationship between the measures Ψ and λ. Consider the dual of $\Psi(A)$, $(\hat{\Psi})(A) = \lambda(A).$). Thus the concepts of opportunity and assurance are duals. However there exists a very important special relationship between these duals, $\Psi(A) \geq \lambda(A)$. In the following we shall look at Ψ and λ for some special cases of μ.

As we previously noted the same piece of information can sometimes be expressed in different ways using a measure representation. Consider the situation in which I know nothing about the value of the variable V other than it must be in the set X. There exists two contrary ways I can express this information. One is optimistic with μ^* and the other pessimistically with μ_*. The optimistic way has $\mu^*(A) = 1$ for all $A \neq \emptyset$ and only $\mu^*(\emptyset) = 0$. The pessimistic method has $\mu_*(A) = 0$ for $A \neq X$ and $\mu_*(X) = 1$. We also emphasize that μ^* and μ_* are duals. We see $\hat{\mu}^*(X) = 1 - \mu^*(\overline{A})$ however since $\mu^*(B) = 1$ for any $B \neq \emptyset$, then $\hat{\mu}^*(A) = 0$ for all $A \neq X$ which is μ^*. As we shall subsequently see that use of λ and Ψ provides unification whereby we have the same information for both.

Consider first μ^* where $\mu^*(A) = 1$ for all $A \neq \emptyset$. In the case $\hat{\mu}^*(A) = 1 - \mu^*(\overline{A}) = 0$ for all $A \neq X$. Here then while $\lambda^*(X) = 1$ we see that for all $A \neq X$

$$\lambda^*(A) = \mu^*(A) \wedge \hat{\mu}^*(A) = 0.$$

Thus in this case while $\mu^*(A) = 1$ for all $A = \emptyset$ we have $\lambda^*(A) = 0$ for all $A \neq X$. Thus this case of complete lack of information about the value of V allows for no assurance about anything except that V lies in its domain X.

Consider now μ_*, we see $\mu_*(A) = 0$ for all $A \neq X$ and hence $\lambda_*(A) = 0$ for $A \neq X$. Thus here again in this case of complete lack of knowledge about a value of V we get $\lambda(A) = 0$. Thus here while $\mu_*(A)$ and $\mu^*(A)$ have different values for most A we have $\lambda^*(A) = \lambda_*(A)$ for all A.

We further observe since Ψ is the dual of λ then in the cases of μ_* and μ^* we have that $\Psi^*(A) = \Psi_*(A)$ and furthermore there are such that $\Psi^*(A) = \Psi_*(A) = 1$ for all $A \neq \emptyset$. Thus here while we have no assurance that V lies in A we have complete opportunity for it to lie in A.

Thus these measures μ^* and μ_* have the same values for λ and Ψ. Since these both correspond to lack of any information we can refer to these as $\lambda_?$ and $\Psi_?$. In the preceding we have shown $\Psi_?(A) = 1$ for all $A \neq \emptyset$ and $\psi_?(\emptyset) = 0$ and $\lambda_?(A) = 0$ for all $A \neq X$. Here we see except for the extremes of X and $\emptyset$ the values of $\Psi_?(A)$ and $\lambda_?(A)$ are as far apart as they can possibly be. Here we see for each A we have an opportunity but no assurance for finding V.

Let us now consider the case where μ is a probability measure. We recall her $\mu(\{x_i\}) = p_i$, $\mu(A) = \sum_{x_i \in A} p_i$ and $\sum_{i=1}^{n} p_i = 1$. Consider now the dual of this measure. In this case

$$\hat{\mu}(A) = 1 - \mu(\overline{A}) = 1 - \sum_{x_i \in A} p_i = \sum_{x_i \in A} p_i = \mu(A).$$

Thus in the case of a probability measure the measure and its dual are the same. We furthermore see it is also equal to the assurance and opportunity measures

$$\lambda(A) = \hat{\mu}(A) \wedge \mu(A) = \mu(A) \wedge \mu(A) = \mu(A),$$
$$\Psi(A) = \hat{\mu}(A) \vee \mu(A) = \mu(A) \vee \mu(A) = \mu(A).$$

Then here we $\Psi(A) = \lambda(A) = \mu(A)$. We note that this fact holds for any self-dual fuzzy measure, some times called participation measures.

This is an important and very special property of the probability distribution. There is no difference between any of the measures. The measures of opportunity, assurance and anticipation are all the same. In the situation $\mu(A)$ is often referred to as the probability of A.

An important special case of probability measure is one in which all $p_j = \frac{1}{n}$. In this case $\mu(A) = \frac{\text{card}(A)}{n}$. In some situations this special case has been used to represent lack of knowledge. However we emphasize this formulation implies some knowledge about V. In particular, it at least assumes that the measure of all the elements $\mu(\{x_j\})$, are the same.

Consider now the measure μ in the case where we know that $V = x^*$. Here we recall $\mu(B) = 1$ if $x^* \in B$ and $\mu(B) = 0$ for $x^* \notin B$. Consider the dual of this $\hat{\mu}(B) = 1 - \mu(\overline{B})$. We see $\mu(\overline{B}) = 1$ if $x^* \notin \overline{B}$ that is if $x^* \notin B$. Also we see that $\mu(\overline{B}) = 0$ if $x^* \notin \overline{B}$ that is if $x^* \in B$. Thus $\hat{\mu}(B) = 1$ if $x^* \in B$. We see that in this case $\hat{\mu}$ and $\mu$are the same and hence

$$\lambda(A) = \hat{\mu}(A) \wedge \mu(A) = \mu(A) = \hat{\mu}(A) \vee \mu(A) = \Psi(A).$$

Thus if $x^* \in A$ we get $\lambda(A) = 1$ and $x \notin A$ we get $\lambda(A) = 0$. Actually this is a special probability measure where $P(x^*) = 1$ and all other are zero, it is a Dirac measure.

We now consider the possibility measure. We recall for these measures $\mu(\{x_i\}) = \pi_i$ and $\mu(A \cup B) = \max(\mu(A), \mu(B))$. Here $\mu(A) = \max_{x_i \in A}(\pi_i)$. We note that at least one $\pi_i = 1$. The dual of a possibility measure $\hat{\mu}(A)$ is defined as $\hat{\mu}(A) = 1 - \mu(\overline{A})$ and hence we get

$$\hat{\mu}(A) = 1 - \max_{x_i \in \overline{A}}(\pi_i).$$

The duals of possibility measures have a very interesting property,

$$\hat{\mu}(A \cap B) = \min(\hat{\mu}(A), \hat{\mu}(B)).$$

Measures having this property are referred to as certainty measures [2]. Since μ and $\hat{\mu}$ are unique pairs we refer to $\hat{\mu}$ as the associated certainty measure. Certainty measures are also referred to as necessity measures

A very important and special relationship exists between a possibility measure and its dual certainty measure. If μ is a possibility measure then for any set A we have $\mu(A) \geq \hat{\mu}(A)$. Let us see that this is true. Here we have $\mu(A) = \max_{x_i \in A}(\pi_i)$ and $\hat{\mu}(A) = 1 - \max_{x_i \in \overline{A}}(\pi_i)$. We note that at least one of the $\pi_i = 1$. If A has one of the x_i with $\pi_i = 1$ then $\mu(A) = 1$ and $\mu(A) \geq \hat{\mu}(A)$. If there exists no element with $\pi_i = 1$ contained in A then at least one such element must be in $\overline{A}$ and hence $\max_{x_i \in \overline{A}}(\pi_i) = 1$ and therefore $\hat{\mu}(A) = 0$ and again we have $\mu(A) \geq \hat{\mu}(A)$. An important implication of this is that for possibility measures we always have

$$\lambda(A) = \hat{\mu}(A) \wedge \mu(A) = \hat{\mu}(A).$$

Thus here the measure of assurance of A is equal to the dual measure of A, the certainty of A. Furthermore in this case of possibility measures we always have

$$\Psi(A) = \hat{\mu}(A) \vee \mu(A) = \mu(A).$$

Thus here $\Psi(A)$ is always the measure of possibility.

We want to emphasize that not all measures have this clear relationship between $\mu(A)$ and $\hat{\mu}(A)$. Thus here in the case of a possibility measure since $\lambda(A) = \hat{\mu}(A)$ then $\Psi(A) = \mu(A) \geq \hat{\mu}(A) = \lambda(A)$. This relationship is greatly used in possibility theory [3].

8 Conclusion

We first looked at the issue of representing information about an uncertain variable using a measure. We focused on some notable measures that can be used. We discussed the role of aggregation functions in the task of combining measures to form, new measures. We looked at this in the framework of multi-source information fusion. We focused on the fusion of probabilistic and possibilistic information and discussed its role in hard-soft information fusion. We looked at some characterizing features associated with measures used to represent uncertain values of variables. We discussed the concepts of assurance and opportunity that play a role in the process of answering questions using information obtained from a measure.

References

1. Beliakov, G., Pradera, A., Calvo, T.: Aggregation Functions: A Guide for Practitioners. Springer, Heidelberg (2007)
2. Dubois, D., Prade, H.: Necessity measures and the resolution principle. IEEE Trans. Syst. Man Cybern. **17**, 474–478 (1987)
3. Dubois, D., Prade, H.: Possibility Theory: An Approach to Computerized Processing of Uncertainty. Plenum Press, New York (1988)

4. Dubois, D., Prade, H.: Formal representations of uncertainty. In: Bouyssou, D., Dubois, D., Pirlot, M., Prade, H. (eds.) Decision Making Process: Concepts and Methods. Wiley, Hoboken (2010)
5. Grabisch, M., Marichal, J.-L., Mesiar, R., Pap, E.: Aggregation Functions. Cambridge University Press, Cambridge (2009)
6. Klement, E.P.: A theory of fuzzy measures: a survey. In: Gupta, M.M., Sanchez, E. (eds.) Fuzzy Information and Decision Processes, pp. 59–66. North-Holland, Amsterdam (1982)
7. Klement, E.P., Mesiar, R., Pap, E.: Triangular Norms. Kluwer Academic Publishers, Dordrecht (2000)
8. Klir, G.J.: Uncertainty and Information. Wiley, New York (2006)
9. Liu, L., Yager, R.R.: Classic works of the Dempster-Shafer theory of belief functions: an introduction. In: Yager, R.R., Liu, L. (eds.) Classic Works of the Dempster-Shafer Theory of Belief Functions, pp. 1–34. Springer, Heidelberg (2008)
10. Llinas, J., Nagi, R., Hall, D.L., Lavery, J.: A multi-disciplinary university research initiative in hard and soft information fusion: Overview, research strategies and initial results. In: Proceedings of the 13th International Conference on Information Fusion (Fusion 2010). Edinburgh, UK, Unpaginated (2010)
11. Shapley, L.S.: A value for n-person games. In: Kuhn, H.W., Tucker, A.W. (eds.) Contributions to Game Theory, pp. 307–317. Princeton University Press, Princeton (1953)
12. Sugeno, M.: Fuzzy measures and fuzzy integrals: a survey. In: Gupta, M.M., Saridis, G.N., Gaines, B.R. (eds.) Fuzzy Automata and Decision Process, pp. 89–102. North-Holland, Amsterdam (1997)
13. Walley, P., Fine, T.: Toward a frequentist theory of upper and lower probability. Ann. Stat. **10**, 741–761 (1982)
14. Wang, Z., Yang, R., Leung, K.S.: Nonlinear Integrals and Their Applications in Data Mining. World Scientific, Singapore (2010)
15. Yager, R.R.: On ordered weighted averaging aggregation operators in multi-criteria decision making. IEEE Trans. Syst. Man Cybern. **18**, 183–190 (1988)
16. Yager, R.R.: On the entropy of fuzzy measures. IEEE Trans. Fuzzy Sets Syst. **8**, 453–461 (2000)
17. Yager, R.R.: Measuring the information and character of a fuzzy measure. In: Proceedings of the Joint 9th IFSA World Congress and the 20th NAFIPS International Conference, pp. 1718–1722. Vancouver (2001)
18. Yager, R.R.: On the cardinality index and attitudinal character of fuzzy measures. Int. J. Gen. Syst. **31**, 303–329 (2002)
19. Yager, R.R.: A measure based approach to the fusion of possibilistic and probabilistic uncertainty. Fuzzy Optim. Decis. Making **10**, 91–113 (2011)
20. Yager, R.R.: Measures of assurance and opportunity in modeling uncertain information. Int. J. Intell. Syst. **27**, 776–796 (2012)
21. Zadeh, L.A.: Fuzzy sets as a basis for a theory of possibility. Fuzzy Sets Syst. **1**, 3–28 (1978)
22. Zadeh, L.A.: From computing with numbers to computing with words–From manipulation of measurements to manipulations of perceptions. IEEE Trans. Circuits Syst. **45**, 105–119 (1999)
23. Zadeh, L.A.: Generalized theory of uncertainty (GTU)-principal concepts and ideas. Comput. Stat. Data Anal. **51**, 15–46 (2006)

Bases and Transforms of Set Functions

Michel Grabisch

Abstract The chapter studies the vector space of set functions on a finite set X, which can be alternatively seen as pseudo-Boolean functions, and including as a special cases games. We present several bases (unanimity games, Walsh and parity functions) and make an emphasis on the Fourier transform. Then we establish the basic duality between bases and invertible linear transform (e.g., the Möbius transform, the Fourier transform and interaction transforms). We apply it to solve the well-known inverse problem in cooperative game theory (find all games with same Shapley value), and to find various equivalent expressions of the Choquet integral.

1 Introduction

Set functions on a finite set X are of fundamental usage in many areas of discrete mathematics, e.g., cooperative game theory [20], combinatorial optimization [11], decision making [14], computer sciences [6], and more generally operations research [16], where in the latter domain, they are more often encountered under the form of pseudo-Boolean functions. Specific domains focus on specific subclasses of set functions, e.g., game theory uses set functions vanishing on the empty set (these are characteristic functions of transferable utility games, which are simply called "games"), decision theory needs games which are monotone with respect to inclusion (called *capacities*), while combinatorial optimization often deals with submodular games.

An interesting feature of set functions and games is that they form a vector space of dimension $2^{|X|}$ ($2^{|X|-1}$ for games). Most often, this feature is ignored, although clearly one can take advantage of the concepts and techniques of linear algebra when dealing with set functions and games. In particular the notion of basis is of importance. The best-known basis is perhaps the basis of unanimity games (this is the

M. Grabisch (✉)
Paris School of Economics, University of Paris I, 106-112, Bd de L'Hôpital,
75013 Paris, France
e-mail: michel.grabisch@univ-paris1.fr

S. Saminger-Platz and R. Mesiar (eds.), *On Logical, Algebraic, and Probabilistic Aspects of Fuzzy Set Theory*, Studies in Fuzziness and Soft Computing 336,
DOI 10.1007/978-3-319-28808-6_13

usual name given in game theory; they are closely related to the incidence functions in combinatorics, see [1]), although each domain has its prefered bases. For example, in computer sciences mostly the basis of parity functions is used, essentially because the encoding of sets is done by $-1, +1$, rather than by 0, 1. The usage of a particular basis induces a particular representation of set functions, viewed as a transform, which is by definition linear and invertible. For example, the representation through the basis of unanimity games is the Möbius transform (or Möbius inverse), widely used in combinatorics and well known in decision making and game theory (known under the name of Harsanyi dividends [17] in the latter domain), while the representation through parity functions is the Fourier transform, which has many applications in computer sciences.

As far as the author can see from the literature, the duality between bases and transforms (i.e., the representation of set functions into a given basis) has never been exploited nor even remarked. A systematic exploitation of this fact can lead to the discovery of new bases and transforms, as well as an easy solution to the so called inverse problem in game theory: find all games having the same Shapley value (and similar ones). Also, it permits to get several different expressions of linear operators on games or set functions, like the Choquet integral.

The aim of this chapter is to bring a survey on the above mentioned issues. Section 2 gives a brief account on set functions and pseudo-Boolean functions. Section 3 describes the best-known bases (unanimity games, Walsh functions and parity functions). Section 4 is about the Fourier transform and its properties, while Sect. 5 explains the fundamental duality between bases and transforms, and gives many examples. Sections 6 and 7 are applications of this duality principle to the solution of the above mentioned inverse problem and to the finding of equivalent expressions of the Choquet integral.

2 Set Functions and Pseudo-Boolean Functions

In the whole chapter, we consider a finite universe X, with $|X| = n$. Occasionally, we will use the notation $[n] = \{1, \dots, n\}$.

A *set function* on X is a mapping $\xi : 2^X \rightarrow \mathbb{R}$. A *game* is a set function v vanishing on the empty set: $v(\varnothing) = 0$.

Clearly, the set $\mathbb{R}^{2^X}$ of set functions on X is a 2^n-dimensional vector space. We introduce on $\mathbb{R}^{2^X}$ the following scalar product:

$$\langle \xi, \xi' \rangle = \frac{1}{2^n} \sum_{S \subseteq X} \xi(S)\xi'(S).$$

There is another vision of set functions, namely the pseudo-Boolean functions [16], noting that any subset A of X can be encoded by its characteristic function 1_A. Formally, a *pseudo-Boolean function* is a mapping $f : \{0, 1\}^n \rightarrow \mathbb{R}$. The equivalence between pseudo-Boolean functions and set functions can be seen through the coding function $1 : 2^X \to \{0, 1\}^n$ defined by $A \mapsto 1_A$, with $1_A(i) = 1$ if and only if $i \in A$. Then

$$\xi_f = f \circ 1, \quad f_\xi = \xi \circ 1^{-1}$$

where ξ_f denotes the set function associated to f, and f_ξ is the pseudo-Boolean function associated to ξ. If follows that the set of pseudo-Boolean functions of n variables is a 2^n-dimensional vector space, with scalar product

$$\langle f, f' \rangle = \frac{1}{2^n} \sum_{x \in \{0,1\}^n} f(x) f'(x).$$

Pseudo-Boolean functions were obtained from set functions by coding subsets by 1 and 0. It is noteworhty that other encodings are possible, for example using $-1, 1$ instead of $0, 1$. We will come back to this encoding later and shows that it is quite useful.

3 Bases of Set Functions and Pseudo-Boolean Functions

3.1 Unanimity Games

Perhaps the best known basis of $\mathbb{R}^{2^X}$ is the basis of the so-called unanimity games. For any nonempty subset $S \subseteq X$, the *unanimity game centered on* S is the game defined by

$$u_S(T) = \begin{cases} 1, & \text{if } T \supseteq S \\ 0, & \text{otherwise} \end{cases}.$$

Defining the set function $u_\varnothing(S) = 1$ for every $S \subseteq X$, it is well known that $\{u_S\}_{S \in 2^X}$ is a basis for $\mathbb{R}^{2^X}$. It is also well known that the coordinates of ξ in this basis are the *Möbius transform* coefficients:

$$\xi = \sum_{S \in 2^X} m^\xi(S) u_S$$

with

$$m^\xi(S) = \sum_{T \subseteq S} (-1)^{|S \setminus T|} \xi(T).$$

The Möbius transform (or *Möbius inverse*) is a well-known tool in combinatorics since the work of [21] (see also [1, 3], etc.).

A drawback of the basis of unanimity games is that it is not orthogonal w.r.t. the above scalar product, as it is easy to see even with $n = 2$:

$$\langle u_{\{1\}}, u_{\{2\}} \rangle = \frac{1}{4}(u_{\{1\}}(\{1\})u_{\{2\}}(\{1\}) + u_{\{1\}}(\{2\})u_{\{2\}}(\{2\}) \\ + u_{\{1\}}(\{1,2\})u_{\{2\}}(\{1,2\}) = \frac{1}{4} \neq 0.$$

In the formalism of pseudo-Boolean functions, unanimity games u_S correspond to monomials $\prod_{i \in S} x_i$. Hence we have

$$f = \sum_{S \subseteq [n]} m^f(S) \prod_{i \in S} x_i$$

where m^f is the Möbius transform of ξ_f (this slight abuse of notation should not be confusing).

3.2 Walsh Functions

Another basis of pseudo-Boolean functions is the basis of *Walsh functions*, which are monomials defined by

$$w_T(x) = \prod_{i \in T}(2x_i - 1) \qquad (T \subseteq [n], x \in \{0,1\}^n)$$

or, in set function notation:

$$w_T(S) = (-1)^{|T \setminus S|} \qquad (S, T \in 2^X). \tag{1}$$

It can be shown that the Walsh functions $\{w_T\}_{T \subseteq [n]}$ form an orthornomal basis of the pseudo-Boolean functions:

$$\langle w_S, w_T \rangle = 1 \text{ iff } S = T, \text{ and 0 otherwise.}$$

It is important to note that letting $z_i = 2x_i - 1 \in \{-1, +1\}$, the Walsh functions reduce to the monomials $\prod_{i \in T} z_i$: hence the Walsh functions are obtained when subsets are encoded by $-1, 1$ instead of $0, 1$. We see that this simple change makes the basis orthonormal. It can be shown that the coordinates of a pseudo-Boolean function into this basis are given by

$$f(x) = \sum_{T \subseteq [n]} \Big(\sum_{S \supseteq T} \frac{m^f(S)}{2^{|S|}} \Big) w_T(x) = \sum_{T \subseteq [n]} \frac{1}{2^{|T|}} I_{\mathrm{B}}^f(T) w_T(x) \tag{2}$$

where I_{B}^{f} is the Banzhaf interaction transform defined in terms of the Möbius transform by

$$I_{\mathrm{B}}^{f}(T) = \sum_{S \supseteq T} \left(\frac{1}{2}\right)^{|S \setminus T|} m^{f}(S) \qquad (T \in 2^{X})$$

(this transform will be properly introduced later).

As a historical remark, we note that the original Walsh functions [26] are rather different in their definition:

$$W_k(x) = (-1)^{\sum_{j=0}^{\infty} k_j x_{j+1}} \qquad (k \in \mathbb{N}_0, x \in [0, 1]), \tag{3}$$

with $k = k_0 + k_1 2 + k_2 2^2 + \cdots k_m 2^m, k_i \in \{0, 1\}$ for all i, and $x = x_1 2^{-1} + x_2 2^{-2} + x_3 2^{-3} + \cdots, x_i \in \{0, 1\}$ for all i, the binary representations of k and x. They form an orthonormal basis of the set of square integrable functions on $[0, 1]$. The connection with our Walsh functions is that the latter have a discretized domain

$$0, \frac{1}{2^n}, \frac{2}{2^n}, \frac{3}{2^n}, \ldots, \frac{2^n - 1}{2^n}$$

of 2^n points, corresponding to the 2^n subsets of $[n]$. More precisely, $w_S(x)$ corresponds to $W_k(x')$ such that S and k have same binary coding, and

$$x'_1 = 1 - x_1, \ldots, x'_n = 1 - x_n, \quad \text{and } x'_j = 0 \text{ for } j > n.$$

3.3 *Parity Functions*

Another family of functions related to the Walsh functions are the parity functions. The *parity function* associated to $S \subseteq [n]$ is the function

$$\chi_S(x) = (-1)^{1_S \cdot x} = (-1)^{\sum_{i \in S} x_i} \qquad (x \in \{0, 1\}^n). \tag{4}$$

Its name comes from the fact that it takes only values -1 and $+1$, depending on whether there is an odd or even number of elements of coordinates of x equal to 1 in S. It expression as a set function is

$$\chi_S(T) = (-1)^{|S \cap T|} \qquad (S, T \in 2^{X}). \tag{5}$$

Up to a recoding by $\varepsilon(1) = 0$ and $\varepsilon(-1) = 1$, the parity functions are the Walsh functions:

$$w_S(z) = \prod_{i \in S} z_i = (-1)^{\sum_{i \in S} \varepsilon(z_i)} = \chi_S(\varepsilon(z)) \qquad (z \in \{-1, 1\}^n).$$

Consequently, they form another orthonormal basis of the vector space of pseudo-Boolean functions. The interest of parity functions is that they lead to the well-known Fourier transform, to which the next section is devoted.

4 The Fourier Transform

In the basis of parity functions, it can be shown that any pseudo-Boolean function f is expressed by

$$f = \sum_{S \subseteq [n]} \widehat{f}(S)\chi_S$$

where the coordinates of f in this basis, denoted by $\widehat{f}(S)$, are given by

$$\widehat{f}(S) = \langle f, \chi_S \rangle = \frac{1}{2^n} \sum_{x \in \{0,1\}^n} (-1)^{1_S \cdot x} f(x) \qquad (S \subseteq [n])$$

or, in terms of set functions,

$$\widehat{\xi}(S) = \frac{1}{2^n} \sum_{T \subseteq [n]} (-1)^{|S \cap T|} \xi(T) \qquad (S \subseteq [n]). \tag{6}$$

The set of coefficients $\widehat{f}(S)$, $S \subseteq [n]$, is called the *Fourier transform* of f, and is widely used in computer sciences (see, e.g., a survey in [6], as well as [19]).

We show now some properties of the Fourier transform, and to this end, we introduce some additional notions and notation. We may consider $x \in \{0, 1\}^n$ as a random variable with uniform distribution. Then, the expected value and the variance of a pseudo-Boolean function f are

$$\mathbb{E}[f] = \frac{1}{2^n} \sum_{x \in \{0,1\}^n} f(x)$$

$$\mathrm{Var}[f] = \mathbb{E}\big[(f - \mathbb{E}[f])^2\big] = \mathbb{E}[f^2] - \mathbb{E}^2[f].$$

The *convolution product* of two pseudo-Boolean functions f, g is defined by

$$(f * g)(x) = \frac{1}{2^n} \sum_{y \in \{0,1\}^n} f(x \oplus y)g(y) \qquad (x \in \{0, 1\}^n)$$

where $\oplus$ denotes the coordinatewise binary addition:

$$1 \oplus 1 = 0 = 0 \oplus 0, \quad 1 \oplus 0 = 0 \oplus 1 = 1.$$

In terms of set functions, we obtain the following expression:

$$(f * g)(S) = \frac{1}{2^n} \sum_{T \subseteq [n]} f(S \Delta T) g(T) \qquad (S \in 2^X).$$

The properties of the Fourier transform are gathered in the next theorem.

Theorem 1 *Let f, g be two pseudo-Boolean functions. The following holds.*

(i) $f(\mathbf{0}) = \sum_{S \subseteq [n]} \widehat{f}(S)$;

(ii) $\widehat{f}(\varnothing) = \mathbb{E}[f]$;

(iii) (Parseval's identity) $\|f\|^2 = \sum_{S \subseteq [n]} \widehat{f}^2(S)$;

(iv) $\sum_{S \in 2^{[n]} \setminus \{\varnothing\}} \widehat{f}^2(S) = \mathrm{Var}[f]$;

(v) *f is constant if and only if $\widehat{f}(S) = 0$ for all $S \neq \varnothing$;*

(vi) $\widehat{(f * g)}(S) = \widehat{f}(S)\widehat{g}(S)$ *for all* $S \in 2^{[n]}$.

The name "Fourier transform" comes from the work of Fourier on the representation of integrable functions. The Fourier transform of a function, viewed as a function of time, gives its frequency representation. Exactly the same results as those in Theorem 1 hold for the original Fourier transform, which explains its name in computer sciences. However, it must be noted that all the properties, except the last one on convolution, are direct consequences of the orthonormality of the basis. To the opinion of the author, this transform should be rather called the *Walsh transform*, since the definition of the Walsh function is an infinite version of the parity function used here, as a comparison of (3) and (4) reveals.

5 Bases and Linear Transforms on Set Functions

The previous sections have introduced various bases on set functions and pseudo-Boolean functions: the unanimity games, the Walsh functions and the parity functions. As a by-product, two fundamental notions in combinatorics and computer sciences have appeared, namely the Möbius transform and the Fourier transform. The name "transform" intuitively means (in particular by reference to well-known transforms used in mathematics for the analysis of real-valued functions: the (original) Fourier transform and the Laplace transform, mainly) a representation in another domain, but equivalent to the original one, i.e., the main desirable characteristic of the transform is that it should be invertible.

Although it comes as an evidence from elementary considerations in linear algebra, there is a duality between bases and linear invertible transforms, which to the knowledge of the author, has never been exploited nor remarked. The aim of this section is to explain this duality and to apply it, first to the known bases and transforms so as to obtain new bases and transforms, and second to some well-known inverse problem and representation of integrals w.r.t. games.

We define a *transform* on the set of set functions on X as a linear invertible mapping $\Psi : \mathbb{R}^{2^X} \to \mathbb{R}^{2^X}$, with $\xi \mapsto \Psi^\xi$. The following discussion is made easier if one consider set functions ξ as row vectors and use matrix notation. To a basis $(b_S)_{S\in 2^X}$, we make correspond the matrix $B = [b_S]$ of row vectors b_S. Hence $\xi = \sum_{S\in 2^X} w_S b_S = wB$ is the expression of ξ in this basis. The following lemma gives the exact equivalence between bases and transforms.

Lemma 1 [10] *For every basis B, there is a (unique) transform Ψ such that for any $\xi \in \mathbb{R}^{2^X}$,*

$$\xi = \sum_{S\in 2^X} \Psi^\xi(S) b_S, \tag{7}$$

whose inverse Ψ^{-1} is given by $\xi \mapsto (\Psi^{-1})^\xi = \sum_{S\in 2^X} \xi(S) b_S = \xi B$.

Conversely, to any transform Ψ corresponds a unique basis B such that (7) holds, given by $b_S = (\Psi^{-1})^{\delta_S}$.

In the above lemma, δ_S denotes the Dirac set function defined by

$$\delta_S(T) = \begin{cases} 1, & \text{if } S = T \\ 0, & \text{otherwise} \end{cases} \quad (S \in 2^X).$$

We apply this result on a number of commonly used bases and transforms.

(i) **The Möbius transform** is, as we have already noticed, related to the basis of unanimity games:

$$\xi(S) = \sum_{T\in 2^X} m^\xi(T) u_T(S) = \sum_{T\subseteq S} m^\xi(T), \quad (S \subseteq X),$$

with

$$m^\xi(S) = \sum_{T\subseteq S} (-1)^{|S\setminus T|} \xi(T).$$

(ii) **The co-Möbius transform** ([15], a.k.a. *commonality function* [23]) is defined by:

$$\check{m}^\xi(S) = \sum_{T\supseteq X\setminus S} (-1)^{n-|T|} \xi(T) = \sum_{T\subseteq S} (-1)^{|T|} \xi(X \setminus T) \quad (S \in 2^X).$$

Its inverse relation is

$$\xi(S) = \sum_{T \subseteq X \setminus S} (-1)^{|T|} \check{m}^{\xi}(T).$$

By Lemma 1, the associated basis is

$$\check{u}_T(S) = \sum_{B \subseteq X \setminus S} (-1)^{|B|} \delta_T(B) = \begin{cases} (-1)^{|T|} & \text{if } S \cap T = \emptyset \\ 0 & \text{otherwise.} \end{cases}$$

(iii) **The (Shapley) interaction transform** [12] is defined by

$$I^{\xi}(S) = \sum_{T \subseteq X \setminus S} \frac{(n-t-s)!t!}{(n-s+1)!} \sum_{L \subseteq S} (-1)^{|S \setminus L|} \xi(T \cup L),$$

and the inverse relation is given by

$$\xi(S) = \sum_{K \subseteq X} \beta^{|K|}_{|S \cap K|} I^{\xi}(K),$$

where

$$\beta^l_k = \sum_{j=0}^{k} \binom{k}{j} B_{l-j} \qquad (k \leq l),$$

and $B_0, B_1, \ldots$ are the Bernoulli numbers. The first values of β^l_k are given in Table 1.
The associated basis $\{b^I_T\}_{T \in 2^X}$ is

$$b^I_T(S) = \beta^{|T|}_{|T \cap S|} \quad (S \in 2^X).$$

Table 1 The coefficients β^l_k

$k \setminus l$	0	1	2	3	4
0	1	$-\frac{1}{2}$	$\frac{1}{6}$	0	$-\frac{1}{30}$
1		$\frac{1}{2}$	$-\frac{1}{3}$	$\frac{1}{6}$	$-\frac{1}{30}$
2			$\frac{1}{6}$	$-\frac{1}{6}$	$\frac{2}{15}$
3				0	$-\frac{1}{30}$
4					$-\frac{1}{30}$

(iv) **The Banzhaf interaction transform** [22] is defined by

$$I_{\rm B}^{\xi}(S) = \left(\frac{1}{2}\right)^{n-s} \sum_{K \subseteq X} (-1)^{|S \setminus K|} \xi(K),$$

with inverse relation

$$(I_{\rm B}^{-1})^{\xi}(S) = \sum_{K \subseteq X} \left(\frac{1}{2}\right)^{k} (-1)^{|K \setminus S|} \xi(K). \tag{8}$$

The associated basis $\{b_T^{I_B}\}_{T \in 2^X}$ is

$$b_T^{I_B}(S) = \sum_{K \subseteq X} \left(\frac{1}{2}\right)^{k} (-1)^{|K \setminus S|} \delta_T(K) = \left(\frac{1}{2}\right)^{|T|} (-1)^{|T \setminus S|}.$$

(v) **The Fourier interaction transform**: as it was already explained in Sect. 4, the Fourier transform, as given by (6), corresponds to the basis of parity functions, given by (5).
The relations between the Fourier transform and the Möbius and Banzhaf transforms are given as follows:

$$\widehat{\xi}(S) = (-1)^{|S|} \sum_{K \supseteq S} \frac{1}{2^k} m^{\xi}(K). \tag{9}$$

$$\widehat{\xi}(S) = \left(\frac{-1}{2}\right)^{s} I_{\rm B}^{\xi}(S). \tag{10}$$

(vi) **The Walsh basis**: This basis is defined by (1). Let us recover the corresponding transform $\xi \mapsto W^{\xi}$ by using Lemma 1, which we already gave in (2). By Lemma 1, the inverse transform is immediate:

$$(W^{-1})^{\xi}(S) = \sum_{T \subseteq X} \xi(T)(-1)^{|T \setminus S|}.$$

The direct transform is obtained by solving the linear system

$$\xi(S) = \sum_{T \subseteq X} W^{\xi}(T)(-1)^{|T \setminus S|} \quad (S \in 2^X),$$

or by simply noticing that $w_T(S) = 2^{|T|} b_T^{I_B}(S)$, which from

$$\xi(S) = \sum_{T \subseteq X} I_B^{\xi}(T) b_T^{I_B}(S) = \sum_{T \subseteq X} W^{\xi}(T) w_T(S)$$

yields the components of W^ξ as

$$W^\xi(T) = \left(\frac{1}{2}\right)^{|T|} I_B^\xi(T) \qquad (T \in 2^X).$$

We recover formula (2). Note that the Fourier and Walsh bases are related as follows:

$$\chi_T(S) = \chi_S(T) = (-1)^{|S \cap T|} = (-1)^{|S \setminus (X \setminus T)|} = w_S(X \setminus T).$$

Also, from (10), we find

$$F^\xi(S) = (-1)^s W^\xi(S) \qquad (S \in 2^X). \tag{11}$$

(vii) **The Yokote basis**: (see [27, 28]) it is a basis of the set of games, which is defined by

$$b_T^Y(S) = \begin{cases} 1, & \text{if } |S \cap T| = 1 \\ 0, & \text{otherwise} \end{cases} \qquad (S \in 2^X \setminus \varnothing). \tag{12}$$

Any game v reads in this basis

$$v = \sum_{T \in 2^X \setminus \varnothing} Y^v(T) b_T^Y \tag{13}$$

where the coordinates $Y^v(S)$ define the *Yokote transform* Y. We give now Y^v in terms of m^v and v, as well as the inverse relations:

$$m^v(S) = |S|(-1)^{|S|+1} \sum_{K \supseteq S} Y^v(K) \qquad (\varnothing \neq S \subseteq X). \tag{14}$$

$$Y^v(S) = (-1)^{|S|+1} \sum_{K \supseteq S} \frac{1}{|K|} m^v(K) \qquad (\varnothing \neq S \subseteq X). \tag{15}$$

$$Y^v(S) = \sum_{L \subseteq X} (-1)^{|S \cap L|+1} \frac{(n-s-l)!(s+l-1)!}{n!} v(L). \tag{16}$$

Table 2 summarizes the correspondence between bases and transforms.

Table 2 Correspondence between bases and transforms

	Transform	Basis
Möbius	$m^{\xi}(S) = \sum_{T \subseteq S} (-1)^{\lvert S \setminus T \rvert} \xi(T)$	$u_T(S) = \begin{cases} 1, & \text{if } S \supseteq T \\ 0, & \text{otherwise} \end{cases}$
co-Möbius	$\check{m}^{\xi}(S) = \sum_{T \supseteq X \setminus S} (-1)^{n-\lvert T \rvert} \xi(T)$	$\check{u}_T(S) = \begin{cases} (-1)^{\lvert T \rvert}, & \text{if } S \cap T = \varnothing \\ 0, & \text{otherwise} \end{cases}$
Conjugate unanimity games	$\overline{U}^{\xi}(S) = (-1)^{\lvert S \rvert + 1} \sum_{T \supseteq X \setminus S} (-1)^{n-\lvert T \rvert} \xi(T)$	$\overline{u_T}(S) = \begin{cases} 1, & \text{if } S \cap T \neq \varnothing \\ 0, & \text{otherwise} \end{cases}$
Shapley interaction	$I^{\xi}(S) = \sum_{K \subseteq X} \frac{\lvert X \setminus (S \cup K) \rvert ! \lvert K \setminus S \rvert !}{(n-s+1)!} (-1)^{\lvert S \setminus K \rvert} \xi(K)$	$b_T^I(S) = \beta_{\lvert T \cap S \rvert}^{\lvert T \rvert}$
Banzhaf interaction	$I_{\mathrm{B}}^{\xi}(S) = \left(\frac{1}{2}\right)^{n-s} \sum_{K \subseteq X} (-1)^{\lvert S \setminus K \rvert} \xi(K)$	$b_T^{I_B}(S) = \left(\frac{1}{2}\right)^{\lvert T \rvert} (-1)^{\lvert T \setminus S \rvert}$
Fourier	$\widehat{\xi}(S) = \frac{1}{2^n} \sum_{K \subseteq X} (-1)^{\lvert S \cap K \rvert} \xi(K)$	$\chi_T(S) = (-1)^{\lvert S \cap T \rvert}$
Walsh	$W^{\xi}(S) = \frac{1}{2^n} \sum_{K \subseteq X} (-1)^{\lvert S \setminus K \rvert} \xi(K)$	$w_T(S) = (-1)^{\lvert T \setminus S \rvert}$
Yokote ($S \neq \varnothing$)	$Y^{v}(S) = \sum_{L \subseteq X} (-1)^{\lvert S \cap L \rvert + 1} \frac{(n-s-l)!(s+l-1)!}{n!} v(L)$	$b_T^Y(S) = \begin{cases} 1, & \text{if } \lvert S \cap T \rvert = 1 \\ 0, & \text{otherwise} \end{cases}$

6 The Inverse Problem for Linear Values

In cooperative game theory, a *linear value* is a linear mapping $\Phi : \mathbb{R}^{2^X} \rightarrow \mathbb{R}^X$ assigning to any game v a n-dim vector $\Phi(v)$, representing a sharing of $v(X)$ among all players (elements of X). For this reason a value most often satisfies efficiency, in the sense that $\sum_{i\in X} \Phi_i(v) = v(X)$. The best-known values are the Shapley value [24] and the Banzhaf value [2]. They are both linear and their definition amounts to considering the respective interaction transforms for singletons, i.e.:

$$\Phi_i^{\mathrm{Sh}}(v) = I^v(\{i\}), \quad \Phi_i^{\mathrm{B}}(v) = I_B^v(\{i\}), \quad (i \in X). \tag{17}$$

Hence, the interaction and Banzhaf interaction transforms can be seen as extensions of these values.

The duality between transforms and bases permits to easily solve the so-called "inverse problem", well-known in game theory (see, e.g., [8, 9, 18, 28]): given a game v on X, find all games v' having the same Shapley value (or any other linear value: Banzhaf, egalitarian, etc.), i.e., $\Phi^{\mathrm{Sh}}(v) = \Phi^{\mathrm{Sh}}(v')$.

Considering a linear value Φ and a game v, finding all games v' such that $\Phi(v) = \Phi(v')$ amounts by linearity to solving $\Phi(v - v') = 0$, i.e., $v - v' \in \ker(\Phi)$. Hence the solution of the inverse problem reduces to finding the kernel of the linear operator Φ.

The kernel is easily found if there exists a transform Ψ extending the linear value Φ, exactly as the interaction transform extends the Shapley value (see (17)). Indeed, the kernel is just the space spanned by the vectors f_S of the corresponding basis with $|S| > 1$. We illustrate this method with the Shapley value. For any game v, its epxression in the basis induced by the interaction transform is:

$$v = \sum_{S\in 2^X} I^v(S) b_S^I = \sum_{i\in X} \Phi_i^{\mathrm{Sh}}(v) b_{\{i\}}^I + \sum_{|S|>1} I^v(S) b_S^I,$$

which implies

$$v \in \ker(\Phi^{\mathrm{Sh}}) \iff v = \sum_{|S|>1} I^v(S) b_S^I$$

i.e.,

$$\ker(\Phi^{\mathrm{Sh}}) = \Big\{ \sum_{|S|>1} \lambda_S b_S^I \mid \lambda_S \in \mathbb{R} \Big\}.$$

In the case where $|X| = 3$, we obtain, using Table 1:

$$v(\emptyset) = \lambda_\emptyset + \frac{1}{6}(\lambda_{12} + \lambda_{13} + \lambda_{23})$$
$$v(1) = \lambda_\emptyset - \frac{1}{3}\lambda_{12} - \frac{1}{3}\lambda_{13} + \frac{1}{6}\lambda_{23} + \frac{1}{6}\lambda_{123}$$

$$
\begin{aligned}
v(2) &= \lambda_\emptyset - \frac{1}{3}\lambda_{12} + \frac{1}{6}\lambda_{13} - \frac{1}{3}\lambda_{23} + \frac{1}{6}\lambda_{123}\\
v(3) &= \lambda_\emptyset + \frac{1}{6}\lambda_{12} - \frac{1}{3}\lambda_{13} - \frac{1}{3}\lambda_{23} + \frac{1}{6}\lambda_{123}\\
v(12) &= \lambda_\emptyset + \frac{1}{6}\lambda_{12} - \frac{1}{3}\lambda_{13} - \frac{1}{3}\lambda_{23} - \frac{1}{6}\lambda_{123}\\
v(13) &= \lambda_\emptyset - \frac{1}{3}\lambda_{12} + \frac{1}{6}\lambda_{13} - \frac{1}{3}\lambda_{23} - \frac{1}{6}\lambda_{123}\\
v(23) &= \lambda_\emptyset - \frac{1}{3}\lambda_{12} - \frac{1}{3}\lambda_{13} + \frac{1}{6}\lambda_{23} - \frac{1}{6}\lambda_{123}\\
v(123) &= \lambda_\emptyset + \frac{1}{6}(\lambda_{12} + \lambda_{13} + \lambda_{23}),
\end{aligned}
$$

where $\lambda_S \in \mathbb{R}$ for every $S \subseteq \{1, 2, 3\}$.

Let us give a second illustrative example with the Banzhaf value. Consider the following problem: *Given a n-dim vector y, find all games v s.t.* $\Phi^{\mathrm{B}}(v) = y$. The set of solutions is simply the set of games of the form

$$v = v_y + w, \text{ with } w \in \ker(\Phi^{\mathrm{B}}),$$

and v_y is any game s.t. $\Phi^{\mathrm{B}}(v_y) = y$. Since the Banzhaf interaction transform generalizes the Banzhaf index, we have

$$\ker(\Phi^{\mathrm{B}}) = \mathrm{Sp}\{b_T^{I_{\mathrm{B}}}, |T| > 1\}$$

with $b_T^{I_{\mathrm{B}}}(S) = (1/2)^{|T|}(-1)^{|T \setminus S|}$, and Sp denotes the space spanned by the vectors. Now, v_y can be obtained as the inverse transform of the game w defined by $w(\{i\}) = y_i$ for all $i \in X$, and $w(S) = 0$ otherwise. This yields by (8):

$$v_y(S) = \frac{1}{2}\Big(\sum_{i \in S} y_i - \sum_{i \notin S} y_i\Big).$$

In the case where there is no known transform which extends the linear value under consideration, a general method is given by [10], consisting in the following steps:

(i) Select a basis $E = \{e_1, \ldots, e_k\}$ of the range $\Phi(\mathbb{R}^{2^X})$.
(ii) Find set functions $b_1, \ldots, b_k \in \mathbb{R}^{2^X}$ such that $\Phi(b_i) = e_i$, $i = 1, \ldots, k$.
(iii) Complete the independent set $\{b_1, \ldots, b_k\}$ to form a basis $B = \{b_1, \ldots, b_{2^n}\}$ of $\mathbb{R}^{2^X}$.
(iv) Compute the coordinates $\epsilon_1^{(j)}, \ldots, \epsilon_k^{(j)}$ of $\Phi(b_j)$ in E for $j = k+1, \ldots, 2^n$.
(v) Compute $b_j^\Phi = b_j - \sum_{i=1}^k \epsilon_i^{(j)} b_i$ for $j = k+1, \ldots, 2^n$.

Finally, the basis of the kernel is $\{b_{k+1}^\Phi, \ldots, b_{2^n}^\Phi\}$.

7 Alternative Expressions of the Choquet Integral

A second application of Lemma 1 is to obtain various equivalent expressions of the Choquet integral. Let v be a game on X, and $f : X \to \mathbb{R}_+$ a real-valued nonnegative mapping. The Choquet integral [5] of f w.r.t. v is defined by

$$\int f \, \mathrm{d}v = \int_0^\infty v(\{x \in X \mid f(x) \geq \alpha\}) \, \mathrm{d}\alpha \qquad (18)$$

which yields in the discrete case ($|X| = n$):

$$\int f \, \mathrm{d}v = \sum_{i=1}^{n} (f_{\sigma(i)} - f_{\sigma(i-1)}) v(\{\sigma(i), \ldots \sigma(n)\})$$

where σ is a permutation on $[n]$ such that $f_{\sigma(1)} \leq f_{\sigma(2)} \leq \cdots \leq f_{\sigma(n)}$, and $f_{\sigma(0)} = 0$.

Remark 1 Usually, the Choquet integral is defined w.r.t. a *capacity*, that is, a game which is monotone with respect to set inclusion. However, it is not possible to extend the definition to arbitrary set functions ξ, i.e., such that $\xi(\varnothing) \neq 0$. Indeed, if $\xi(\varnothing) \neq 0$, it is easily seen from (18) that the integral becomes unbounded.

The Choquet integral is positively homogeneous but not additive in general, i.e., $\int (f + g) \, \mathrm{d}v \neq \int f \, \mathrm{d}v + \int g \, \mathrm{d}v$. However, an interesting feature of this integral is that it is linear w.r.t. the game:

$$\int f \, \mathrm{d}(v + \alpha v') = \int f \, \mathrm{d}v + \alpha \int f \, \mathrm{d}v'.$$

Expressing v in some basis, it is then possible to get the expression of the Choquet integral w.r.t. the corresponding transform.

Let Ψ be a linear invertible transform, and $\{b_A^\Psi\}_{A \in 2^X}$ be the corresponding basis of set functions. Due to Remark 1, one has to be careful because many bases are composed of set functions which are not games. Therefore, some adaptation is necessary. From a basis $\{b_A^\Psi\}_{A \in 2^X}$, we build a basis of games $\{b_A'^\Psi\}_{A \in 2^X \setminus \{\varnothing\}}$ as follows:

$$b_S'^\Psi(T) = \begin{cases} b_S^\Psi(T), & \text{if } T \neq \varnothing \\ 0, \text{otherwise} & \end{cases} \qquad (S \in 2^X \setminus \{\varnothing\}).$$

Then, from the linearity of the integral, for every $f \in \mathbb{R}^X$ and every game v,

$$\int f \, \mathrm{d}v = \int f \, \mathrm{d}\Big(\sum_{\varnothing \neq A \subseteq X} \Psi^v(A) b_A'^\Psi \Big) = \sum_{\varnothing \neq A \subseteq X} \Psi^v(A) \int f \, \mathrm{d}b_A'^\Psi.$$

It is therefore sufficient to compute $\int f \, \mathrm{d}b_A'^{\Psi}$ for every $A \subseteq X$, $A \neq \varnothing$. One obtains the following expressions, for the main bases:

- For the Möbius transform:

$$\int f \, \mathrm{d}u_A = \bigwedge_{i \in A} f_i \tag{19}$$

- For the co-Möbius transform:

$$\int f \, \mathrm{d}\check{u}'_A = (-1)^{|A|+1} \bigvee_{i \in A} f_i .$$

- For the Fourier transform:

$$\int f \, \mathrm{d}\chi'_A = f_{\sigma(n)} + 2 \sum_{j=1}^{|A|} (-1)^j f_{i_j}$$

  with $A = \{i_1, \ldots, i_{|A|}\}$ and $f_{i_1} \geqslant \cdots \geqslant f_{i_{|A|}}$.

Equation (19) is well-known and was first proved by [4] (also by [25]), extending a result of [7]).

There is no simple expression for the case of the interaction transform, although such an expression exists and has been obtained through a different method [13]:

$$\int f \, \mathrm{d}v = \sum_{A \subseteq X} \Bigg(\sum_{K \subseteq X \setminus A} B_{|K|} I^{v^+}(A \cup K) \Bigg) \bigwedge_{i \in A} f_i$$
$$+ \sum_{\varnothing \neq A \in 2^X} (-1)^{|A|+1} \Bigg(\sum_{K \subseteq X \setminus A} B_{|K|} I^{v^-}(A \cup K) \Bigg) \bigvee_{i \in A} f_i \tag{20}$$

where

$$I^{v^+}(A) = \begin{cases} I^v(A), & \text{if } I^v(A) > 0 \\ 0, & \text{otherwise} \end{cases},$$

$$I^{v^-}(A) = \begin{cases} I^v(A), & \text{if } I^v(A) < 0 \\ 0, & \text{otherwise} \end{cases} \qquad (A \in 2^X).$$

References

1. Aigner, M.: Combinatorial Theory. Springer (1979)
2. Banzhaf, J.: Weighted voting doesn't work: a mathematical analysis. Rutgers Law Rev. **19**, 317–343 (1965)
3. Berge, C.: Principles of Combinatorics. Academic Press (1971)

4. Chateauneuf, A., Jaffray, J.-Y.: Some characterizations of lower probabilities and other monotone capacities through the use of Möbius inversion. Math. Soc. Sci. **17**, 263–283 (1989)
5. Choquet, G.: Theory of capacities. Annales de l'Institut Fourier **5**, 131–295 (1953)
6. de Wolf, R.: A brief introduction to Fourier analysis on the Boolean cube. Theory Comput. Lib. Graduate Surv. **1**, 1–20 (2008)
7. Dempster, A.P.: Upper and lower probabilities induced by a multivalued mapping. Ann. Math. Stat. **38**, 325–339 (1967)
8. Dragan, I.: The potential basis and the weighted Shapley value. Libertas Mathematica **11**, 139–150 (1991)
9. Dragan, I.: The least square values and the Shapley value for cooperative TU games. TOP **14**, 61–73 (2006)
10. Faigle, U., Grabisch, M.: Bases and linear transforms of TU-games and cooperation systems. Int. J. Game Theory, to appear
11. Fujishige, S.: Submodular functions and optimization. In: Annals of Discrete Mathematics, vol. 58, 2nd edn. Elsevier, Amsterdam (2005)
12. Grabisch, M.: k-Order additive discrete fuzzy measures and their representation. Fuzzy Sets Syst. **92**, 167–189 (1997)
13. Grabisch, M., Labreuche, Ch.: The symmetric and asymmetric Choquet integrals on finite spaces for decision making. Stat. Pap. **43**, 37–52 (2002)
14. Grabisch, M., Labreuche, Ch.: A decade of application of the Choquet and Sugeno integrals in multi-criteria decision aid. Ann. Oper. Res. **175**, 247–286 (2010). doi:10.1007/s10479-009-0655-8
15. Grabisch, M., Marichal, J.-L., Roubens, M.: Equivalent representations of set functions. Math. Oper. Res. **25**(2), 157–178 (2000)
16. Hammer, P.L., Rudeanu, S.: Boolean Methods in Operations Research and Related Areas. Springer (1968)
17. Harsanyi, J.C.: A simplified bargaining model for the n-person cooperative game. Int. Econ. Rev. **4**, 194–220 (1963)
18. Kleinberg, N.L., Weiss, J.H.: Equivalent n-person games and the null space of the Shapley value. Math. Oper. Res. **10**(2), 233–243 (1985)
19. O'Donnell, R.: Analysis of Boolean functions, draft 2.0, ch. 1–3 (2007). http://www.cs.cmu.edu/~odonnell11/boolean-analysis
20. Peleg, B., Sudhölter, P.: Introduction to the Theory of Cooperative Games. Kluwer Academic Publisher (2003)
21. Rota, G.C.: On the foundations of combinatorial theory I. Theory of Möbius functions. Zeitschrift für Wahrscheinlichkeitstheorie und Verwandte Gebiete. **2**, 340–368 (1964)
22. Roubens, M.: Interaction between criteria and definition of weights in MCDA problems. In: 44th Meeting of the European Working Group "Multicriteria Aid for Decisions". Brussels, Belgium (1996)
23. Shafer, G.: A Mathematical Theory of Evidence. Princeton University Press (1976)
24. Shapley, L.S.: A value for n-person games. In: Kuhn, H.W., Tucker, A.W. (eds.) Contributions to the Theory of Games, Vol. II, number 28 in Annals of Mathematics Studies, pp. 307–317. Princeton University Press (1953)
25. Walley, P.: Coherent lower (and upper) probabilities. Technical Report 22, University of Warwick, Coventry (1981)
26. Walsh, J.: A closed set of normal orthogonal functions. Am. J. Math. **45**, 5–24 (1923)
27. Yokote, K.: Weak addition invariance and axiomatization of the weighted Shapley value. Int. J. Game Theory **44**, 275–293 (2015)
28. Yokote, K., Funaki, Y., Kamijo, Y.: Linear basis to the Shapley value. Technical report, Waseda Economic Working Paper Series (2013)

Conditioning for Boolean Subsets, Indicator Functions and Fuzzy Subsets

Siegfried Weber

Abstract This chapter deals with measure-free conditioning. It starts with the mean value based definition of conditional fuzzy subsets which again gives a fuzzy subset. Applying this general construction to indicator functions, it is proved that these conditionals form an MV-algebra and that this is isomorphic to the already known MV-algebra of the interval based conditional Boolean subsets. In the following, the problem of iteration is completely solved with the result that there are exactly two types of iteration, called the blurred resp. the sharper one, which remain in the corresponding MV-algebras. Moreover, the general concept of conditional operators plays a significant role. Finally, the problem of extending an uncertainty measure is discussed.

1 Introduction

The first step in "measure-free conditioning" consists in the construction of *conditional events* "a given b" as well-defined elements of some structured set such that the (unconditional) events a are described by the conditional events "a given the *sure* event". In a second step, the uncertainty of such conditional events is expressed by elements of the real unit interval as values of a suitable measure.

Whereas the author's last papers [19–21] treated problems mainly from the second step, the present paper concentrates on the first step. In the introduction of our joint paper [11] with Ulrich Höhle we observed that "... the iteration of measure-free conditioning is still an open problem." In [11] we gave a partial solution for the quite general situation of events from a Girard algebra. In the present paper we will give a complete solution to this problem for the classical situation of events from a Boolean algebra $\mathbb{B}$ of subsets of an universe Ω, where conditioning operators, as introduced in [11], will play an essential role.

S. Weber (✉)
Institut Für Mathematik, Universität Mainz, FB 08, Mainz, Germany
e-mail: sweber@mathematik.uni-mainz.de

S. Saminger-Platz and R. Mesiar (eds.), *On Logical, Algebraic, and Probabilistic Aspects of Fuzzy Set Theory*, Studies in Fuzziness and Soft Computing 336,
DOI 10.1007/978-3-319-28808-6_14

Another motivation for the present paper came from several talks at the Linz Seminar about "a new axiomatic approach", where the conditional events are introduced by three-valued "generalized indicator functions"

$$(*) \qquad \mathbf{1}_{A|B} = 1 \cdot \mathbf{1}_{A \cap B} + u \cdot \mathbf{1}_{B^c} + 0 \cdot \mathbf{1}_{A^c \cap B}\,, \quad A, B \in \mathbb{B}, \quad \text{with some} \quad u \in [0, 1]\,,$$

where the "undetermined value" u is not a constant number but is considered as value $u = t(A \mid B)$ of a "conditional uncertainty measure", see [5, 6]. Therefore, this interesting approach is not measure-free. Contrary to this, in the present paper the measure-free version of the right side in $(*)$ will be considered, i.e. where u is a constant. We will see that $u = (0 \mid 0)$ should be a self-complemented element of the MV-chain [0, 1] and, therefore, $u = \frac{1}{2}$. It will be shown that these *conditional indicator functions* form an MV-algebra, where the underlying partial ordering results to be equivalent to the interval ordering used for the interval based definition of *conditional Boolean subsets*

$$(CB) \qquad (A \parallel B) = [A \cap B\,,\ B \to A] \quad \text{for} \quad A, B \in \mathbb{B},$$

denoting this set by $\tilde{\mathbb{B}}$. This lattice-interval approach has been treated in detail in a lot of publications, see e.g. [9, 11, 17] and the references therein. In this approach the "conditional Boolean event" $(A \parallel B)$ is defined as the set of *all* possible conditional candidates between the Boolean events "A and B" resp. "if B then A" as the extreme candidates. In contrast to this approach, it is well-known that there is no reasonable way to define a conditional as *some* Boolean subset $(A \mid B)$ between these two extreme Boolean subsets. On the other hand, this alternative approach really is applicable to indicator functions. It seems that this approach has not yet treated systematically. This will be done in the present paper, including the problem of iteration.

More concretely, we start with the general construction, introduced in [18], of *conditional fuzzy sets* defined pointwisely by

$$(CF) \qquad (\varphi \mid \psi) = C(\varphi \wedge \psi\,,\ \psi \to \varphi) \in \mathbb{F} \quad \text{for fuzzy subsets} \quad \varphi, \psi \in \mathbb{F} = [0, 1]^{\Omega}\,,$$

based on a *mean value function* C on the unit interval [0, 1] which is *compatible* with the complement in [0, 1]. The properties of a mean value function C are very natural in order to choose *some* fuzzy subset *between* the extreme fuzzy subsets. The additional property of compatibility guarantees that C generates a conditioning operator $\mid$ on $\mathbb{F}$. Now, applying this construction to indicator functions, we prove that the subset $\mathbb{F}_1$ of the *conditional indicator functions*

$$(CI) \qquad (\mathbf{1}_A \mid \mathbf{1}_B) = 1 \cdot \mathbf{1}_{A \cap B} + \frac{1}{2} \cdot \mathbf{1}_{B^c} + 0 \cdot \mathbf{1}_{A^c \cap B}$$

is an MV-algebra and closed under iteration if and only if $C = C_1$ given by $C_1(0, \frac{1}{2}) = \frac{1}{2}$ or $C = C_2$ given by $C_2(0, \frac{1}{2}) = 0$. Because of these values we call the first type "blurred iteration" and the second type "sharper iteration". Furthermore, we prove that any conditioning operator | on $\mathbb{F}_1$ is generated by exactly one of these two compatible mean value functions.

Moreover, it follows that $i : \mathbb{F}_1 \to \tilde{\mathbb{B}}$ is an MV-algebra isomorphism between the two types (*CI*) resp. (*CB*) of conditionals. We use this result to obtain two corresponding conditioning operators $\tilde{|}$ on $\tilde{\mathbb{B}}$ by

$$((A_1 \parallel B_1) \,\tilde{|}\, (A_2 \parallel B_2)) = i \circ ((\mathbf{1}_{A_1} \mid \mathbf{1}_{B_1}) \mid (\mathbf{1}_{A_2} \mid \mathbf{1}_{B_2}))\ .$$

Particularly, it follows that the conditional Boolean subsets $(A \parallel B) = (\{A\} \,\tilde{|}\, \{B\})$ are recovered as values of the conditioning operators $\tilde{|}$ applied to singletons. This approach to the iteration process in $\tilde{\mathbb{B}}$ leads to the same result as in [11] although is very different, but it shows that the two special compatible mean value functions on $\tilde{\mathbb{B}}$ given in [11] are the only two and are obtained as

$$\tilde{C}_k(\, i \circ \varphi\ ,\ i \circ \psi\ ) = i \circ C_k(\varphi, \psi) \quad \text{for} \quad \varphi, \psi \in \mathbb{F}_1\ ,\ \varphi \leq \psi\ ,\ k = 1, 2\ .$$

The main results of the present paper were presented in the author's talk at the 34th Linz Seminar on Fuzzy Set Theory, 2013.

The paper is organized as follows. In Sect. 2 we put together the basic prerequisites, referring to MV-algebras, mean value functions, conditioning operators and the conditional fuzzy subsets (*CF*) in $\mathbb{F}$. Section 3 treats the conditional indicator functions (*CI*) in $\mathbb{F}_1$ which form an MV-algebra (Theorem 1). Section 4 treats the conditional Boolean subsets (*CB*) in $\tilde{\mathbb{B}}$ (Theorem 2) and the MV-algebra isomorphism between $\mathbb{F}_1$ and $\tilde{\mathbb{B}}$ (Theorem 3). Section 5 is dedicated to the iteration processes in both $\mathbb{F}_1$ (Theorem 4) and $\tilde{\mathbb{B}}$ (Theorem 5). Finally, in Sect. 6 we present the basic topics from the second step of measure-free conditioning, i.e. referring to uncertainty measures: the general Definition 5, the Theorem 6 for $\mathbb{F}_1$ and $\tilde{\mathbb{B}}$, some examples and remarks, including the Remark 9 for the alternative approach (∗).

2 Basic Definitions and Results

The basic structure needed in the paper is that of an MV-algebra which we introduce as follows.

Definition 1 *(MV-algebras)*

(i) A set $\mathbb{L}$ is called a *residuated lattice* if it is equipped with the two structures of a bounded lattice $(\mathbb{L}, \leq, \wedge, \vee)$ with universal upper (resp. lower) bound 1 (resp. 0) and a commutative monoid (semigroup with 1 as unit) $(\mathbb{L}, \sqcap)$, such that there exist all residuals $b \to a$ given by the residuation property

$$c \sqcap b\ \leq\ a \quad \Longleftrightarrow \quad c\ \leq\ b \to a.$$

For each $b \in \mathbb{L}$ a *residual complement* can be defined by

$$b' = b \to 0.$$

(ii) Furthermore, $\mathbb{L}$ is called an *MV-algebra* if the lattice-join can be expressed by residuals by means of the additional property

$$\text{(MV)} \qquad b \vee a = (b \to a) \to a.$$

For each $a, b \in \mathbb{L}$ a dual semigroup operation (with 0 as unit) can be defined by

$$a \sqcup b = (a' \sqcap b')'.$$

Sometimes we will use the notation $(\mathbb{L}, \sqcap, ', \sqcup)$ instead of $\mathbb{L}$, dealing with explicitely the MV-operations in this order.

The values $a \sqcup b$ of the dual operation in (ii) will be interpreted as "unions". If the "intersection" satisfies $a \sqcap b = 0$ we call them "disjoint unions" and write $a \dot{\sqcup} b$, for short.

In previous papers, e.g. in [21], the author used the name *commutative residuated lattice ordered semigroup with zero* from [1] rather than the name *residuated lattice*, see [15], the references therein and the book [8].

The operations denoted here by $\sqcap, ', \sqcup$ play the role of the MV-algebra operations $\cdot, ^-, +$ originally used by Chang in [3] resp. $\odot, \neg, \oplus$ used e.g. in the book [4] as general reference to MV-algebras. See also the biographical Remark 6.5 in [21].

In previous papers we needed the structure of a *Girard algebra*, which is a residuated lattice where the residual complementation is idempotent, i.e. a structure "between" residuated lattices and MV-algebras, see [11, 21] and the biographical Remark 6.2 therein.

Remark 1 (Boolean algebras) It is well known that an MV-algebra $\mathbb{L}$ is a *Boolean algebra* if and only if $\sqcap = \wedge$.

In the present paper we will often tacitly use the following well-known properties:

Proposition 1 (Additional properties in MV-algebras)

(i) The residual complement has the involution property

$$b'' = b.$$

(ii) The residuals can be expressed as

$$b \to a = b' \sqcup a = b' \dot{\sqcup} (a \wedge b).$$

(iii) *The lattice-meet can be expressed by the semigroup operations and the residual by means of the divisibility property*

$$b \wedge a = b \sqcap (b \to a) = b \sqcap (b' \sqcup a).$$

The following well-known result is the starting point of the present paper.

Lemma 1 (The MV-algebra of fuzzy subsets) *The set*

$$\mathbb{F} = [0, 1]^{\Omega} = \{\varphi : \Omega \to [0, 1]\}$$

of all fuzzy subsets φ *of an universe* Ω *has the structure of an MV-algebra, inherited pointwisely from the standard MV-(algebra-)chain* $([0, 1], \odot, \neg, \oplus)$ *for the values* $\varphi(\omega)$ *by*

$$a \odot b = (a + b - 1) \vee 0\,, \quad \neg a = 1 - a\,, \quad a \oplus b = (a + b) \wedge 1\,,$$

where $\odot$, $\neg$, $\oplus$ *denote (here and in the following) the MV-operations "intersection", "complement", "union" in* $[0, 1]$ *resp.* $\mathbb{F}$.

In the following we deal with the basics about mean value functions which will play a crucial role in our presentation of conditioning.

Definition 2 *(Mean value functions on MV-algebras)* A *compatible mean value function* C on an MV-algebra $\mathbb{L} = (\mathbb{L}, \sqcap, ', \sqcup)$ is

(i) a *mean value function* on $\mathbb{L}$, i.e. a map C defined for all $a, b \in \mathbb{L}$ with $a \leq b$ and values in $\mathbb{L}$, which is isotone in both arguments and idempotent,
(ii) which is *compatible with the complement* in $\mathbb{L}$, i.e. satisfying $(C(a, b))' = C(b', a')$.

Remark 2 *(Compatible mean value functions)* The additional property that a mean value function C on an MV-algebra $\mathbb{L}$ is compatible implies that $(C(0, 1))' = C(0, 1)$, i.e. in $\mathbb{L}$ has to exist a self-complemented element which can serve as value $C(0, 1)$.

Example 1 *(Compatible mean value functions on [0,1])* The standard MV-chain $[0, 1]$ has $\frac{1}{2}$ as the (unique) self-complemented element and admits a lot of compatible mean value functions C_k. In the following we deal with four of these to which we will refer in the following sections. Particularly, the first two will be essential in Sect. 5, where the special value $C_k(0, \frac{1}{2})$ will play an important role. We denote by

(i) $C_1(a, b) = \begin{Bmatrix} b \text{ if } & b < \frac{1}{2} \\ \frac{1}{2} \text{ if } & a \leq \frac{1}{2} \leq b \\ a \text{ if } & \frac{1}{2} < a \end{Bmatrix}$ with $C_1(0, \frac{1}{2}) = \frac{1}{2}$,

(ii) $C_2(a, b) = \begin{Bmatrix} \frac{a}{1+a-b} & \text{if } (a, b) \neq (0, 1) \\ \frac{1}{2} & \text{if } (a, b) = (0, 1) \end{Bmatrix}$ with $C_2(0, \frac{1}{2}) = 0$,

(iii) $C_3(a, b) = \frac{a+b}{2}$ with $C_3(0, \frac{1}{2}) = \frac{1}{4}$,

(iv) $C_4(a, b) = \frac{b}{1+b-a}$ with $C_4(0, \frac{1}{2}) = \frac{1}{3}$.

In contrast to these four, the following mean value function, needed in Sect. 6, is not compatible:

(v) $C_5(a, b) = (1 - t) \cdot a + t \cdot b$ for some $t \in [0, 1]$, $t \neq \frac{1}{2}$.

In the last part of this section, we present one concept of conditioning, first the general definition introduced in [18], then the connection with the axiomatic approach via conditioning operators introduced in [10] and, finally, the application to fuzzy subsets. For more details see also [11], where this concept is extended to the more general structure of Girard algebras.

Definition 3 *(Conditional elements in an MV-algebra)* Let $\mathbb{L}$ be an MV-algebra with a self-complemented element and let C be any compatible mean value function on $\mathbb{L}$. Then *conditional elements* "a given b" are defined by

$$(C) \qquad (a \mid b) = C(a \wedge b\,,\; b \to a) \in \mathbb{L} \quad \text{for} \quad a, b \in \mathbb{L}.$$

Proposition 2 (Conditional elements and conditioning operator in an MV-algebra) *Let $\mathbb{L} = (\mathbb{L}, \sqcap, ', \sqcup)$ be an MV-algebra with a self-complemented element. Then:*

(i) Each compatible mean value function C on $\mathbb{L}$ "generates" a conditioning operator $\mid$ on $\mathbb{L}$, i.e. a binary operation $\mid: \mathbb{L} \times \mathbb{L} \longrightarrow \mathbb{L}$ with values $(a \mid b)$ given by (C), which satisfies

(C1) $(a \mid 1) = a$,
(C2) $(a \mid b) = (a \wedge b \mid b)$,
(C3) $a_1 \leq a_2 \;\Rightarrow\; (a_1 \mid b) \leq (a_2 \mid b)$,
(C4) $b_1 \leq b_2$ *and* $a \wedge b_2 \leq a \wedge b_1 \;\Rightarrow\; (a \mid b_2) \leq (a \mid b_1)$,
(C5) $(a \mid b)' = (a' \sqcap b \mid b)$.

Particularly, conditioning only with lower bound 0 and upper bound 1 in $\mathbb{L}$ leads to

$$(0 \mid 1) = C(0, 0) = 0\,, \quad (0 \mid 0)' = (0 \mid 0) = C(0, 1)\,, \quad (1 \mid 1) = C(1, 1) = 1\,.$$

(ii) The compatible mean value function C can be recovered by

$$C(a, b) = (a \mid b \to a) \;\; \textit{for} \;\; a \leq b.$$

(iii) Vice versa, given any conditioning operator $\mid$ on $\mathbb{L}$, i.e. fulfilling the axioms (C1) – (C5) from (i), then there exists a compatible mean value function C, given by (ii) if and only if $\mid$ satisfies the additional condition

$$(cmvf) \qquad a_1 \leq a_2 \;\Rightarrow\; (a_1 \wedge b \mid b \to a_1) \leq (a_2 \wedge b \mid b \to a_2).$$

Corollary 1 (Conditional fuzzy subsets) *Let* $\mathbb{F} = [0, 1]^{\Omega}$ *be the MV-algebra of fuzzy subsets of a universe* Ω. *Then each compatible mean value function* C *on* $[0, 1]$ *leads to a conditioning operator* $|$ *on* $\mathbb{F}$, *where its values are the "conditional fuzzy subsets" given pointwise by*

$$(CF) \qquad (\varphi \mid \psi) = C(\varphi \wedge \psi \, , \; \psi \to \varphi) \in \mathbb{F} \;\; for \;\; \varphi, \psi \in \mathbb{F}.$$

3 The MV-Subalgebra $\mathbb{F}_1$ of Conditional Indicator Functions

Notation 1 Let $\mathbb{B} = (\mathbb{B}, \cap, ^c, \cup)$ be a Boolean (MV-)algebra of (crisp) subsets of the universe Ω. Then we will denote by

$$\mathbb{F}_0 = \{\mathbf{1}_A : A \in \mathbb{B}\} \;\; \text{resp.} \;\; \mathbb{F}_1 = \{(\mathbf{1}_A \mid \mathbf{1}_B) : \mathbf{1}_A, \mathbf{1}_B \in \mathbb{F}_0\}$$

the subsets of $\mathbb{F}$ containing all *indicator functions* resp. all *conditional indicator functions*.

Remark 3 It is well known that $\mathbb{F}_0$ is a Boolean (MV-)subalgebra of $\mathbb{F}$ such that $\mathbb{F}_0 \cong \mathbb{B}$ is a Boolean algebra isomorphism:

$$\mathbf{1}_A \odot \mathbf{1}_B = \mathbf{1}_{A \cap B} \, , \quad \neg \mathbf{1}_A = \mathbf{1}_{A^c} \, , \quad \mathbf{1}_A \oplus \mathbf{1}_B = \mathbf{1}_{A \cup B} \, .$$

Applying the general construction (*CF*) of conditioning fuzzy subsets in $\mathbb{F}$ to indicator functions requires only the properties $C(0, 0) = 0, C(1, 1) = 1$ of any mean value function C on $[0, 1]$ and $C(0, 1) = \frac{1}{2}$ of any compatible one and, therefore, leads directly to the following

Proposition 3 (Conditional indicator functions) *The elements of* $\mathbb{F}_1$ *can be written as*

$$(CI) \qquad (\mathbf{1}_A \mid \mathbf{1}_B) = 1 \cdot \mathbf{1}_{A \cap B} + \frac{1}{2} \cdot \mathbf{1}_{B^c} + 0 \cdot \mathbf{1}_{A^c \cap B} = \mathbf{1}_{A \cap B} + \frac{1}{2} \cdot \mathbf{1}_{B^c} \, .$$

These "conditional indicator functions" do not depend on the choice of the compatible mean value function C *on* $[0, 1]$, *in other words, only the values* $(1 \mid 1) = 1$, $(0 \mid 1) = 0$ *and* $(1 \mid 0) = (0 \mid 0) = \frac{1}{2}$ *are needed.*

It follows from (*CI*) that the conditioning operator $|$ on $\mathbb{F}$ cannot be restricted to a conditioning operator on $\mathbb{F}_0$. The question, if this is possible on $\mathbb{F}_1$, will be answered in Sect. 5. As a first step in this direction we need the following

Theorem 1 (The MV-subalgebra of conditional indicator functions) *The subset* $\mathbb{F}_1$ *of all conditional indicator functions is an MV-subalgebra of* $\mathbb{F}$ *with respect to*

(i) $(\mathbf{1}_{A_1} \mid \mathbf{1}_{B_1}) \odot (\mathbf{1}_{A_2} \mid \mathbf{1}_{B_2}) = (\mathbf{1}_D \mid \mathbf{1}_E)$ *with* $D = A_1 \cap B_1 \cap A_2 \cap B_2$ *and* $E = D \cup (A_1^c \cap B_1) \cup (A_2^c \cap B_2) \cup (B_1^c \cap B_2^c)$

(ii) $\neg(\mathbf{1}_A \mid \mathbf{1}_B) = (\mathbf{1}_{A^c} \mid \mathbf{1}_B)$,

(iii) $(\mathbf{1}_{A_1} \mid \mathbf{1}_{B_1}) \oplus (\mathbf{1}_{A_2} \mid \mathbf{1}_{B_2}) = (\mathbf{1}_F \mid \mathbf{1}_G)$ *with* $F = (A_1 \cap B_1) \cup (A_2 \cap B_2) \cup (B_1^c \cap B_2^c)$ *and* $G = F \cup (B_1 \cap B_2)$.

Clearly, $\mathbb{F}_0$ *is a Boolean (MV-)subalgebra of* $\mathbb{F}_1$.

Proof (i) is obtained by applying the conditionals $(\cdot \mid \cdot)$ in the form (CI) and the values $a \odot b$ for $a, b \in \{0, \frac{1}{2}, 1\}$:

$$\begin{aligned}
&(\mathbf{1}_{A_1} \mid \mathbf{1}_{B_1}) \odot (\mathbf{1}_{A_2} \mid \mathbf{1}_{B_2}) \\
&= (1 \cdot \mathbf{1}_{A_1 \cap B_1} + \tfrac{1}{2} \cdot \mathbf{1}_{B_1^c} + 0 \cdot \mathbf{1}_{A_1^c \cap B_1}) \odot (1 \cdot \mathbf{1}_{A_2 \cap B_2} + \tfrac{1}{2} \cdot \mathbf{1}_{B_2^c} + 0 \cdot \mathbf{1}_{A_2^c \cap B_2}) \\
&= 1 \cdot \mathbf{1}_{A_1 \cap B_1 \cap A_2 \cap B_2} + \tfrac{1}{2} \cdot \mathbf{1}_{(A_1 \cap B_1 \cap B_2^c) \cup (B_1^c \cap A_2 \cap B_2)} + 0 \cdot \mathbf{1}_{(A_1^c \cap B_1) \cup (A_2^c \cap B_2) \cup (B_1^c \cap B_2^c)} .
\end{aligned}$$

From this we obtain D directly and $E = ((A_1 \cap B_1 \cap B_2^c) \cup (B_1^c \cap A_2 \cap B_2))^c$ can be rewritten into the given form, where $D \subseteq E$.
Formula (ii) follows directly:

$$\begin{aligned}
\neg(\mathbf{1}_A \mid \mathbf{1}_B) &= \neg(1 \cdot \mathbf{1}_{A \cap B} + \tfrac{1}{2} \cdot \mathbf{1}_{B^c} + 0 \cdot \mathbf{1}_{A^c \cap B}) \\
&= 0 \cdot \mathbf{1}_{A \cap B} + \tfrac{1}{2} \cdot \mathbf{1}_{B^c} + 1 \cdot \mathbf{1}_{A^c \cap B} = (\mathbf{1}_{A^c} \mid \mathbf{1}_B).
\end{aligned}$$

In analogy to (i) it follows (iii):

$$\begin{aligned}
&(\mathbf{1}_{A_1} \mid \mathbf{1}_{B_1}) \oplus (\mathbf{1}_{A_2} \mid \mathbf{1}_{B_2}) \\
&= 1 \cdot \mathbf{1}_{(A_1 \cap B_1) \cup (A_2 \cap B_2) \cup (B_1^c \cap B_2^c)} + \tfrac{1}{2} \cdot \mathbf{1}_{(A_1^c \cap B_1 \cap B_2^c) \cup (B_1^c \cap A_2^c \cap B_2)} + 0 \cdot \mathbf{1}_{A_1^c \cap B_1 \cap A_2^c \cap B_2} .
\end{aligned}$$

From this we obtain F directly and $G = ((A_1^c \cap B_1 \cap B_2^c) \cup (B_1^c \cap A_2^c \cap B_2))^c$ can be rewritten into the given form, where $F \subseteq G$. □

Remark 4 (Measure-free conditioning) The form (CI) has a long history, for which we refer to the book [9] and the references therein, where instead of $\frac{1}{2}$ often the value u but also other symbols as e.g. **?** in [7] are used for this third "undefined" or "undetermined" value. Some proposals for connecting these, sometimes named "generalized indicator functions", were given, very different among themselves and from those in the preceding theorem. But all proposals correspond to "measure-free conditioning".

A completely different concept was proposed in [5], where the undetermined value u is not a constant, i.e. is no longer the same number for all "conditional events". This concept will briefly be discussed in Remark 9.

4 The Isomorphism Between $\mathbb{F}_1$ and the MV-Algebra $\tilde{\mathbb{B}}$ of Conditional Boolean Subsets

In Sect. 2, a conditional element $(a \mid b)$ of elements a, b in an MV-algebra $\mathbb{L}$ is introduced as *some* mean value $C(a \wedge b, b \to a)$ between the "natural conditional candidates" $a \wedge b$ and $b \to a$. But this requires that $\mathbb{L}$ has a self-complemented element.

Therefore, this "mean value approach" is not applicable for a Boolean algebra $\mathbb{L}$. A solution is to take *all* candidates between the two mentioned above. This leads to the "interval approach" of the following

Definition 4 *(Conditional Boolean events)* The *conditional Boolean event* "*a* given *b*" of two *(crisp) events a, b* in any Boolean algebra $\mathbb{L} = (\mathbb{L}, \wedge, ', \vee)$ is defined as the lattice interval

$$(a \parallel b) = [\, a \wedge b \,,\ b \to a \,] = [\, a \wedge b \,,\ b' \vee a \,].$$

The set of all conditional Boolean events of events in $\mathbb{L}$ will be denoted by $\tilde{\mathbb{L}}$.

The author get to know this construction during the talk [12], for details see the book [9] and the references therein. An extension to MV-algebras was presented in the author's talk [16], for details see [17, 18]. In [11] this concept has been extended to Girard algebras.

Remark 5 The conditional events are in a one-to-one correspondence to intervals via

$$[a, c] = (a \parallel c' \vee a).$$

Therefore, in the following we will alternatively take the conditional events or the intervals as elements of $\tilde{\mathbb{L}}$.

Lemma 2 (The MV-algebra of conditional Boolean events) *Let* $\tilde{\mathbb{L}}$ *be the set of conditional Boolean events from the preceding definition. Then the following assertions hold:*

(I1) $(a \parallel 1) = [a, a] = \{a\}$, *i.e.* $\tilde{\mathbb{L}}$ *extends* $\mathbb{L}$,
(I2) $(a \parallel b) = (a \wedge b \parallel b)$.

Furthermore, $\tilde{\mathbb{L}}$ *is a lattice with respect to the partial ordering for intervals*

$$[a, c] \leq [d, f] \text{ if and only if } a \leq d\ ,\ c \leq f$$

where, therefore, $(1 \parallel 1)$ *is the upper resp.* $(0 \parallel 1)$ *the lower universal bound. The monotonicity properties follow from the lattice structure:*

(I3) $a_1 \le a_2 \Rightarrow (a_1 \parallel b) \le (a_2 \parallel b)$,
(I4) $b_1 \le b_2$ *and* $a \wedge b_2 \le a \wedge b_1 \Rightarrow (a \parallel b_2) \le (a \parallel b_1)$.

Finally, $\tilde{\mathbb{L}}$ *is an MV-algebra with respect to*

$$(a \parallel b) \sqcap (c \parallel d) = (a \wedge b \wedge c \wedge d \parallel (a \wedge b \wedge c \wedge d) \vee (a' \wedge b) \vee (c' \wedge d) \vee (b' \wedge d')),$$

(I5) $(a \parallel b)' = (a' \parallel b)$,

$$(a \parallel b) \sqcup (c \parallel d) = ((a \wedge b) \vee (c \wedge d) \vee (b' \wedge d') \parallel (a \wedge b) \vee (c \wedge d) \vee (b' \wedge d') \vee (b \wedge d)).$$

Proof The result was proved in [11] more generally for any MV-algebra $\mathbb{L}$, but only in terms of intervals. Rewriting these intervals into conditional events lead to the given formulae for the MV-algebra operations as follows.

In order to obtain the first result apply, in this order, the definition of $(\cdot \parallel \cdot)$, the definition of $\sqcap$ from Theorem 2.3 in [11], the preceding remark and known properties in a Boolean algebra:

$$\begin{aligned}
&(a \parallel b) \sqcap (c \parallel d) = [\, a \wedge b \,,\, b' \vee a \,] \sqcap [\, c \wedge d \,,\, d' \vee c \,] \\
&= [\, (a \wedge b) \wedge (c \wedge d) \,,\, ((a \wedge b) \wedge (d' \vee c)) \vee ((b' \vee a) \wedge (c \wedge d)) \,] \\
&= (\, a \wedge b \wedge c \wedge d \parallel (\, (a' \vee b' \vee (d \wedge c')) \wedge ((b \wedge a') \vee c' \vee d') \,) \vee (a \wedge b \wedge c \wedge d) \,) \\
&= (\, a \wedge b \wedge c \wedge d \parallel (\, (a' \wedge b) \vee (c' \wedge d) \vee (b' \wedge d') \vee (a \wedge b \wedge c \wedge d) \,).
\end{aligned}$$

By analogous steps (I5) follows:

$$\begin{aligned}
(a \parallel b)' &= [\, a \wedge b \,,\, b' \vee a \,]' = [\, (b' \vee a)' \,,\, (a \wedge b)' \,] = [\, b \wedge a' \,,\, a' \vee b' \,] \\
&= (\, b \wedge a' \parallel (a' \vee b')' \vee (b \wedge a') \,) = (b \wedge a' \parallel b) = (a' \parallel b).
\end{aligned}$$

Analogously and using also the former formulae and in the final step the property $(I2)$ the third formula follows:

$$\begin{aligned}
&(a \parallel b) \sqcup (c \parallel d) = (\, (a \parallel b)' \sqcap (c \parallel d)' \,)' = (\, (a' \parallel b) \sqcap (c' \parallel d) \,)' \\
&= (\, a' \wedge b \wedge c' \wedge d \parallel (a' \wedge b \wedge c' \wedge d) \vee (a \wedge b) \vee (c \wedge d) \vee (b' \wedge d') \,)' \\
&= (\, a \vee b' \vee c \vee d' \parallel (a' \wedge b \wedge c' \wedge d) \vee (a \wedge b) \vee (c \wedge d) \vee (b' \wedge d') \,) \\
&= (\, (a \wedge b) \vee (c \wedge d) \vee (b' \wedge d') \parallel (b \wedge d) \vee (a \wedge b) \vee (c \wedge d) \vee (b' \wedge d') \,). \quad \square
\end{aligned}$$

The properties $(I2) - (I5)$ for the interval based conditionals $(a \parallel b)$ are completely analogous to $(C2) - (C5)$ for the mean value based conditionals $(a \mid b)$, whereas the difference between $(I1)$ and $(C1)$ means simply that the original elements are embedded into the conditionals in the former construction and are special conditionals in the latter one.

The preceding result was already proved in [9], mainly in Theorem 3 of Sect. 3, which can be seen after rewriting the operations given there.

As already mentioned within the proof, in [11] the result was generalized to each MV-algebra $\mathbb{L}$ where the set $\tilde{\mathbb{L}}$ results to be a Girard algebra.

Structures $\mathbb{L}$ which are more general than an MV-algebra can also be extended to the set $\tilde{\mathbb{L}}$ of intervals. On the one hand, the Main Theorem 3.1 from [11] extends a Girard algebra $\mathbb{L}$ to a Girard algebra $\tilde{\mathbb{L}}$. On the other hand, Theorem 15 from [15] deals with an "interval-valued residuated lattice" (IVRL) $\tilde{\mathbb{L}}$, where the underlying "base lattice" $\mathbb{L}$ results to be a residuated lattice. The particular case for $\alpha = 0$ in Theorem 15 from [15] corresponds to the Main Theorem 3.1 from [11]. But in both situations, intervals cannot be rewritten into conditional events.

Now let $\mathbb{B} = (\mathbb{B}, \cap, ^c, \cup)$ be a Boolean (MV-)algebra of (crisp) subsets of the universe Ω and let $\tilde{\mathbb{B}}$ be the corresponding set of *conditional Boolean subsets*

$$(CB) \qquad (A \parallel B) = [A \cap B\,,\ B \to A] = [A \cap B\,,\ B^c \cup A] \quad \text{for} \quad A, B \in \mathbb{B}.$$

Then the result of the preceding lemma can be rewritten into the following

Theorem 2 (The MV-algebra of conditional Boolean subsets) *For a Boolean algebra $\mathbb{B}$ of subsets of the universe Ω, the set $\tilde{\mathbb{B}}$ is an MV-algebra with respect to*

(i) $(A_1 \parallel B_1) \sqcap (A_2 \parallel B_2) = (D \parallel E)$,
(ii) $(A \parallel B)' = (A^c \parallel B)$,
(iii) $(A_1 \parallel B_1) \,\bar{\sqcup}\, (A_2 \parallel B_2) = (F \parallel G)$,

where D, E resp. F, G result to be the same as for the MV-algebra $\mathbb{F}_1$ in Sect. 3.

As a summary of the preceding theorem and the above mentioned corresponding result in Sect. 3 we obtain explicitely the following (compare with our Remarks 2.5 and 3.3 in [11])

Theorem 3 (The MV-algebra isomorphism between $\mathbb{F}_1$ and $\tilde{\mathbb{B}}$) *It follows that $\mathbb{F}_1 \cong \tilde{\mathbb{B}}$ is an MV-algebra isomorphism between the two MV-algebras of "conditionals", i.e. of*

(CI) *conditional indicator functions* $(\boldsymbol{1}_A \mid \boldsymbol{1}_B) = 1 \cdot \boldsymbol{1}_{A \cap B} + \frac{1}{2} \cdot \boldsymbol{1}_{B^c} + 0 \cdot \boldsymbol{1}_{A^c \cap B} \in \mathbb{F}_1$ *and*
(CB) *conditional Boolean subsets* $(A \parallel B) = [\,A \cap B\,,\ B^c \cup A\,] \in \tilde{\mathbb{B}}$,

for $A, B \in \mathbb{B}$, where both partial orderings are equivalent:

$$(\boldsymbol{1}_{A_1} \mid \boldsymbol{1}_{B_1}) \le (\boldsymbol{1}_{A_2} \mid \boldsymbol{1}_{B_2}) \Leftrightarrow \left\{ \begin{array}{c} A_1 \cap B_1 \subseteq A_2 \cap B_2 \\ B_1 \to A_1 \subseteq B_2 \to A_2 \end{array} \right\} \Leftrightarrow (A_1 \parallel B_1) \le (A_2 \parallel B_2).$$

Furthermore, the MV-algebra isomorphism $\mathbb{F}_1 \cong \tilde{\mathbb{B}}$ extends the Boolean algebra isomorphism $\mathbb{F}_0 \cong \mathbb{B}$ because of $(\boldsymbol{1}_A \mid \boldsymbol{1}_\Omega) = \boldsymbol{1}_A \in \mathbb{F}_0$ and $(A \parallel \Omega) = \{A\}$, $A \in \mathbb{B}$.

Proof Only the assertion referring to the orderings has not yet proved explicitely. In order to establish the first equivalence, it follows from the left side that $A_1 \cap B_1 \subseteq A_2 \cap B_2$ (from "$1 \leq 1$") and $A_2^c \cap B_2 \subseteq A_1^c \cap B_1$ (from "$0 \leq 0$") and, therefore, also $B_1 \to A_1 \subseteq B_2 \to A_2$, i.e. the direction "$\Rightarrow$" is proved. The other direction in the first equivalence can easily be obtained by checking the possible cases. The second equivalence is precisely the definition of the ordering for intervals. □

For the author it was a little surprising when he realized that the (very natural) ordering in $\mathbb{F}_1$ is equivalent to the interval ordering in $\tilde{\mathbb{B}}$ in such a direct way, that this can be seen as a motivation for considering the interval ordering as the adequate ordering.

5 Iteration of Conditioning

The conditional indicator functions are fuzzy subsets and, therefore, the general construction (CF) from Sect. 2 can be applied and leads to the following

Lemma 3 (Iteration of conditional indicator functions) *The iterated conditional indicator functions have the form*

$$((\boldsymbol{1}_{A_1} \mid \boldsymbol{1}_{B_1}) \mid (\boldsymbol{1}_{A_2} \mid \boldsymbol{1}_{B_2})) = 1 \cdot \boldsymbol{1}_{A_1 \cap B_1 \cap A_2 \cap B_2} + (1-c) \cdot \boldsymbol{1}_{(A_1 \cup B_1^c) \cap B_2^c}$$
$$+ \frac{1}{2} \cdot \boldsymbol{1}_{(B_1^c \cup A_2^c) \cap B_2} + c \cdot \boldsymbol{1}_{A_1^c \cap B_1 \cap B_2^c} + 0 \cdot \boldsymbol{1}_{A_1^c \cap B_1 \cap A_2 \cap B_2}$$

with $c = (0 \mid \frac{1}{2}) = C(0, \frac{1}{2}) \in [0, \frac{1}{2}]$ *and, therefore,* $1 - c = (\frac{1}{2} \mid \frac{1}{2}) = C(\frac{1}{2}, 1) \in [\frac{1}{2}, 1]$.

Proof The announced formula follows from the values

$$\begin{aligned}
&(1 \mid 1) = C(1,1) = 1\,, && (1 \mid \tfrac{1}{2}) = (\tfrac{1}{2} \mid \tfrac{1}{2}) = C(\tfrac{1}{2}, 1) = 1 - c\,,\\
&(1 \mid 0) = (\tfrac{1}{2} \mid 0) = (0 \mid 0) = C(0,1) = \tfrac{1}{2}\,, && (\tfrac{1}{2} \mid 1) = C(\tfrac{1}{2}, \tfrac{1}{2}) = \tfrac{1}{2}\,,\\
&(0 \mid \tfrac{1}{2}) = C(0, \tfrac{1}{2}) = c\,, && (0 \mid 1) = C(0,0) = 0\,.
\end{aligned}$$
□

As a direct consequence of the preceding lemma we obtain the following

Theorem 4 (Iteration within $\mathbb{F}_1$) *Iteration of conditional indicator functions remains in $\mathbb{F}_1$, i.e. the conditioning operator $\mid$ on $\mathbb{F}$ can be restricted to a conditioning operator on $\mathbb{F}_1$,*
if and only if $(0 \mid \frac{1}{2}) = C(0, \frac{1}{2}) \in \{0, \frac{1}{2}\}$.
Therefore, denoting by C_1 resp. C_2 any compatible mean value function on $[0, 1]$ *with*

$$C_1(0, \frac{1}{2}) = \frac{1}{2} \quad \textit{resp.} \quad C_2(0, \frac{1}{2}) = 0\,,$$

these two mean value functions lead to the following two types of iteration, namely the

$$\text{“blurred iteration” for } C_1: \quad ((\mathbf{1}_{A_1} \mid \mathbf{1}_{B_1}) \mid (\mathbf{1}_{A_2} \mid \mathbf{1}_{B_2})) = (\mathbf{1}_{A_1} \mid \mathbf{1}_{B_1 \cap A_2 \cap B_2}),$$

$$\text{“sharper iteration” for } C_2: \quad ((\mathbf{1}_{A_1} \mid \mathbf{1}_{B_1}) \mid (\mathbf{1}_{A_2} \mid \mathbf{1}_{B_2})) = (\mathbf{1}_{A_1 \cup B_1^c} \mid \mathbf{1}_{(B_1 \cap A_2) \cup B_2^c}).$$

Moreover, these two types of iterations can be obtained from <u>*any*</u> *conditioning operator* $\mid$ *on* $\mathbb{F}_1$*, i.e. fulfilling the conditions* $(C1) - (C5)$*, with the additional property* $(0 \mid \frac{1}{2}) = \frac{1}{2}$ *for the blurred resp.* $(0 \mid \frac{1}{2}) = 0$ *for the sharper iteration.*

Proof The formulae follow from the preceding lemma and the property $(C2)$. On the one hand, the value $c = \frac{1}{2}$ leads to the blurred iteration

$$(\mathbf{1}_{A_1 \cap B_1 \cap A_2 \cap B_2} \mid \mathbf{1}_{B_1 \cap A_2 \cap B_2}) = (\mathbf{1}_{A_1} \mid \mathbf{1}_{B_1 \cap A_2 \cap B_2}).$$

On the other hand, the value $c = 0$ leads to the sharper iteration

$$(\mathbf{1}_{(A_1 \cap B_1 \cap A_2) \cup (A_1 \cap B_2^c) \cup (B_1^c \cap B_2^c)} \mid \mathbf{1}_{(B_1 \cap A_2) \cup B_2^c}) = (\mathbf{1}_{A_1 \cup B_1^c} \mid \mathbf{1}_{(B_1 \cap A_2) \cup B_2^c}).$$

Finally, we have to prove the additional condition *(cmvf)* for any conditioning operator $\mid$ on $\mathbb{F}_1$. For that purpose, let $\varphi_1 = (\mathbf{1}_{A_1} \mid \mathbf{1}_{B_1})$, $\varphi_2 = (\mathbf{1}_{A_2} \mid \mathbf{1}_{B_2})$, $\psi = (\mathbf{1}_A \mid \mathbf{1}_B)$ with $\varphi_1 \leq \varphi_2$. Then

$$\begin{aligned}
&(\varphi_1 \wedge \psi \mid \psi \to \varphi_1) \\
&= (1 \cdot \mathbf{1}_{A_1 \cap B_1 \cap A \cap B} + \tfrac{1}{2} \cdot \mathbf{1}_{(A_1 \cap B^c) \cup (B_1^c \cap A) \cup (B_1^c \cap B^c)} + 0 \cdot \mathbf{1}_{(A_1^c \cap B_1) \cup (A^c \cap B)} \mid \\
&1 \cdot \mathbf{1}_{(A^c \cap B) \cup (A_1 \cap B_1) \cup (B_1^c \cap B^c)} + \tfrac{1}{2} \cdot \mathbf{1}_{(A_1^c \cap B_1 \cap B^c) \cup (B_1^c \cap A \cap B)} + 0 \cdot \mathbf{1}_{A_1^c \cap B_1 \cap A \cap B}).
\end{aligned}$$

For $(0 \mid \frac{1}{2}) = \frac{1}{2}$ this is equal to

$$\begin{aligned}
(\mathbf{1}_{A_1 \cap B_1 \cap A \cap B} \mid \mathbf{1}_{(A_1 \cap B_1 \cap B) \cup (A^c \cap B)}) &\leq (\mathbf{1}_{A_2 \cap B_2 \cap A \cap B} \mid \mathbf{1}_{(A_2 \cap B_2 \cap B) \cup (A^c \cap B)}) \\
&= (\varphi_2 \wedge \psi \mid \psi \to \varphi_2),
\end{aligned}$$

where "$\leq$" follows from

$$\begin{aligned}
&\underline{A_1 \cap B_1} \cap A \cap B \subseteq \underline{A_2 \cap B_2} \cap A \cap B \quad \text{and} \\
&((A_1 \cap B_1 \cap B) \cup (A^c \cap B)) \to (A_1 \cap B_1 \cap A \cap B) = B \to A
\end{aligned}$$

which does not depend on k = 1.

For $(0 \mid \frac{1}{2}) = 0$ it follows analogously

$$\begin{aligned}
&(\mathbf{1}_{(A_1 \cup B_1^c) \cap A \cap B} \mid \mathbf{1}_{((A_1 \cup B_1^c \cup A^c) \cap B) \cup (A_1^c \cap B_1 \cap B^c)}) \\
&\leq (\mathbf{1}_{(A_2 \cup B_2^c) \cap A \cap B} \mid \mathbf{1}_{((A_2 \cup B_2^c \cup A^c) \cap B) \cup (A_2^c \cap B_2 \cap B^c)})
\end{aligned}$$

because of

$$(A_1 \cup B_1^c) \cap A \cap B = \underline{(B_1 \to A_1)} \cap A \cap B \subseteq \ldots \quad \text{and}$$

$$(((A_1 \cup B_1^c \cup A^c) \cap B) \cup (A_1^c \cap B_1 \cap B^c)) \to ((A_1 \cup B_1^c) \cap A \cap B)$$
$$= (\underline{(B_1 \to A_1)} \cap B^c) \cup (A \cap B) \subseteq \ldots$$

Let us observe that for the blurred iteration only the part $A_1 \cap B_1 \subseteq A_2 \cap B_2$ of the inequality $\varphi_1 \leq \varphi_2$ was needed, whereas for the sharper iteration only the other part $B_1 \to A_1 \subseteq B_2 \to A_2$ was needed. □

In the preceding theorem and in the following, the second type we call "sharper iteration" because the conditionals with the undetermined value as condition have the "sharp values" $(0 \mid \frac{1}{2}) = 0$ and $(1 \mid \frac{1}{2}) = (\frac{1}{2} \mid \frac{1}{2}) = 1$, whereas the first type we call "blurred iteration" because all these conditionals have the "blurred value" $\frac{1}{2}$.

In the following two corollaries we will specify the result of the preceding theorem for two special situations, which will be taken up and briefly discussed at the end of this section.

Corollary 2 (Iteration conditioned on an indicator function) *In the special case of crisp indicator function in the condition, both types of iteration lead to the same result, i.e. for*

$$((\mathbf{1}_{A_1} \mid \mathbf{1}_{B_1}) \mid \mathbf{1}_{A_2}) = (\mathbf{1}_{A_1} \mid \mathbf{1}_{B_1 \cap A_2}) \,.$$

Corollary 3 (Iteration with equal indicator function as conditions) *In the special case of equal indicator function in the conditions, both types of iteration lead to different results, i.e. for the*

$$\textit{blurred iteration:} \quad ((\mathbf{1}_{A_1} \mid \mathbf{1}_B) \mid (\mathbf{1}_{A_2} \mid \mathbf{1}_B)) = (\mathbf{1}_{A_1} \mid \mathbf{1}_{A_2 \cap B}) \,,$$

$$\textit{sharper iteration:} \quad ((\mathbf{1}_{A_1} \mid \mathbf{1}_B) \mid (\mathbf{1}_{A_2} \mid \mathbf{1}_B)) = (\mathbf{1}_{A_1 \cup B^c} \mid \mathbf{1}_{A_2 \cup B^c}) \,.$$

Now we will prove the result of an iteration process for conditional Boolean subsets using essentially the isomorphism between these and the conditional indicator functions. Although the result can also be obtained as a particular case of the more general situation of Sect. 5 in [11], the formulation of the following theorem and its proof are very different from [11] and show cleary the analogy of both iteration processes and that, furthermore, also in the following setting there are exactly two types of iteration.

Theorem 5 (Iteration within $\tilde{\mathbb{B}}$) *Let* $((\mathbf{1}_{A_1} \mid \mathbf{1}_{B_1}) \mid (\mathbf{1}_{A_2} \mid \mathbf{1}_{B_2})) \in \mathbb{F}_1$ *be any of the two types of iteration of conditional indicator functions from the preceding theorem and let* $i : \mathbb{F}_1 \longrightarrow \tilde{\mathbb{B}}$ *be the isomorphism from Sect. 4. Then there exist two conditioning operators* $\tilde{\rceil}$ *on* $\tilde{\mathbb{B}}$ *given by*

$$((A_1 \parallel B_1) \,\tilde{\rceil}\, (A_2 \parallel B_2)) = i \circ ((\mathbf{1}_{A_1} \mid \mathbf{1}_{B_1}) \mid (\mathbf{1}_{A_2} \mid \mathbf{1}_{B_2})) \,,$$

which result to be the corresponding blurred resp. sharper iteration in $\tilde{\mathbb{B}}$ *:*

$$((A_1 \parallel B_1) \tilde{|} (A_2 \parallel B_2)) = (A_1 \parallel B_1 \cap A_2 \cap B_2) \;\; for \;\; (0 \mid \frac{1}{2}) = C_1(0, \frac{1}{2}) = \frac{1}{2} ,$$

$$((A_1 \parallel B_1) \tilde{|} (A_2 \parallel B_2)) = (A_1 \cup B_1^c \parallel (B_1 \cap A_2) \cup B_2^c) \;\; for \;\; (0 \mid \frac{1}{2}) = C_2(0, \frac{1}{2}) = 0 .$$

Moreover, the two conditioning operators $\tilde{|}$ *are generated by two compatible mean value functions* $\tilde{C}_k$ *on* $\tilde{\mathbb{B}}$ *given by*

$$\tilde{C}_k(i \circ \varphi \, , \, i \circ \psi \,) = i \circ C_k(\varphi, \psi) \;\; for \;\; \varphi, \psi \in \mathbb{F}_1 \, , \, \varphi \leq \psi \, , \, k = 1, 2 \, ,$$

which result to be

$$\tilde{C}_1([\alpha, \beta] \, , \, [\gamma, \delta] \,) = [\alpha, \delta] \;\; for \;\; C_1 \, ,$$

$$\tilde{C}_2([\alpha, \beta] \, , \, [\gamma, \delta] \,) = [\beta \cap \gamma \, , \, \beta \cup \gamma] \;\; for \;\; C_2 \, ,$$

written the arguments of $\tilde{C}_k$ *as intervals* $[\alpha, \beta] \leq [\gamma, \delta]$ *of elements* $\alpha \subseteq \beta, \gamma \subseteq \delta$ *in* $\mathbb{B}$.

Proof The assertion that both types of $\tilde{|}$ are conditioning operators on $\tilde{\mathbb{B}}$ generated by some compatible mean value function follows immediately from the facts that both types of $|$ are conditioning operators on $\mathbb{F}_1$ fulfilling (*cmvf*) and the isomorphism i between these two MV-algebras. Naturally, it is needed that i is an isomorphism with respect to the partial orderings, the lattice operations and the MV-algebra operations. This will be illustrated in the following only for two properties, namely for $(C1)$:

$$((A_1 \parallel B_1) \tilde{|} (\Omega \parallel \Omega)) = i \circ ((\mathbf{1}_{A_1} \mid \mathbf{1}_{B_1}) \mid (\mathbf{1}_{\Omega} \mid . \mathbf{1}_{\Omega})) = i \circ (\mathbf{1}_{A_1} \mid \mathbf{1}_{B_1}) = (A_1 \parallel B_1),$$

and for $(C5)$:

$$\begin{aligned} ((A_1 \parallel B_1) \tilde{|} (A_2 \parallel B_2))' &= (i \circ ((\mathbf{1}_{A_1} \mid \mathbf{1}_{B_1}) \mid (\mathbf{1}_{A_2} \mid \mathbf{1}_{B_2})))' \\ &= i \circ (\neg((\mathbf{1}_{A_1} \mid \mathbf{1}_{B_1}) \mid (\mathbf{1}_{A_2} \mid \mathbf{1}_{B_2}))) \\ &= i \circ (\neg(\mathbf{1}_{A_1} \mid \mathbf{1}_{B_1}) \odot (\mathbf{1}_{A_2} \mid \mathbf{1}_{B_2}) \mid (\mathbf{1}_{A_2} \mid \mathbf{1}_{B_2})) \\ &= ((A_1 \parallel B_1)' \sqcap (A_2 \parallel B_2) \tilde{|} (A_2 \parallel B_2)). \end{aligned}$$

Therefore, it follows by the definitions of $\tilde{|}$ and $|$ that $\tilde{C}_k(i \circ \varphi \, , \, i \circ \psi \,) = i \circ C_k(\varphi, \psi)$ for

$$i \circ \varphi = (A_1 \parallel B_1) \wedge (A_2 \parallel B_2) = [\alpha, \beta] \;\leq\; i \circ \psi = (A_2 \parallel B_2) \rightarrow (A_1 \parallel B_1) = [\gamma, \delta].$$

Vice versa, given $\varphi, \psi \in \mathbb{F}_1$ with $\varphi \leq \psi$, then there exist $[\alpha, \beta] \leq [\gamma, \delta]$ in $\tilde{\mathbb{B}}$ such that

$$i \circ \varphi = [\alpha, \beta] \,, \; i \circ \psi = [\gamma, \delta].$$

Finally, the last two formulae are obtained if we start with

$$\begin{aligned}\tilde{C}_k([\alpha, \beta] \,, \; [\gamma, \delta]) &= \tilde{C}_k((\alpha \parallel \beta^c \cup \alpha) \,, \; (\gamma \parallel \delta^c \cup \gamma)) = \ldots \\ &= i \circ ((\mathbf{1}_\alpha \mid \mathbf{1}_{\beta^c \cup \alpha}) \mid (\mathbf{1}_{\alpha \cup (\beta \cap \gamma^c) \cup \delta^c} \mid \mathbf{1}_{\alpha \cup (\beta \cap \gamma^c) \cup (\beta^c \cap \gamma) \cup \delta^c}))\end{aligned}$$

and apply the formulae for the blurred and sharper iteration from the preceding theorem. □

In [10] the formula for the blurred iteration has been proposed as one concrete example, in [11] the formulae for both types of iteration appear as concrete examples. But here we have shown that there do not exist more. We will compare these two types of iteration in the following

Remark 6 (Comparison between blurred and sharper iteration in $\tilde{\mathbb{B}}$) Let us write the results of the two types of iteration from the preceding theorem as intervals:

$$(\,(A_1 \parallel B_1) \,\tilde{]}\, (A_2 \parallel B_2)\,) = [\,\alpha_k \,, \; \beta_k\,] \,,$$

where $k = 1$ resp. $k = 2$ correspond to the blurred resp. sharper iteration. Then the following assertions can be established as an exercise. It follows that

(i) in general, $\alpha_1 \subseteq \alpha_2 \subseteq \beta_2 \subseteq \beta_1$, i.e. $[\alpha_2, \beta_2] \subseteq [\alpha_1, \beta_1]$,
(ii) particularly, $[\alpha_2, \beta_2] = [\alpha_1, \beta_1]$ if and only if $B_2 = \Omega$, where $[\alpha_k, \beta_k] = (A_1 \parallel B_1 \cap A_2)$,
(iii) $\alpha_2 = \beta_2$ if and only if $B_2 \subseteq B_1 \cap A_2$, where $\{\alpha_2\} = (B_1 \to A_1 \parallel \Omega) \subseteq (A_1 \parallel B_2) = [\alpha_1, \beta_1]$.

Part (i) of the preceding remark shows that the two types of iteration lead to conditional Boolean subsets which are not comparable with respect to the partial ordering $\leq$ but with respect to the set-inclusion $\subseteq$ in $\tilde{\mathbb{B}}$. Moreover, this relation $\subseteq$ can be interpreted as "sharper than" resp. $\supseteq$ as "more blurred than", because a sharper interval has less Boolean subsets than a more blurred interval. In this sense, $[\alpha_2, \beta_2]$ as result of the sharper iteration is "sharper than" $[\alpha_1, \beta_1]$ as result of the blurred iteration.

In the former papers [10, 11, 20, 21] we used the identification of an interval $[\alpha, \beta]$ with the ordered pair (α, β) of its endpoints α, β in an MV-algebra $\mathbb{L}$. Therefore, the singletons $\{\alpha\} = [\alpha, \alpha]$ are identified with the pairs (α, α) and we will refer to as elements of the "diagonal" of $\tilde{\mathbb{L}}$ in both cases. This will be used in the following

Remark and Notation 1 *(Conditioning operator on $\tilde{\mathbb{B}}$ restricted to its diagonal)* The conditioning operator $\tilde{]}$ on $\tilde{\mathbb{B}}$ from the preceding theorem applied to elements of the diagonal of $\tilde{\mathbb{B}}$ permits to recover the conditional Boolean subsets via the property

$$(D) \quad (\{A\} \,\tilde{|}\, \{B\}) = (A \parallel B).$$

Particularly, it follows that $(\{\Omega\}\tilde{|}\{\Omega\}) = (\Omega \parallel \Omega) = \{\Omega\}$ and $(\{\emptyset\}\tilde{|}\{\Omega\}) = (\emptyset \parallel \Omega) = \{\emptyset\}$ are in the diagonal, but not the self-complemented element in $\tilde{\mathbb{B}}$: $(\{\emptyset\}\tilde{|}\{\emptyset\}) = (\emptyset \parallel \emptyset) = \mathbb{B}$.

Moreover, property (D) permits to rewrite the iterates of conditional Boolean subsets from the preceding theorem also into the form

$$((\{A_1\}\tilde{|}\{B_1\}) \,\tilde{|}\, (\{A_2\}\tilde{|}\{B_2\})) \quad \text{using the conditioning operator } \tilde{|} \text{ on } \tilde{\mathbb{B}},$$

which result to be completely analogous to the iterates of conditional indicator functions

$$((\mathbf{1}_{A_1} \mid \mathbf{1}_{B_1}) \mid (\mathbf{1}_{A_2} \mid \mathbf{1}_{B_2})) \quad \text{using the conditioning operator } \mid \text{ on } \mathbb{F}_1.$$

Also the results of both iteration procedures are completely analogous.
Therefore, for both we use for short the following

formal notation of iteration: $((A_1 \mid B_1) \mid (A_2 \mid B_2))$ using here | as formal symbol.

Particularly, $(\{A\}\tilde{|}\{\Omega\}) = \{A\}$ and $(\mathbf{1}_A \mid \mathbf{1}_\Omega) = \mathbf{1}_A$ both are written for short as $(A \mid \Omega) = A$.

Now, using the preceding notation, we can rewrite the results of the two corollaries from the beginning of this section not only as iterates in $\mathbb{F}_1$ but at the same time as iterates in $\tilde{\mathbb{B}}$.

Remark 7 (Blurred vs. sharper iterations) For the blurred iterations the following three special iterations lead to the same result.

(i) $((A \mid B) \mid D) = (A \mid B \cap D)$,
(ii) $(A \mid (B \mid D)) = (A \mid B \cap D)$,
(iii) $((A \mid D) \mid (B \mid D)) = (A \mid B \cap D)$.

For the sharper iterations property (i) remains valid, but the other two lead to the following different results.

(ii)* $(A \mid (B \mid D)) = (A \mid D \to B)$,
(iii)* $((A \mid D) \mid (B \mid D)) = (D \to A \mid D \to B)$.

These tree special iterations were treated and discussed by many authors but mostly in an ad hoc manner.

One of the references in which the first two are systematically discussed is [7] at the end of Sect. 2.4. There the authors justified property (i) by the "convention **?** | 0 = **?** = **?** | 1". This corresponds precisely to the two properties $C(0, 1) = \frac{1}{2} =$

$C(\frac{1}{2}, \frac{1}{2})$ valid for any compatible mean value function C on [0, 1]. Calabrese ([2]) took (i) as definition for $((A \mid B) \mid D)$.
In the following the authors made out clearly the two possible meanings for $(A \mid (B \mid D))$, the first justified by "$1 \mid ? = ? = 0 \mid ?$" which corresponds to C_1 and leads to property (ii) and the second by "$1 \mid ? = 1$ and $0 \mid ? = 0$" which corresponds to C_2 and leads to property (ii)* taken by Calabrese as definition for $(A \mid (B \mid D))$.
Based on these arguments the authors of [7] proposed (but without additional justification) two possible meanings for the general iteration $((A \mid B) \mid (D \mid E))$. As first definition for this they proposed $(A \mid B \cap D \cap E)$ and called it "associative definition" due to the equality $((A \mid B) \mid D) = (A \mid (B \mid D))$. This is indeed our blurred iteration corresponding to the choice C_1. But as second definition they proposed $(A \mid B \cap (E \to D))$ which is quite different from our sharper iteration $(B \to A \mid E \to (B \cap D))$ we derived from the choice C_2.
For the third special iteration, Copeland (see e.g. the book [13]) took property (iii) as definition for $((A \mid D) \mid (B \mid D))$ which corresponds to our blurred iteration. It seems that property (iii)*, which corresponds to the sharper iteration, has not been considered in the literature.

Finally, let us observe that the iteration process proposed in [9], Sect. 8.1, is completely different because its result does not remain in $\tilde{\mathbb{B}}$ but is an interval of elements of $\tilde{\mathbb{B}}$.

6 Uncertainty Measures of Conditionals

Definition 5 *(Uncertainty measures on MV-algebras)* Let $\mathbb{L} = (\mathbb{L}, \sqcap, ', \sqcup)$ be an MV-algebra. Then we use the following notations.

(i) A function $m : \mathbb{L} \to [0, 1]$ will be called an *uncertainty measure* if it satisfies the following conditions:

(M1) $m(0) = 0, \quad m(1) = 1$ (boundary conditions),
(M2) $a \le b \quad \Rightarrow \quad m(a) \le m(b)$ (isotonicity).

(ii) An uncertainty measure m on $\mathbb{L}$ will be called, resp.

(M3) *compatible (with the complement)* if $m(a') = 1 - m(a)$,
(A) *additive* if $m(a \dot{\sqcup} b) = m(a) + m(b)$ for all disjoint unions.

As in the preceding sections we use also here the same symbols 0 resp. 1 for the universal lower resp. upper bound in the lattice $\mathbb{L}$ as well as for the real numbers in the unit interval [0, 1].

The additivity of measures on MV-algebras has a clear meaning in analogy to the additivity on Boolean algebras, for some details we refer to the biographical Remark 6.7 in [21] and the references therein, particularly [14]. In contrast to this, the additivity in more general structures is not so clear, for the case of Girard algebras we refer to [21], its biographical Remark 6.8 and the references therein. Particularly, in the process of extending the additivity on a finite MV-algebra $\mathbb{L}$ to $\tilde{\mathbb{L}}$, there appears the compatible mean value function M from the example (iv) in Sect. 2, see also one of the following examples.

Returning now to the MV-algebras of conditionals, we will present the corresponding results in the following

Theorem 6 (Uncertainty measures on $\tilde{\mathbb{B}}$ and on $\mathbb{F}_1$) *Let* $\mathbb{B} = (\mathbb{B}, \cap, ^c, \cup)$ *be a Boolean algebra of (crisp) subsets of the universe* Ω *and let* $\tilde{\mathbb{B}}$ *resp.* $\mathbb{F}_1$ *be the corresponding MV-algebras of conditional Boolean subsets* $(A \parallel B)$ *resp. conditional indicator functions* $(\boldsymbol{1}_A \mid \boldsymbol{1}_B)$. *Then:*

(i) Any uncertainty measure μ *on* $\mathbb{B}$ *can be extended to uncertainty measures*

$$\tilde{\mu} \text{ on } \tilde{\mathbb{B}}, \quad \text{given by} \quad \tilde{\mu}(A \parallel B) = M(\mu(A \cap B)\,,\ \mu(B^c \,\dot{\cup}\, (A \cap B)))\,,$$

which are "generated by" mean value functions M *on* $[0, 1]$, *resp.*

$$m \text{ on } \mathbb{F}_1, \quad \text{given by} \quad m = \tilde{\mu} \circ i\,, \quad \text{i.e. } m(\boldsymbol{1}_A \mid \boldsymbol{1}_B) = \tilde{\mu}(A \parallel B)\,,$$

with the isomorphism $i : \mathbb{F}_1 \longrightarrow \tilde{\mathbb{B}}$ *from Sect. 4. Particularly, it follows that*

$$m(\boldsymbol{1}_A) = m(\boldsymbol{1}_A \mid \boldsymbol{1}_\Omega) = \tilde{\mu}(A \parallel \Omega) = \mu(A) \quad \text{for all } A \in \mathbb{B}.$$

(ii) If, furthermore, M *is a compatible mean value function and* μ *is a compatible (uncertainty) measure, then* $\tilde{\mu}$ *and* m *are compatible measures. Moreover, it follows that*

$$m(\boldsymbol{1}_A \mid \boldsymbol{1}_B) = \tilde{\mu}(A \parallel B) = \frac{1}{2} \quad \text{for} \quad \mu(B) = 0,$$
$$\text{particularly} \quad m(\boldsymbol{1}_\emptyset \mid \boldsymbol{1}_\emptyset) = \tilde{\mu}(\emptyset \parallel \emptyset) = \frac{1}{2}.$$

(iii) An additive (probability) measure μ *has a unique extension to an additive measure* $\tilde{\mu}$ *resp.* m *given by*

$$m(\boldsymbol{1}_A \mid \boldsymbol{1}_B) = \tilde{\mu}(A \parallel B) = \mu(A \cap B) + \frac{1}{2} \cdot \mu(B^c) = \int_\Omega (\boldsymbol{1}_A \mid \boldsymbol{1}_B)\, d\mu\,,$$

where $\tilde{\mu}$ *is generated by*

$$M(x, y) = \frac{x + y}{2}\,.$$

from the example (iii) in Sect. 2.

Proof (i) The boundary conditions ($M1$) follow directly:

$$m(\mathbf{1}_\emptyset \mid \mathbf{1}_\Omega) = \tilde{\mu}(\emptyset \parallel \Omega) = M(\mu(\emptyset), \mu(\emptyset)) = 0\,,$$
$$m(\mathbf{1}_\Omega \mid \mathbf{1}_\Omega) = \tilde{\mu}(\Omega \parallel \Omega) = M(\mu(\Omega), \mu(\Omega)) = 1.$$

The isotonicity condition ($M2$) follows from the isotonicity of μ and M and the isomorphism i.
(ii) Also the compatibility condition is transfered from M and μ to $\tilde{\mu}$ and m:

$$\begin{aligned} m(\neg(\mathbf{1}_A \mid \mathbf{1}_B)) &= \tilde{\mu}((A \parallel B)') = \tilde{\mu}(A^c \parallel B) \\ &= M(\mu(A^c \cap B), \mu(B^c \cup A^c)) \\ &= 1 - M(1 - \mu(B^c \cup A^c), 1 - \mu(A^c \cap B)) \\ &= 1 - M(\mu(A \cap B), \mu(B^c \cup A)) \\ &= 1 - \tilde{\mu}(A \parallel B) = 1 - m(\mathbf{1}_A \mid \mathbf{1}_B). \end{aligned}$$

(iii) was proved in [10] for $\tilde{\mu}$, in a more general context. □

Remark 8 (The classical conditional probability) Another extension of an additive measure μ on $\mathbb{B}$ is generated by the compatible mean value function M from the example (ii) in Sect. 2, i.e. given by

$$M(x, y) = \frac{x}{1 + x - y} \quad \text{for} \quad (x, y) \neq (0, 1), \quad M(0, 1) = \frac{1}{2}\,,$$

and leads to the classical "conditional probability"

$$m(\mathbf{1}_A \mid \mathbf{1}_B) = \tilde{\mu}(A \parallel B) = \mu_B(A) \quad \text{for } \mu(B) > 0, \quad \text{where} \quad \mu_B(A) = \frac{\mu(A \cap B)}{\mu(B)}\,.$$

Particularly, it follows that $\tilde{\mu}(B \parallel B) = 1$ for $\mu(B) > 0$ <u>and</u> $\tilde{\mu}(B \parallel B) = \frac{1}{2}$ for $\mu(B) = 0$.
By part (ii) of the preceding theorem this extension is a compatible measure. By part (iii) it is not additive, but only "levelwise additive":

$$\begin{aligned} m(\mathbf{1}_{A_1} \dot{\oplus} \mathbf{1}_{A_2} \mid \mathbf{1}_B) &= \tilde{\mu}(A_1 \dot{\cup} A_2 \parallel B) \\ &= \tilde{\mu}(A_1 \parallel B) + \tilde{\mu}(A_2 \parallel B) \quad \text{(but only) for } \mu(B) > 0. \end{aligned}$$

Example 2 (A further compatible measure extension) As already mentioned at the beginning of this section, a further compatible measure extension $\tilde{\mu}$ of an additive measure μ on $\mathbb{B}$ is generated by the compatible mean value function M from the example (iv) in Sect. 2, i.e. given by

$$M(x, y) = \frac{y}{1 + y - x}\,,$$

and has, therefore, the following form

$$m(\mathbf{1}_A \mid \mathbf{1}_B) = \tilde{\mu}(A \parallel B) = \frac{\mu(A \cap B) + \mu(B^c)}{1 + \mu(B^c)} .$$

But the extensions $\tilde{\mu}$ resp. m are neither additive nor levelwise additive.

Example 3 (A non-compatible measure extension) Extensions $\tilde{\mu}$ resp. m of an additive measure μ on $\mathbb{B}$ which are neither compatible (therefore not additive) nor levelwise additive are those, where $\tilde{\mu}$ is generated by the non-compatible mean value function M from the example (v) in Sect. 2, i.e. given by

$$M(x, y) = (1 - t) \cdot x + t \cdot y \quad \text{with some} \quad t \in [0, 1],\ t \neq \frac{1}{2} .$$

These have the following form

$$m(\mathbf{1}_A \mid \mathbf{1}_B) = \tilde{\mu}(A \parallel B) = \mu(A \cap B) + t \cdot \mu(B^c) .$$

Remark 9 (The alternative model of PACS) The definition (CI) of conditional indicator functions $(\mathbf{1}_A \mid \mathbf{1}_B)$ in Sect. 3 requieres only the values $C(0, 0) = 0$, $C(1, 1) = 1$ of any mean value function C as "determined truth values" and $C(0, 1) = \frac{1}{2}$ of any compatible C as "undetermined truth value". If we drop the assumption of compatibility and set $C(0, 1) = t \neq \frac{1}{2}$, we are led to the two expressions

$$1 \cdot \mathbf{1}_{A\cap B} + t \cdot \mathbf{1}_{B^c} + 0 \cdot \mathbf{1}_{A^c \cap B} \quad \text{for} \quad (\mathbf{1}_A \mid \mathbf{1}_B) ,$$

$$1 \cdot \mathbf{1}_{A\cap B} + (1 - t) \cdot \mathbf{1}_{B^c} + 0 \cdot \mathbf{1}_{A^c \cap B} \quad \text{for} \quad \neg(\mathbf{1}_{A^c} \mid \mathbf{1}_B) ,$$

which should be equal because of the property $(C5)$ in $\mathbb{F}_1$, i.e. $\neg(\mathbf{1}_A \mid \mathbf{1}_B) = (\mathbf{1}_{A^c} \mid \mathbf{1}_B)$. But this property, generally accepted as a reasonable one, implies that t cannot be a constant.

Therefore, instead of (CI) we are led to the alternative expression

$$(*) \qquad \mathbf{1}_{A|B} = 1 \cdot \mathbf{1}_{A\cap B} + t(A \mid B) \cdot \mathbf{1}_{B^c} + 0 \cdot \mathbf{1}_{A^c \cap B} \quad \text{with some} \quad t(A \mid B) \in [0, 1] ,$$

where we used the notation $\mathbf{1}_{A|B}$ from [5] instead of our notation $(\mathbf{1}_A \mid \mathbf{1}_B)$. Really, in spite of the very similar expressions (CI) and $(*)$, the difference between them is essential. While (CI) is a "measure-free" definition of conditionals, the approach based on $(*)$ requires some knowledge or assumptions for the undetermined values $t(A \mid B)$. For instance, if we assume that the (reasonable) property $1 - \mathbf{1}_{A|B} = \mathbf{1}_{A^c|B}$ holds then, applying $(*)$ to both $\mathbf{1}_{A|B}$ and $\mathbf{1}_{A^c|B}$, it follows that $t(A^c \mid B) = 1 - t(A \mid B)$.

Indeed, several authors take $(*)$ as definition for a "truth-value of the conditional event $A \mid B$" and show that $t(A \mid B)$ fulfills the axioms of a general conditional probability, see the above mentioned paper [5] and the references therein. In [6], models of "Partial Algebraic Conditional Spaces" over a Boolean algebra $\mathbb{B}$ of subsets of the universe Ω are axiomatically introduced and it is shown that they are in a one-to-one correspondence to "De Finetti-Popper conditional probabilities".

Now, if μ is a probability (measure) on the Boolean algebra $\mathbb{B}$ then it follows from $(*)$ that

$$(**) \qquad m(\mathbf{1}_{A|B}) = \int_{\Omega} \mathbf{1}_{A|B}\, d\mu = \mu(A \cap B) + t(A, B) \cdot \mu(B^c)\,,$$

compare with the result of the preceding example. On the other hand, it follows from $(**)$ for $0 < \mu(B) < 1$ that

$$m(\mathbf{1}_{A|B}) = \frac{\mu(A \cap B)}{\mu(B)} \quad \text{if and only if} \quad t(A \mid B) = \frac{\mu(A \cap B)}{\mu(B)}\,,$$

i.e. in this setting we have the classical conditional probability $t(A \mid B) = \mu_B(A) = m(\mathbf{1}_{A|B})$.

7 Conclusions

All authors treating measure-free conditioning agree that a conditional element interpreted as "a given b" for the (unconditional) elements a, b from some structured set $\mathbb{L}$ should be defined using the elements interpreted as "a and b" resp. "b implies a". For an MV-algebra $\mathbb{L}$ there are two reasonable ways:

In the interval approach, a conditional element is defined as

$$(a \parallel b) = [a \wedge b\,,\ b \rightarrow a] \in \tilde{\mathbb{L}} \quad \text{for} \quad a, b \in \mathbb{L},$$

i.e. as the set of *all* elements between $a \wedge b$ and $b \rightarrow a$. Particularly, it follows that $(0 \parallel 0) = \mathbb{L}$.

In the mean value approach, a conditional element is defined as

$$(a \mid b) = C(a \wedge b\,,\ b \rightarrow a) \in \mathbb{L} \quad \text{for} \quad a, b \in \mathbb{L},$$

i.e. as *some* element between $a \wedge b$ and $b \rightarrow a$. It is crucial that this approach requires that $\mathbb{L}$ should have a self-complemented element u which results to be $(0 \mid 0) = u$. Therefore, for a Boolean algebra $\mathbb{L}$ only the first approach is applicable. But for a Boolean algebra $\mathbb{L} = \mathbb{B}$ of subsets of a universe Ω, we can use the one-to-one correspondence between Boolean subsets $A \in \mathbb{B}$ and its indicator functions $\mathbf{1}_A$ as special fuzzy subsets and then apply the second approach. The present paper contains three main results:

- The conditionals $(\mathbf{1}_A \mid \mathbf{1}_B)$ form an MV-subalgebra $\mathbb{F}_1$ of the MV-algebra $\mathbb{F}$ of fuzzy subsets, see Theorem 1.
- There is an MV-algebra isomorphism between $\mathbb{F}_1$ and $\tilde{\mathbb{B}}$, see Theorem 3.
- There are exactly two types of iterations within $\mathbb{F}_1$ and, because of the isomorphism, also within $\tilde{\mathbb{B}}$, see Theorems 4 resp. 5.

Acknowledgments I am very grateful to Peter Klement, as organizator and motor of *his* Linz Seminars where I received a lot of stimulations since my first participation in 1983, and as friend from our common time working together and in private occations.

References

1. Birkhoff, G.: Lattice Theory. Providence, RI (1960)
2. Calabrese, P.: An algebraic synthesis of the foundations of logic and probability. Inf. Sci. **42**, 187–237 (1987)
3. Chang, C.C.: Algebraic analysis of many valued logics. Trans. Am. Math. Soc. **88**, 467–490 (1958)
4. Cignoli, R.L.O., D'Ottaviano, I.M.L., Mundici, D.: Algebraic Foundations of Many-valued Reasoning. Kluwer Academic Publishers, Dordrecht (2000)
5. Coletti, G., Scozzafava, R.: From conditional events to conditional measures: a new axiomatic approach. Ann. Math. Artif. Intell. **32**, 373–392 (2001)
6. Di Nola, A., Scozzafava, R.: Partial algebraic conditional spaces. Int. J. Uncertainty, Fuzziness Knowl.-Based Syst. **12**, 781–789 (2004)
7. Dubois, D., Prade, H.: Conditioning, non-monotonic logic and non-standard uncertainty models. In: Goodman, I.R., Gupta, M.M., Nguyen, H.T., Rogers, G.S. (eds.) Conditional Logic in Expert Systems, pp. 115–158. North-Holland, Amsterdam (1991)
8. Galatos, N., Jipsen, P., Kowalski, T., Ono, H.: Residuated lattices: an algebraic glimpse at substructural logics. Studies in Logic and the Foundations of Mathematics 151, Elsevier (2007)
9. Goodman, I.R., Nguyen, H.T., Walker, E.A.: Conditional Inference and Logic for Intelligent Systems—A Theory of Measure-free Conditioning. North-Holland, Amsterdam (1991)
10. Höhle, U., Weber, S.: Uncertainty measures, realizations and entropies. In: Goutsias, J., Mahler, R.P.S., Nguyen, H.T. (eds.) Random Sets: Theory and Applications, pp. 259–295. Springer, Berlin (1997)
11. Höhle, U., Weber, S.: On conditioning operators. In: Höhle, U., Rodabaugh, S. (eds.) Mathematics of Fuzzy Sets—Logic, Topology and Measure Theory, pp. 653–673. Kluwer Academic Publishers, Dordrecht (1999)
12. Nguyen, H.T.: On representation and combinability of uncertainty. In: Abstracts of the 2nd IFSA Congress, Tokyo (1987)
13. Pfanzagl, J.: Theory of Measurement, 2nd edn. Physica, Würzburg, Wien (1971)
14. Riečan, B., Mundici, D.: Probability on MV-algebras. In: Pap, E. (ed.) Handbook of Measure Theory, vol. 2, pp. 869–909. North-Holland, Amsterdam (2002)
15. Van Gasse, B., Cornelis, C., Deschrijver, G., Kerre, E.E.: A characterization of interval-valued residuated lattices. Int. J. Approximate Reasoning **49**, 478–487 (2008)
16. Weber, S.: On conditional measures and events. In: Abstracts of the 9th Linz Seminar on Fuzzy Set Theory (1987)
17. Weber, S.: Conditioning on MV-algebras and additive measures. I. Fuzzy Sets Syst. **92**, 241–250 (1997)
18. Weber, S.: Conditioning on MV-algebras and additive measures, further results. In: Dubois, D., Prade, H., Klement, E.P. (eds.) Fuzzy Sets, Logics and Reasoning about Knowledge, pp. 175–199. Kluwer Academic Publishers, Dordrecht (1999)

19. Weber, S.: Uncertainty measures—problems concerning additivity. Fuzzy Sets Syst. **160**, 371–383 (2009)
20. Weber, S.: A complete characterization of all weakly additive measures and of all valuations on the canonical extension of any finite MV-chain. Fuzzy Sets Syst. **161**, 1350–1367 (2010)
21. Weber, S.: Measure-free conditioning and extensions of additive measures on finite MV-algebras. Fuzzy Sets Syst. **161**, 2479–2504 (2010)

Multivalued Functions Integration: from Additive to Arbitrary Non-negative Set Function

Endre Pap

Abstract It is given a short overview of some integrals of multifunctions based on additive measures, as strong, Aumann and Aumann-Gould integrals. It is considered also a multi-valued Choquet integral based on a multisubmeasure. Then it is introduced a set-valued Gould type integral of multifunctions with values in the family of all nonempty bounded subsets of a real Banach space X and with respect to an arbitrary non-negative set function. There are given some basic properties of the integrable multifunctions, and some continuity properties of the multimeasure induced by set-valued integral.

1 Introduction

Theory of multifunctions, i.e., set-valued maps, correspondences, etc., is important field of investigations as theoretical and practical applications [1, 2, 4, 5, 9–11, 20, 21, 32, 42, 55, 60]. It allows one to take into account the multiplicity of possible choices, the lack of information and/or the uncertainty in a lot of situations ranging from Optimal Control to Economic Theory, see [3, 31, 60]. In particular, measurable multifunctions, i.e., set-valued random variables, random sets, are investigated in probability and statistics, with many applications, see first papers [38, 54]. Various types of integrals for multifunctions have many applications in mathematical economics, theory of control, probabilities. Integrals of multifunctions can be used as an aggregation tool when dealing with a large amount of information fusing and with data mining problems such as programming and classification. In processes of subjective evaluation, for instance, the integral of a multifunction can be a tool in synthetic evaluation of the quality of a given object, when the score function may

E. Pap (✉)
Singidunum University, Danijelova 32, 11000 Belgrade, Serbia
e-mail: epap@singidunum.ac.rs

E. Pap
Óbuda University, Becsi út 92, Budapest 1034, Hungary

S. Saminger-Platz and R. Mesiar (eds.), *On Logical, Algebraic, and Probabilistic Aspects of Fuzzy Set Theory*, Studies in Fuzziness and Soft Computing 336,
DOI 10.1007/978-3-319-28808-6_15

be set-valued i.e., for each quality factor there exists a multiple score or a set of estimations.

A generalization of the classical integrals is related to their extension on non-additive measures. For non-additive integrals see [8, 37, 39, 46, 47, 50]. Integrals of multifunctions have been defined in different ways: Aumann method [4, 7, 10], by the Rådström-Hörmander [53] embedding theorem [17], via Pettis method [11, 19, 58, 59], using (as in Dunford [20]) sequences of multifunctions [13, 16, 42], using finite or countable sums that generalize the Riemann sums [5, 6, 10, 33], via Choquet or Sugeno integrals [14, 15, 41, 45, 49, 58–63], as extension of pseudo-integral [28]. For survey on multivalued integrals of Aumann and Debreu types see [55]. The Gould integral was defined in [27] via finite sums for real functions relative to a finitely additive vector measure μ. Different generalizations of the Gould integral were introduced and studied in [22–25, 51, 56] for real functions and in [52] for multifunctions relative to a finitely additive vector measure (via Aumann method). In this paper we define and study another set-valued Gould type integral of multifunctions (taking values in the space of all nonempty bounded subsets of a real Banach space) based on an arbitrary non-negative set function. We define the integral as a limit of a net of finite integral sums. Many important properties of the integral work with an arbitrary non-negative set function μ without supplementary conditions on μ.

This paper is organized as follows. In Sect. 2 we collect some facts about families of subsets of Banach space, set functions and general multimeasures. Section 3 deals with the integration of strongly measurable multifunctions and, more particularly, with those that can be approximated by simple measurable multifunctions in the sense of the Hausdorff distance, the Aumann integral, whose construction is based on integrable selections and which has become the most popular set-valued integral, and Aumann-Gould integral. In Sect. 4 is presented a multi-valued Choquet integral based on multisubmeasure. In Sect. 5 we define a Gould type integral for multifunctions relative to an arbitrary non-negative set function and point out some relationships between integrability and total measurability. There are given some basic properties of the integrable multifunctions, and some continuity properties of the multimeasure induced by set-valued integral. Section 6 is for the conclusion.

2 Preliminaries

2.1 *Families of Subsets of Banach Space*

Let T be an abstract nonempty set, $\mathscr{P}(T)$ the family of all subsets of T, $\mathscr{A}$ an algebra of subsets of T, $(X, \|\cdot\|)$ a real Banach space with the metric d induced by its norm, $\mathscr{P}_0(X)$ the family of all nonempty subsets of X, $\mathscr{P}_b(X)$ the family of all nonempty bounded subsets of X, $\mathscr{P}_{bc}(X)$ the family of all nonempty bounded convex subsets of X, $\mathscr{C}_b(X)$ the family of all nonempty bounded closed subsets of X, $\mathscr{C}_{bc}(X)$ the family of all nonempty bounded closed convex subsets of X and $\mathcal{K}_c(X)$ the family of all nonempty compact convex subsets of X.

For every $A, B \in \mathcal{P}_0(X)$ and every $\alpha \in \mathbb{R}$, let

$$A + B = \{x + y \mid x \in A, y \in B\},$$
$$\alpha A = \{\alpha x \mid x \in A\}.$$

We denote by $\overline{A}$ the closure of A with respect to the topology induced by the norm of X. Let $\mp$ be the Minkowski addition on $\mathcal{P}_0(X)$, i.e.,

$$A \mp B = \overline{A + B} \quad (A, B \in \mathcal{P}_0(X)).$$

Let h be the Hausdorff metric given by

$$h(A, B) = \max\left(\sup_{x \in A} d(x, B), \sup_{x \in B} d(x, A)\right) \quad (A, B \in \mathcal{P}_0(X)),$$

and $d(x, B) = \inf_{y \in B} d(x, y)$. $(\mathcal{C}_b(X), h)$ and $(\mathcal{K}_c(X), h)$ are complete metric spaces, see [32]. We denote $|A| = h(A, \{0\})$, for every $A \in \mathcal{P}_0(X)$, where 0 is the origin of X.

Relations of the Hausdorff metric h with respect to the operations $+$ and $\mp$ are given by the following inequalities. If $A, B, C, D, A_i, B_i \in \mathcal{P}_0(X)$, for every $i \in \{1, \dots, n\}$ and $n \in \mathbb{N}$, then

$$h\left(\sum_{i=1}^{n} A_i, \sum_{i=1}^{n} B_i\right) \leqslant \sum_{i=1}^{n} h(A_i, B_i),$$

$$h(\alpha A \mp \beta B, \gamma A \mp \delta B) \leqslant |\alpha - \gamma| \cdot |A| + |\beta - \delta| \cdot |B| \quad (\alpha, \beta, \gamma, \delta \in \mathbb{R}).$$

Definition 1 A *partition* of T is a finite family $\pi = \{A_i \mid i = 1, \dots, n\} \subset \mathcal{A}$ such that $A_i \cap A_j = \emptyset, i \neq j$ and $\bigcup_{i=1}^{n} A_i = T$.

Then, as usual, we have that for two partitions $\pi = \{A_i \mid i = 1, \dots n\}$ and $\pi' = \{B_j \mid j = 1, \dots m\}$ of T we say that π' is *finer than* π, denoted $\pi' \geqslant \pi$, if for every $j = 1, \dots, m$, there exists $i_j = 1, \dots, n$, so that $B_j \subseteq A_{i_j}$. *The common refinement* of two partitions $\pi = \{A_i \mid i = 1, \dots n\}$ and $\pi' = \{B_j \mid j = 1, \dots m\}$ is the partition

$$\pi \wedge \pi' = \{A_i \cap B_j \mid i = 1, \dots, n; j = 1, \dots, m\}.$$

We denote by $\mathcal{P}$ the class of all partitions of T and if $A \in \mathcal{A}$ is fixed, by $\mathcal{P}_A$ we denote the class of all partitions of A.

2.2 Set Functions

We have by [46].

Definition 2 Let $\mu : \mathscr{A} \to [0, \infty]$ be a non-negative set function with $\mu(\emptyset) = 0$. We say that μ is:

(i) *monotone* if $\mu(A) \leqslant \mu(B)$, for every $A, B \in \mathscr{A}$, with $A \subseteq B$;
(ii) *subadditive* if $\mu(A \cup B) \leqslant \mu(A) + \mu(B)$, for every $A, B \in \mathscr{A}$, with $A \cap B = \emptyset$;
(iii) *submeasure* if μ is monotone and subadditive;
(iv) *σ-subadditive* if $\mu(A) \leqslant \sum_{n=1}^{\infty} \mu(A_n)$, for every sequence of pairwise disjoint sets $(A_n)_{n \in \mathbb{N}} \subset \mathscr{A}$, with $A = \bigcup_{n=1}^{\infty} A_n \in \mathscr{A}$;
(v) *finitely additive* if $\mu(A \cup B) = \mu(A) + \mu(B)$ for every disjoint $A, B \in \mathscr{A}$;
(vi) *measure* if $\lim_{n \to \infty} \sum_{k=1}^{n} \mu(A_k) = \mu(A)$, for every sequence of pairwise disjoint sets $(A_n)_{n \in \mathbb{N}} \subset \mathscr{A}$, with $A = \bigcup_{n=1}^{\infty} A_n \in \mathscr{A}$;
(vii) *increasing convergent* if $\lim_{n \to \infty} \mu(A_n) = \mu(A)$, for every increasing sequence of sets $(A_n)_{n \in \mathbb{N}} \subset \mathscr{A}$ (i.e. $A_n \subset A_{n+1}$, for every $n \in \mathbb{N}$), with $\bigcup_{n=1}^{\infty} A_n = A \in \mathscr{A}$ (denoted by $A_n \nearrow A$);
(viii) *decreasing convergent* if $\lim_{n \to \infty} \mu(A_n) = \mu(A)$, for every decreasing sequence of sets $(A_n)_{n \in \mathbb{N}} \subset \mathscr{A}$, i.e. $A_{n+1} \subset A_n$, for every $n \in \mathbb{N}$, with $\bigcap_{n=1}^{\infty} A_n = A \in \mathscr{A}$ (denoted by $A_n \searrow A$), and $\mu(A_1) < \infty$;
(ix) *order-continuous* (shortly, *o-continuous*) if $\lim_{n \to \infty} \mu(A_n) = 0$, for every decreasing sequence of sets $(A_n)_{n \in \mathbb{N}} \subset \mathscr{A}$, with $A_n \searrow \emptyset$.;
(x) *exhaustive* if $\lim_{n \to \infty} \mu(A_n) = 0$, for every sequence of pairwise disjoint sets $(A_n)_{n \in \mathbb{N}} \subset \mathscr{A}$.

We introduce two types of variations of set function based on [46].

Definition 3 We consider the following set functions associated to an arbitrary set function $\mu : \mathscr{A} \to [0, \infty]$ with $\mu(\emptyset) = 0$:

(i) $\overline{\mu}$ (*the disjoint variation* of μ) defined, for every $A \in \mathscr{A}$, by

$$\overline{\mu}(A) = \sup \left\{ \sum_{i=1}^{n} \mu(B_i) \right\},$$

where the supremum is considered over all finite partitions $\{B_i\}_{i=1}^{n}$ of A, with $B_i \in \mathscr{A}$, for every $i \in \{1, \ldots, n\}$. μ is said to be *of finite variation* if $\overline{\mu}(T) < \infty$.

(ii) $\widetilde{\mu}$ defined, for every $A \subseteq T$, by

$$\widetilde{\mu}(A) = \inf\{\overline{\mu}(B) \mid A \subseteq B, B \in \mathscr{A}\}.$$

The properties of set functions $\overline{\mu}$ and $\tilde{\mu}$ are investigated in details in [46]. If we deal with $\mathscr{K}_c(X)$-valued multifunctions, then the Minkowski addition $\overline{+}$ becomes the usual addition of sets $+$, i.e., $A + B = \{x + y | x \in A, y \in B\}$).

Definition 4 Let $\mu : \mathscr{A} \to [0, \infty]$ be an arbitrary set function. We say that *a property* (P) *holds* μ *-almost everywhere* (briefly, μ-ae) if there exists $A \in \mathscr{P}(T)$, with $\widetilde{\mu}(A) = 0$, such that the property (P) is valid on $T \backslash A$.

Definition 5 Let $\nu, \mu : \mathscr{A} \to [0, \infty]$ be two arbitrary set functions. We say that ν *is absolutely continuous with respect to* μ if for every $\varepsilon > 0$, there is $\delta > 0$, such that for any $A \in \mathscr{A}$, with $\overline{\mu}(A) < \delta$, we have $\nu(A) < \varepsilon$ (denoted by $\nu \ll \mu$).

Definition 6 Let $f, f_n : T \to \mathbb{R}$, $n \in \mathbb{N}$. The sequence $(f_n)_{n \in \mathbb{N}}$ is said to be:

(i) *convergent in* μ*-measure* if for every $\varepsilon > 0$, $\{t \in T; |f_n(t) - f(t)| > \varepsilon\} \in \mathscr{A}$ and $\lim\limits_{n \to \infty} \mu(\{t \in T; |f_n(t) - f(t)| > \varepsilon\}) = 0$ (denoted by $f_n \stackrel{\mu}{\to} f$),
(ii) *convergent almost everywhere* if there exists $A \in \mathscr{A}$, with $\mu(A) = 0$, such that $\lim\limits_{n \to \infty} f_n(t) = f(t)$ for every $t \in T \backslash A$ (denoted by $f_n \stackrel{ae}{\to} f$).

For more details on convergences see [40].

2.3 General Multimeasures

The theory of set-valued measures (multimeasures) whose values are subsets of some Banach space, under additivity condition, was investigated in [2, 12, 18, 26]. Some of the motivations were the applications to mathematical economics and to statistics. To extend the notion of σ-additivity, there are at least three possible definitions depending on the series summability concept considered in the space of closed sets, see [30].

Here we shall consider general multimeasures with some additional properties. We have by [22].

Definition 7 Let $M : \mathscr{A} \to \mathscr{P}_0(X)$ be a set multifunction, with $M(\emptyset) = \{0\}$. M is said to be:

(i) *monotone* if $M(A) \subseteq M(B)$, for every $A, B \in \mathscr{A}$, with $A \subseteq B$;
(ii) *additive multimeasure* if $M(A \cup B) = M(A) + M(B), \forall A, B \in \mathscr{A}, A \cap B = \emptyset$;
(iii) *absolutely continuous with respect to* μ if for every $\varepsilon > 0$, there is $\delta > 0$ such that for every $A \in \mathscr{A}$ with $\overline{\mu}(A) < \delta$, we have $|M(A)| < \varepsilon$ (denoted by $M \ll \mu$);

(iv) *increasing convergent* if $\lim_{n\to\infty} h(M(A_n), M(A)) = 0$, for every increasing sequence of sets $(A_n)_{n\in\mathbb{N}} \subset \mathscr{A}$, with $\bigcup_{n=0}^{\infty} A_n = A \in \mathscr{A}$;

(v) *decreasing convergent* if $\lim_{n\to\infty} h(M(A_n), M(A)) = 0$, for every decreasing sequence of sets $(A_n)_{n\in\mathbb{N}} \subset \mathscr{A}$, with $\bigcap_{n=0}^{\infty} A_n = A \in \mathscr{A}$, and $M(A_1) < \infty$;

(vi) *order-continuous* (shortly, *o-continuous*) if $\lim_{n\to\infty} |M(A_n)| = 0$, for every decreasing sequence of sets $(A_n)_{n\in\mathbb{N}} \subset \mathscr{A}$, with $A_n \searrow \emptyset$;

(vii) *exhaustive* if $\lim_{n\to\infty} |M(A_n)| = 0$, for every sequence of pairwise disjoint sets $(A_n)_{n\in\mathbb{N}} \subset \mathscr{A}$;

(viii) *h-multimeasure* if $\lim_{n\to\infty} h\left(\sum_{k=0}^{n} M(A_k), M(A)\right) = 0$, for every sequence of pairwise disjoint sets $(A_n)_{n\in\mathbb{N}} \subset \mathscr{A}$, with $A = \bigcup_{n=0}^{\infty} A_n \in \mathscr{A}$;

(ix) *of finite variation* if $\overline{M}(T) < \infty$, where $\overline{M}$ is defined for every $A \in \mathscr{A}$ by

$$\overline{M}(A) = \sup\left(\sum_{i=1}^{n} |M(B_i)|\right),$$

where the supremum is considered over all finite partitions $\{B_i\}_{i=1}^{n}$ of A, with $B_i \in \mathscr{A}, \forall i \in \{1, \ldots, n\}$;

(x) *null-additive* if for every $A, B \in \mathscr{A}$ with $M(A) = \{0\}$ we have $M(A \cup B) = M(B)$;

(xi) *weakly null-additive* if for every $A, B \in \mathscr{A}$ with $M(A) = M(B) = \{0\}$ we have $M(A \cup B) = \{0\}$.

Remark 1 The theory of random sets is closely related to multimeasures, especially when X is finite dimensional, see [35, 43, 44]. There are lot of important probabilistic and statistical applications, e.g., to Stereology and to Image Processing.

3 Some Multivalued Additive Integrals

3.1 Strong Integral

Let $(T, \mathscr{A}, p)$ be a probability space. A multifunction is a map, defined on T, whose values are subsets of some given set. In this section, we shall restrict our attention to the space $\mathscr{C}_b(X)$ of closed bounded subsets of X, endowed with the topology τ_H generated by the Hausdorff metric h. Further, we consider the Borel σ-field $\mathscr{B}(\mathscr{C}_b(X), \tau_H)$ generated by the τ_H-open subsets of $\mathscr{C}_b(X)$. A multifunction $F : T \to \mathscr{C}_b(X)$ is said to be strongly $\mathscr{A}$-measurable (or simply, strongly measurable) if, for every member $\mathscr{W}$ of $\mathscr{B}(\mathscr{C}_b, \tau_H)$, one has $F^{-1}(\mathscr{W}) \in \mathscr{A}$. Multifunctions that enjoy some measurability property are also called "random sets".

We start by defining the set-valued integral of a simple multifunction, i.e., of a multifunction F assuming only a finite number of values, see [30].

Definition 8 Let $\{A_1, \ldots, A_k\}$ be an $\mathscr{A}$-measurable partition of T and let $F : T \to \mathscr{C}_b(X)$ be a measurable multifunction taking on the value $K_i \in \mathscr{C}_b$ for any $t \in A_i$ $(i = 1, \ldots, k)$, i.e., $F = \sum_{i=1}^{n} \mathbf{1}_{A_i} K_i$, where $\mathbf{1}_{A_i}$ denotes the characteristic function of A_i. Then the integral (or expectation) of F is the member of $\mathscr{C}_b$ defined by

$$E(F) = \overline{\sum_{i=1}^{n} p(A_i) K_i} \tag{1}$$

When X is finite dimensional, the closure operation is not necessary. Given a subspace $\mathscr{C}'$ of $\mathscr{C}_b$, we denote by $\mathscr{L}^1(\mathscr{C}', \mathscr{A})$ the class of strongly $\mathscr{A}$-measurable multifunctions with values in $\mathscr{C}'$ such that $E||F|| <$, where E is the integral given by (1), and by $\mathscr{S}(\mathscr{C}', \mathscr{A})$ the subclass of $\mathscr{L}^1(\mathscr{C}', \mathscr{A})$ whose members are strongly $\mathscr{A}$-measurable simple multifunctions.

Now, we construct the integral of a strongly measurable multifunction $F : T \to \mathscr{C}'$, where $\mathscr{C}'$ is a τ_H-separable subspace of $\mathscr{C}_{cb}$. Without restriction, we assume that $\mathscr{C}'$ is τ_H-closed, and stable under the Minkowski addition and multiplication by positive scalar.

Definition 9 For a strongly $\mathscr{A}$-measurable simple multifunction $F : T \to \mathscr{C}'$, we define the map $\Phi : \mathscr{S}(\mathscr{C}', \mathscr{A}) \to \mathscr{C}'$ by

$$\Phi(F) = E(F) \tag{2}$$

where $E(F)$ is defined by (1).

Now, we can extend the integral on the space of multifunctions whose members can be approximated by simple functions. We denote by $\mathscr{L}_h^1(\mathscr{C}, \mathscr{A})$ the subclass of $\mathscr{L}^1(\mathscr{C}', \mathscr{A})$ of those F that can be approximated by simple multifunctions, i.e., such that one can find a sequence $(F_n)_{n \in \mathrm{N}}$ in $\mathscr{S}(\mathscr{C}', \mathscr{A})$ such that

$$\lim_{n \to \infty} h(F(t), F_n(t)) = 0 \;\; p\text{-}a.e.$$

We have by [29].

Theorem 1 *The map $\Phi : \mathscr{S}(\mathscr{C}', \mathscr{A}) \to \mathscr{C}'$ given by (2) is extended to a map $\tilde{\Phi}$ from $\mathscr{L}^1(\mathscr{C}', \mathscr{A})$ into $\mathscr{C}'$ (called strong integral) with the following properties*

(i) $\tilde{\Phi}(F + G) = \overline{\tilde{\Phi}(F) + \tilde{\Phi}(G)}$.
(ii) $\tilde{\Phi}(aF) = a\tilde{\Phi}(F)$, *for* $a \geqslant 0$.
(iii) $h(\tilde{\Phi}(F), \tilde{\Phi}(G)) \leqslant \tilde{\Phi}(h(F, G))$, *specially* $||E(F)|| \leqslant E||F||$.

Remark 2 The preceding approach for defining the set-valued integral is explicit, in that it starts with simple multifunctions and uses only elementary operations such as the Minkowski addition and the scalar multiplication. An alternative approach for constructing the set-valued conditional expectation for a compact convex multifunction is given in [17].

3.2 Aumann Integral

Let $(T, \mathscr{A}, p)$ be a probability space and $\mathscr{B}$ be a sub-σ-field of $\mathscr{A}$. By $L^0(T, \mathscr{B}, p; X)$ we denote the space of all (classes of) measurable functions from $(T, \mathscr{B})$ into $(X, \mathscr{B}(X))$. For every $F \in \mathscr{M}(\mathscr{C}(X))$ (the set of closed valued multifunctions) we define

$$\mathscr{S}(F, \mathscr{A}) = \left\{ f \in L^0(T, \mathscr{B}, p; X) \mid f(t) \in F(t), \text{ for } p\text{-almost every } t \in dom(F) \right\}.$$

$L^0(T, \mathscr{A}, p; X)$ endowed with the topology of convergence in probability is a metrizable topological vector space, provided one identify two functions that coincide p-almost everywhere. Since a sequence converging in probability admits an almost everywhere converging subsequence, then for any sub-σ-field $\mathscr{B}$ of $\mathscr{A}$, the set $\mathscr{S}(F, \mathscr{B})$ is closed in $L^0(T, \mathscr{B}, p; X)$. We denote by $L^1(T, \mathscr{A}, p; X)$ the subspace of $L^0(T, \mathscr{A}, p; X)$, whose members are Bochner integrable. Given a sub-σ-field $\mathscr{B}$ of $\mathscr{A}$ and a multifunction F, we define the following $L^1(X)$-closed subset of $L^1(T, \mathscr{A}, p; X)$

$$\mathscr{S}^1(F, \mathscr{B}) = \left\{ f \in L^1(T, \mathscr{A}, p; X) \mid f(t) \in F(t), \text{ for } p\text{-almost every } t \in dom(F) \right\}.$$

The following notion of integral for multifunctions was introduced by Aumann [4].

Definition 10 For any measurable multifunction F and any sub-σ-field $\mathscr{B}$ of $\mathscr{A}$, the set-valued (Aumann) integral of F over T, with respect to $\mathscr{B}$, is denoted by $I(F, \mathscr{B})$ and defined by

$$I(F, \mathscr{B}) = \left\{ \int_T f \, dp \mid f \in \mathscr{S}^1(F, \mathscr{B}) \right\}.$$

The relation between strong and Aumann integral is given in the following theorem.

Theorem 2 *Let $F \in \mathscr{M}(\mathscr{C}(X))$. If we assume in addition that F is integrable, then the following property hold. If F is integrably bounded and takes on its values in a τ_H-separable subspace of $\mathscr{C}_{bc}(X)$, then one has*

$$\tilde{\Phi}(F) = \overline{I(F)}$$

where $\tilde{\Phi}(F)$ denotes the integral as defined in Sect. 3.1 (Theorem 1).

3.3 Aumann-Gould Type Integral

The Gould integral was introduced in [27] using finite sums for real functions with respect to a finitely additive vector measure μ. If μ is a countably additive vector measure of finite variation, then the Gould integral coincides with the Dunford integral and the Gelfand-Pettis integral. Different generalizations of the Gould integral were introduced and studied in [22–25, 51, 56]. We consider in this section a Gould integral of multifunction with respect to a finitely additive multimeasure.

Definition 11 If $\mu : \mathscr{A} \to X$ is a finitely additive measure, we say that μ is a selector of a multimeasure $M : \mathscr{A} \to \mathscr{P}_0(X)$ if $\mu(A) \in M(A)$, for every $A \in \mathscr{A}$. We denote by S_M the family of all selectors of M.

We have by [52].

Definition 12 Let $M : \mathscr{A} \to \mathscr{P}_{wkc}(X)$ (nonempty weakly compact convex subsets of X) be a finitely additive multimeasure of bounded variation and $f : T \to \mathbb{R}$ a bounded function. We say that f is Gould integrable with respect to M if there exists a nonempty weakly compact convex subset of X, denoted by $\int_T f\,dM$, which satisfies the condition that for every $\varepsilon > 0$ there exists a partition π_ε of T such that, for every partition $\pi = \{E_i\}_{i=0}^n$ finer than π_ε and for every choice of $s_i \in E_i$, we have

$$h\left(\sum_{i=0}^{n} f(s_i)M(E_i), \int_T f\,dM\right) < \varepsilon.$$

The set $\int_T f\,dM$ is called the M-Gould integral of f.

Definition 13 If $A \in \mathscr{A}$, the set

$$\left\{(G)\int_A f\,d\mu \mid \mu \in S_M^f\right\}$$

is called the Aumann-Gould integral of the function $f : T \to \mathbb{R}$ on A, where $(G)\int f\,d\mu$ denotes the Gould integral of f with respect to the vector measure μ. We denote it by $(AG)\int_A f\,dM$. If $(AG)\int_A f\,dM \neq \emptyset$, we say that f is Aumann-Gould integrable on A.

Theorem 3 *If M is an additive multimeasure, $(T, \mathscr{A}, \overline{M})$ is a complete measurable space and the Banach space X has the Radon-Nikodym property, then any M-Gould-integrable function f is Aumann-Gould-integrable and, moreover,*

$$\int_A f\,dM = \overline{(AG)\int_A f\,dM}, \quad A \in \mathscr{A}.$$

4 A Multi-valued Choquet Integral Based on a Multisubmeasure

Jang et al. [34] have introduced a multi-valued Choquet integral for multifunction taking values in the class of all closed, nonempty sets of the interval $[0, \infty[$ based on a nonnegative monotone set function, for which the selectors are real Choquet integrals.

We present in this section another multivalued Choquet integral, this time for nonnegative functions with respect to multisubmeasures taking values in $\mathcal{K}_c([0, \infty[)$. We can associate to $M : \mathcal{A} \to \mathcal{K}_c([0, \infty[)$ two nonnegative set functions: $\mu_1(A) = \inf M(A)$, and $\mu_2(A) = \sup M(A)$, for every $A \in \mathcal{A}$, such that $M(A) = [\mu_1(A), \mu_2(A)]$, for every $A \in \mathcal{A}$, and set of selectors for M is nonempty since contains the submeasures μ_1 and μ_2. Let be $f : T \to [0, \infty[$ a measurable function. We denote by $S_M^C(f)$ the set of all monotone set functions $\mu : \mathcal{A} \to [0, \infty[$ which are selectors of the multisubmeasure M with respect to which the function f is Choquet integrable. Using a similar procedure as in Sect. 3.3 for Aumann-Gould integral we introduce the following notion by [57].

Definition 14 Let $M : \mathcal{A} \to \mathcal{K}_c([0, \infty[)$ be a multisubmeasure, $f : T \to [0, \infty[$ a measurable function and $A \in \mathcal{A}$. The Aumann-Choquet integral (shortly (AC)-integral) of f on A with respect to the multisubmeasure M is the set

$$(AC) \int_A f \, dM = \left\{ (C) \int_A f \, d\mu \mid \mu \in S_M^C(f) \right\},$$

where $(C) \int_A f \, d\mu$ is the usual Choquet integral. A function f is said to be Aumann-Choquet integrable on A if $(AC) \int_A f \, dM \neq \emptyset$.

We denote by $\mathcal{L}_{AC}(M)$ the set of all Aumann-Choquet integrable functions with respect to M on A.

Theorem 4 *The multifunction*

$$N(A) = (AC) \int_A f \, dM \quad (A \in \mathcal{A}),$$

has the following properties.

(i) N is a monotone set multifunction.
(ii) If $f \in \mathcal{L}_{AC}(M)$ and μ is weakly null-additive, then the multifunction defined by N is weakly null-additive.
(iii) If $f \in \mathcal{L}_{AC}(M)$ and μ is null-additive, then the multifunction defined by N is null-additive.

5 Gould Type Integral of Multifunctions Based on Arbitrary Nonegative Set Function

We present here some recent results without proofs, which can be found in [48].

5.1 Definition and Basic Properties

In this section we introduce a Gould type set-valued integral for $\mathcal{P}_b(X)$-valued multifunctions with respect to a non-negative set function and point out some relationships between total measurability and integrability. Suppose $\mu : \mathcal{A} \to [0, \infty[$ is a non-negative set function, with $\mu(\emptyset) = 0$. We introduce total measurability and Gould type integrability for multifunctions relative to a set function.

Definition 15 Let $F : T \to \mathcal{P}_0(X)$ be a multifunction. F is said to be $\widetilde{\mu}$-totally measurable (on T) if for every $\varepsilon > 0$ there exists a partition $\pi_\varepsilon = \{A_i \mid i = 0, \ldots, n\}$ of T such that $\widetilde{\mu}(A_0) < \varepsilon$, and $\sup\limits_{t,s \in A_i} h(F(t), F(s)) < \varepsilon$, for every $i \in \{1, \ldots, n\}$. F is said to be $\widetilde{\mu}$-totally measurable on $B \in \mathcal{A}$ if the restriction $F|_B$ of F to B is $\widetilde{\mu}$-totally measurable on $(B, \mathcal{A}_B, \mu_B)$.

Remark 3 (i) If F is $\widetilde{\mu}$-totally measurable on T, then F is $\widetilde{\mu}$-totally measurable on every $A \in \mathcal{A}$.
(ii) Let $F : T \to \mathcal{C}_c(\mathbb{R})$, $F(t) = [f(t), g(t)]$, $\forall t \in T$, where $f, g : T \to \mathbb{R}$ are real functions, such that $f \leqslant g$. Then F is $\widetilde{\mu}$-totally measurable if and only if f and g are both $\widetilde{\mu}$-totally measurable. This result follows by the equality

$$h([x, y], [z, t]) = \max\{|x - z|, |y - t|\} \quad (x, y, z, t \in \mathbb{R}, x \leqslant y, z \leqslant t).$$

For every multifunction $F : T \to \mathcal{P}_b(X)$ we denote

$$\sigma_{F,\mu}(\pi) = \sum_{i=1}^{n} F(t_i)\mu(A_i) = \mu(A_1)F(t_1) + \cdots + \mu(A_n)F(t_n)$$

for every partition $\pi = \{A_i \mid i = 1, \ldots, n\}$ of T and every $t_i \in A_i, i \in \{1, \ldots, n\}$. If there is no doubt, we shall denote $\sigma_{F,\mu}(\pi)$ shortly by $\sigma(\pi)$.

Definition 16 A multifunction $F : T \to \mathcal{P}_b(X)$ is said to be μ-integrable (on T) if the net $(\sigma(\pi))_{\pi \in (\mathcal{P}, \leqslant)}$ is convergent in $(\mathcal{C}_b(X), h)$, where $\mathcal{P}$, the set of all partitions of T, is ordered by the order relation "$\leqslant$" given in Sect. 2.1. If $(\sigma(\pi))_{\pi \in (\mathcal{P}, \leqslant)}$ is convergent, then its limit is called *the integral of F on T* with respect to μ, denoted by

$$(G) \int_T F \, d\mu.$$

F is said to be *μ-integrable on $B \in \mathscr{A}$* if the restriction $F|_B$ of F to B is μ-integrable on $(B, \mathscr{A}_B, \mu_B)$.

We easily obtain by the Definition 16.

Proposition 1 *(i) If $(G) \int_T F\, d\mu$ exists, it is unique.*
(ii) F is μ-integrable on T if and only if there exists a set $I \in \mathscr{C}_b(X)$ such that for every $\varepsilon > 0$, there exists a partition π_ε of T, such that for every other partition of T, $\pi = \{A_i \mid i = 1 \dots, n\}$, with $\pi \geqslant \pi_\varepsilon$ and every choice of points $t_i \in A_i$, $i \in \{1, \dots, n\}$, we have $h(\sigma(\pi), I) < \varepsilon$.
(iii) If $F : T \to \mathscr{K}_c(X)$ is $\mathscr{K}_c(X)$-valued, then $(G) \int_T F\, d\mu \in \mathscr{K}_c(X)$.
(iv) If $\mu(T) = 0$, then F is μ-integrable on T and $(G) \int_T F\, d\mu = \{0\}$.

A multifunction $F : T \to \mathscr{P}_0(X)$ is called *bounded* if there exists $C \geqslant 0$ such that $|F(t)| \leqslant C$, $\forall t \in T$. If $F : T \to \mathscr{P}_0(X)$ is bounded, then F is $\mathscr{P}_b(X)$-valued. The converse is not true.

Example 1 Let $F : [0, \infty[\to \mathscr{P}_b([0, \infty[)$ be defined by $F(t) = [t, t+1]$ for every $t \in [0, \infty[$. Then F is $\mathscr{P}_b([0, \infty[)$-valued, but F is not bounded since $\sup\limits_{t \in [0,\infty[} |F(t)| = \infty$.

5.2 Gould μ-Integrability of Multifunction

Theorem 5 *Suppose $F : T \to \mathscr{P}_b(X)$ is a μ-integrable multifunction. Then*

(i) for every $A \in \mathscr{A}$, F is μ-integrable on A;
(ii) F is μ-integrable on $A \in \mathscr{A}$ if and only if $F\mathbf{1}_A$ is μ-integrable on T.

In the following, we present some results concerning the relation between $\widetilde{\mu}$-total measurability and μ-integrability.

Theorem 6 *Suppose $\mu : T \to [0, \infty[$ is finitely additive. If $F : T \to \mathscr{P}_b(X)$ is a bounded $\widetilde{\mu}$-totally measurable multifunction, then F is μ-integrable.*

Example 2 (i) Suppose μ is finitely additive. If $F(t) = E \in \mathscr{C}_{bc}(X)$, for every $t \in T$, then F is μ-integrable and

$$(G) \int_T F\, d\mu = \mu(T)E.$$

(ii) Generally, if $F = \sum\limits_{i=1}^{n} E_i \cdot \mathbf{1}_{A_i}$, $E_i \in \mathscr{C}_{bc}(X)$, $i \in \{1, \dots, n\}$, $\{A_i\}_{i=1}^n \subset \mathscr{A}$ is a partition of T and $\mathbf{1}_{A_i}$ is the characteristic function of A_i, for every $i \in \{1, \dots, n\}$, then F is μ-integrable and

$$(G) \int_T F\, d\mu = \sum_{i=1}^{n} E_i \mu(A_i).$$

(iii) Let $F : T \to \mathscr{P}_{kc}(\mathbb{R})$ be defined by $F(t) = [f_1(t), f_2(t)]$, $\forall t \in T$, where $f_1, f_2 : T \to \mathbb{R}$ are real functions such that $f_1 \leqslant f_2$. Then F is Gould μ-integrable if and only if f_1 and f_2 are both Gould μ-integrable as usual functions. In this case we obtain

$$(G) \int_T F\, d\mu = \left[\int_T f_1\, d\mu, \int_T f_2\, d\mu \right].$$

Particularly, if $f_1 = f_2 = f$, then

$$(G) \int_T F\, d\mu = \left\{ \int_T f\, d\mu \right\}.$$

We have by [46].

Definition 17 If $\mu : \mathscr{A} \to [0, \infty[$ is a non-negative set function, with $\mu(\emptyset) = 0$, then a set $A \in \mathscr{A}$ is said to be *an atom* of μ if $\mu(A) > 0$ and for every $B \in \mathscr{A}$, with $B \subseteq A$, we have $\mu(B) = 0$ or $\mu(A \setminus B) = 0$.

Theorem 7 *Let $\mu : \mathscr{A} \to [0, \infty[$ be a submeasure of finite variation and $F : T \to \mathscr{P}_b(X)$ a $\widetilde{\mu}$-totally measurable bounded multifunction. Then F is Gould μ-integrable on every atom of μ.*

5.3 Further Properties of Integrable Multifunctions

In this section we list some general properties of the integral introduced in Definition 16, based on [48]. Suppose $\mu : \mathscr{A} \to [0, \infty[$ is a non-negative set function, with $\mu(\emptyset) = 0$.

Theorem 8 *Let $F, G : T \to \mathscr{P}_b(X)$ be Gould μ-integrable multifunctions. Then we have the following statements.*

(i) $F + G$ is Gould μ-integrable and

$$(G) \int_T (F + G)\, d\mu = (G) \int_T F\, d\mu \mp (G) \int_T G\, d\mu.$$

(ii) For every $\alpha \in \mathbb{R}$ a multifunction αF is Gould μ-integrable on T and

$$(G) \int_T \alpha F\, d\mu = \alpha \cdot (G) \int_T F d\mu.$$

(iii) For every $\alpha \geqslant 0$ a multifunction F is Gould $\alpha\mu$-integrable on T and

$$(G)\int_T F\,d(\alpha\mu) = \alpha \cdot (G)\int_T F\,d\mu.$$

(iv) For $F(t) \subseteq G(t)$, for every $t \in T$, we have

$$\int_T F\,d\mu \subseteq \int_T G\,d\mu.$$

(v) Related Hausdorff metric we have

$$h\left((G)\int_T F\,d\mu, (G)\int_T G\,d\mu\right) \leqslant \sup_{t\in T} h(F(t), G(t)) \cdot \overline{\mu}(T).$$

(vi) Specially, related to absolute value we obtain

$$\left|(G)\int_T F\,d\mu\right| \leqslant \sup_{t\in T} |F(t)| \cdot \overline{\mu}(T).$$

Theorem 9 *Suppose $\mu_1, \mu_2 : \mathscr{A} \to [0, \infty[$ are non-negative set functions, with $\mu_1(\emptyset) = \mu_2(\emptyset) = 0$. Then we have the following statements.*

(i) If $F : T \to \mathscr{P}_{bc}(X)$ is both μ_1-integrable and μ_2-integrable and $\mu : \mathscr{A} \to [0, \infty[$ is defined by $\mu(A) = \mu_1(A) + \mu_2(A)$, for every $A \in \mathscr{A}$, then F is μ-integrable and

$$(G)\int_T F\,d(\mu_1 + \mu_2) = (G)\int_T F d\mu_1 \mp (G)\int_T F\,d\mu_2.$$

(ii) Let $\mu_1 \leqslant \mu_2$, and $F : T \to \mathscr{P}_b(\mathbb{R})$ be defined by $F(t) = [0, f(t)]$, where $f : T \to [0, \infty[$ is a real function. If F is both μ_1-integrable and μ_2-integrable, then

$$(G)\int_T F\,d\mu_1 \subseteq (G)\int_T F\,d\mu_2.$$

5.4 Set Multifunction Induced by Set-Valued Integral

Suppose $\mu : \mathscr{A} \to [0, \infty[$ is a non-negative set function with $\mu(\emptyset) = 0$ and $F : T \to \mathscr{P}_b(X)$ is a μ-integrable multifunction. In this section, we present several properties concerning the set multifunction $M : \mathscr{A} \to \mathscr{C}_b(X)$ defined by

$$M(A) = (G)\int_A F d\mu, \quad (A \in \mathscr{A}). \tag{3}$$

Theorem 10 *Let be $B, C \in \mathscr{A}$, with $B \cap C = \emptyset$. If $F : T \to \mathscr{P}_b(X)$ is μ-integrable both on B and on C, then F is μ-integrable on $B \cup C$ and moreover*

$$M(A \cup B) = (G) \int_{B \cup C} F\, d\mu = (G) \int_B F\, d\mu \dot{+} (G) \int_C F\, d\mu = M(A) \dot{+} M(B).$$

We list some basic properties of set multifunction M defined by (3).

Theorem 11 *Let $F : T \to \mathscr{P}_b(X)$ be a μ-integrable multifunction and the set multifunction $M : \mathscr{A} \to \mathscr{C}_b(X)$, defined by $M(A) = (G) \int_A F\, d\mu$, for every $A \in \mathscr{A}$. Then the following hold*

(i) M is an additive multimeasure.
(ii) $M \ll \mu$.
(iii) If μ is of finite variation, then M is of finite variation.
(iv) If $\overline{\mu}$ is o-continuous (exhaustive respectively), then M is also o-continuous (exhaustive respectively).

For monotone set function μ we obtain the following theorem.

Theorem 12 *Suppose $\mu : \mathscr{A} \to [0, \infty[$ is monotone. Let $F, G : T \to \mathscr{P}_b(X)$ be bounded multifunctions on T such that F is μ-integrable on T and $F = G$ μ-ae. Then G is μ-integrable on T and $(G) \int_T F\, d\mu = (G) \int_T G\, d\mu$.*

Theorem 13 *Suppose $\mu : \mathscr{A} \to [0, \infty[$ is a submeasure of finite variation. Let $F : T \to \mathscr{P}_b(X)$ be a bounded μ-integrable multifunction and $M : \mathscr{A} \to \mathscr{C}_b(X)$ defined by $M(A) = (G) \int_A F d\mu$, $\forall A \in \mathscr{A}$. Then the following properties hold*

(i) If μ is o-continuous (increasing convergent, decreasing convergent respectively), then the same is M.
(ii) If μ is σ-subadditive, then M is an h-multimeasure.

For special multifunctions (interval valued) we can prove some additional properties.

Theorem 14 *Suppose $\mu : \mathscr{A} \to [0, \infty[$ is monotone. Let $F : T \to \mathscr{P}_b(\mathbb{R})$ be defined by $F(t) = [0, f(t)]$, $\forall t \in T$, where $f : T \to [0, \infty[$ is a real function and let $B, C \in \mathscr{A}$ be such that $B \subseteq C$. If F is μ-integrable both on B and on C, then*

$$(G) \int_B F d\mu \subseteq (G) \int_C F d\mu.$$

Theorem 15 *Let $F : T \to \mathscr{P}_b(\mathbb{R})$ be a μ-integrable multifunction defined by $F(t) = [0, f(t)]$, $\forall t \in T$, where $f : T \to [0, \infty[$ is a bounded $\widetilde{\mu}$-totally measurable function. If μ is finitely additive and of finite variation, then*

$$\overline{M}(A) = (G) \int_A f\, d\mu = |M(A)| \quad (A \in \mathscr{A}).$$

6 Conclusion

It is given a short overview of basic integrals of multifunctions with respect to additive set functions, as strong integral, Aumann integral, Aumann-Gould integral. Then it is presented an extension of Choquet integral based on multisubmeasure. It is introduced a set-valued Gould type integral of $\mathscr{P}_b(X)$-valued multifunctions with respect to an arbitrary non-negative set function. There are presented important properties of this integral. In our future works we shall investigate the relationships among this set-valued Gould type integral and other set-valued integrals of Dunford, Pettis, Choquet, Sugeno, Birkhoff, McShane, Henstock-Kurzweil. In this spirit it will be investigated the extension of universal integral, introduced in [36], for multifunctions.

Acknowledgments This research was supported by the grant MNPRS 174009 and by the project "Mathematical models of intelligent systems and their applications" which was supported by the Provincial Secretariat for Science and Technological Development of Vojvodina.

References

1. Apreutesei, G.: Cauchy nets and convergent nets on semilinear topological spaces. Topology Appl. **159**, 2922–2931 (2012)
2. Artstein, Z.: Set-valued measures. Trans. Amer. Math. Soc. **165** (1972)
3. Aubin, J.P., Frankowska, H.: Set-Valued Analysis. Birkhäuser, Boston (1990)
4. Aumann, R.J.: Integrals of set-valued maps. J. Math. Anal. Appl. **12**, 1–12 (1965)
5. Boccuto, A., Sambucini, A.R.: A McShane integral for multifunctions. J. Concr. Appl. Math. **2**(4), 307–325 (2004)
6. Bongiorno, B., Pfeffer, W.F., Thomson, B.S.: A full descriptive definition of the gage integral. Canad. Math. Bull. **39**(4), 390–401 (1996)
7. Brink, H.E., Maritz, P.: Integration of multifunctions with respect to a multimeasure. Glasnik Math. **35**, 313–334 (2000)
8. Cao, Y.: Aggregating multiple classification results using Choquet integral for financial distress early warning. Expert Syst. Appl. **38**(7), 8285–8292 (2011)
9. Caponetti, D., Di Piazza, L., Kadets, V.: Description of the limit set of Henstock-Kurzweil integral sums of vector-valued functions. J. Math. Anal. Appl. **421**, 1151–1162 (2015)
10. Cascales, B., Rodriguez, J.: Birkhoff integral for multi-valued functions. J. Math. Anal. Appl. **297**, 540–560 (2004)
11. Cascales, B., Kadets, V., Rodriguez, J.: The Pettis integral for multivalued functions via single-valued functions. J. Math. Anal. Appl. **332**, 1–10 (2007)
12. Costé, A.: Sur les multimesures valeurs fermes bornes d'un espace de Banach. C.R. Acad. Sci. Paris **280**, 567–570 (1975)
13. Croitoru, A.: An integral for multifunctions with respect to a multimeasure. An. Şt. Univ. "Al. I. Cuza" Iaşi **49**, 95–106 (2003)
14. Croitoru, A.: Fuzzy integral of measurable multifunctions. Iran. J. Fuzzy Syst. **9**(4), 133–140 (2012)
15. Croitoru, A.: Strong integral of multifunctions relative to a monotone measure. Fuzzy Sets Syst. **244**, 20–33 (2014)
16. Croitoru, A., Godet-Thobie, C.: Set-valued integration in seminorm. I. Annals of University of Craiova, Mathematics and Computer Science Series, vol. 33, pp. 16–25 (2006)
17. Debreu, G.: Integration of correspondences. In: Proceedings 5th Berkely Symposium on Mathematical Statistics and Probability II, Part. I, pp. 351–372 (1967)

18. Debreu, G., Schmeidler, D.: The Radon-Nikodym derivative of a correspondence. In: Proceedings of the Sixth Berkeley Symposium of Mathematical Statistics and Probability (1971)
19. Di Piazza, L., Musial, K.: Set-valued Kurzweil-Henstock-Pettis integral. Set-Valued Anal. **13**, 167–179 (2005)
20. Dunford, N., Schwartz, J.: Linear Operators I. General Theory. Interscience, New York (1958)
21. Frankowska, H.: An open mapping principle for set-valued map. J. Math. Anal. Appl. **127**, 172–180 (1987)
22. Gavriluţ, A.: On some properties of the Gould type integral with respect to a multisubmeasure. An. Şt. Univ. "Al. I. Cuza" Iaşi **52**(1), 177–194 (2006)
23. Gavriluţ, A.: Fuzzy Gould integrability on atoms. Iran. J. Fuzy Syst. **8**(3), 113–124 (2011)
24. Gavriluţ, A.: The general Gould type integral with respect to a multisubmeasure. Math. Slovaca **60**(3), 289–318 (2010)
25. Gavriluţ, A., Iosif, A.E., Croitoru, A.: The Gould integral in Banach lattices. Positivity **19**, 65–82 (2015)
26. Godet-Thobie, C.: Some results about multimeasures and their selections. In: Measure Theory at Oberwolfach 1979, Lecture Notes in Mathematics, vol. 794, pp. 112–116. Springer (1980)
27. Gould, G.G.: Integration over vector-valued measures. Proc. London Math. Soc. **15**, 193–205 (1965)
28. Grbić, T., Štajner-Papuga, I., Štrboja, M.: An approach to pseudo-integration of set-valued functions. Inf. Sci. **181**(11), 2278–2292 (2011)
29. Hess, C.: Conditional expectation and martingales of random sets. J. Pattern Recognit. **32**, 1543–1567 (1999)
30. Hess, C.: Set-valued integration and set-valued probability theory: an overview. in [47], 617–673
31. Hildebrand, W.: Core and Equilibria of a Large Economy. Princeton University Press (1974)
32. Hu, S., Papageorgiou, N.S.: Handbook of Multivalued Analysis, vol. I. Kluwer Academic Publishers, Dordrecht (1997)
33. Hukuhara, M.: Integration des applications mesurables dont la valuer est un compact convexe. Funkcialaj Ekvacioj **10**, 205–223 (1967)
34. Jang, L.C., Kwon, J.S.: On the representation of Choquet integrals of set-valued functions, and null sets. Fuzzy Sets Syst. **112**, 233–239 (2000)
35. Kendall, M.G., Moran, P.A.P.: Geometrical Probability. Charles Griffin, London (1963)
36. Klement, E.P., Mesiar, R., Pap, E.: A universal integral as common frame for Choquet and Sugeno integral. IEEE Trans. Fuzzy Syst. **18**(1), 178–187 (2000)
37. Klement, E.P., Mesiar, R., Li, J., Pap, E.: Integrals based on monotone set functions. Fuzzy Sets Syst. **281**, 88–102 (2015)
38. Kudo, H.: Dependent experiments and sufficient statistics. Natur. Sci. Rep. Ochanomizu Univ. **4**, 151–163 (1954)
39. Kuncová, K., Malý, J.: Non-absolutely convergent integrals in metric spaces. J. Math. Anal. Appl. **401**, 578–600 (2013)
40. Li, J., Mesiar, R., Pap, E., Klement, E.P.: Convergence theorems for monotone measures. Fuzzy Sets Syst. **281**, 103–127 (2015)
41. Liu, W.L., Song, X.Q., Zhang, Q.Z., Zhang, S.B.: (T) Fuzzy integral of multi-dimensional function with respect to multi-valued measure. Iran. J. Fuzzy Syst. **9**(3), 111–126 (2012)
42. Martellotti, A., Sambucini, A.R.: A Radon-Nikodym theorem for multimeasures. Atti Sem. Mat. Fis. Univ. Modena **42**, 579–599 (1994)
43. Matheron, G.: Random Sets and Integral Geometry. Wiley (1975)
44. Molchanov, I.S.: Limit Theorems for Unions of Random Closed Sets. Lecture Notes in Mathematics, vol. 1561. Springer (1993)
45. Narukawa, Y., Torra, V.: Multidimensional generalized fuzzy integral. Fuzzy Sets Syst. **160**, 802–815 (2009)
46. Pap, E.: Null-Additive Set Functions. Kluwer Academic Publishers, Dordrecht (1995)
47. Pap, E.: Handbook of Measure Theory. Elsevier (2002)

48. Pap, E., Gavriluţ, A., Croitoru, A.: Gould type integral of multifunctions relative to non-negative set functions (submitted)
49. Park, C.K.: Set-valued Choquet-Pettis integrals. Korean J. Math. **20**(4), 381–393 (2012)
50. Pham, T.D., Brandl, M., Nguyen, N.D., Nguyen, T.V.: Fuzzy measure of multiple risk factors in the prediction of osteoporotic fractures. In: Proceedings of the 9-th WSEAS International Conference on Fuzzy Systems (FS'08), pp. 171–177 (2008)
51. Precupanu, A., Croitoru, A.: A Gould type integral with respect to a multimeasure, I/I. An. Şt. Univ. "Al. I. Cuza" Iaşi **48**, 165–200/ 49 (2003); 183–207 (2002)
52. Precupanu, A.M., Satco, B.: The Aumann-Gould integral. Medit. J. Math. **5**, 429–441 (2008)
53. Rådström, H.: An embedding theorem for spaces of convex sets. Proc. Amer. Math. Soc. **3**, 151–158 (1952)
54. Robbins, H.E.: On the measure of random set I. Ann. Math. Stat. **15**, 70–74 (1944)
55. Sambucini, A.R.: A survey on multivalued integration. Atti Sem. Mat. Fis. Univ. Modena **50**, 53–63 (2002)
56. Satco, B.: A Vitali type theorem for the set-valued Gould integral. An. Şt. Univ. "Al. I. Cuza" Iaşi **51**, 191–200 (2005)
57. Sofian-Boca, F-N.: A multi-valued Choquet integral with respect to a multisubmeasure. An. Şt. Univ. "Al. I. Cuza" Iaşi **61**, 1, 129–152 (2015)
58. Stamate, C.: Vector fuzzy integral. Recent Advances in Neural Network. Fuzzy Systems and Evolutionary Computing, pp. 221–224 (2010)
59. Stamate, C., Croitoru, A.: Non-linear integrals, properties and relationships. Recent Advances in Telecommunications, Signals and Systems (Proceedings of NOLASC'13), WSEAS Press, 118–123 (2013)
60. Tversky, A., Kahneman, D.: Advances in prospect theory: commulative representation of uncertainty. J. Risk Uncertain. **5**, 297–323 (1992)
61. Zhang, D., Wang, Z.: On set-valued fuzzy integrals. Fuzzy Sets Syst. **56**, 237–241 (1993)
62. Zhang, D., Guo, C.: Fuzzy integrals of set-valued mappings and fuzzy mappings. Fuzzy Sets Syst. **75**, 103–109 (1995)
63. Zhang, D., Guo, C., Liu, D.: Set-valued Choquet integrals revisited. Fuzzy Sets Syst. **147**, 475–485 (2004)

Author Index

S. Saminger-Platz and R. Mesiar (eds.), *On Logical, Algebraic, and Probabilistic Aspects of Fuzzy Set Theory*, Studies in Fuzziness and Soft Computing 336,
DOI 10.1007/978-3-319-28808-6